Reliability of Geotechnical Structures
in ISO2394

T0199598

Reliability of Geotechnical Structures in ISO2394

Editors

K.K. Phoon
Department of Civil and Environmental Engineering,
National University of Singapore

J.V. Retief
Department of Civil Engineering, Stellenbosch University,
South Africa

CRC Press
Taylor & Francis Group
Boca Raton London New York

CRC Press is an imprint of the
Taylor & Francis Group, an **informa** business

A BALKEMA BOOK

Cover photo: Abel Erasmus Pass, Limpopo, South Africa, courtesy of Sue Keartland, Johannesburg, South Africa.

CRC Press
Taylor & Francis Group
6000 Broken Sound Parkway NW, Suite 300
Boca Raton, FL 33487-2742

First issued in paperback 2020

© 2007 by Taylor & Francis Group, LLC
CRC Press is an imprint of Taylor & Francis Group, an Informa business

No claim to original U.S. Government works

ISBN 13: 978-0-367-57446-8 (pbk)
ISBN 13: 978-1-138-02911-8 (hbk)

Typeset by MPS Limited, Chennai, India

Library of Congress Cataloging-in-Publication Data

Names: Phoon, Kok-Kwang, editor. | Retief, J. V. (Johannes Verster), 1940– editor.
Title: Reliability of geotechnical structures in ISO2394 / editors, K.K. Phoon, National University of Singapore, J.V. Retief, Department of Civil Engineering, Stellenbosch University, Matieland, South Africa.
Description: Leiden, The Netherlands : CRC Press/Balkema, [2016] | Includes bibliographical references and index.
Identifiers: LCCN 2016020462 (print) | LCCN 2016036243 (ebook) | ISBN 9781138029118 (hardcover : alk. paper) | ISBN 9781315643892 (eBook PDF) | ISBN 9781315643892 (ebook)
Subjects: LCSH: Geotechnical engineering—Standards. | Earthwork—Reliability.
Classification: LCC TA705 .R3926 2016 (print) | LCC TA705 (ebook) | DDC 624.1/891—dc23
LC record available at https://lccn.loc.gov/2016020462

Visit the Taylor & Francis Web site at
http://www.taylorandfrancis.com

and the CRC Press Web site at
http://www.crcpress.com

Table of contents

Preface

The central role of ISO2394 in providing a common basis of reliability principles for structural design standards is affirmed by the fourteen ISO Standards for which it serves as a normative reference and ten ISO member states who have adopted it as a national standard; together with extensive citations in the related literature. The key departure of the current ISO2394:2015 from previous versions is the introduction of risk and risk-informed decision making as the fundamental basis for the regulation and standardization of safety and reliability of structures. From a geotechnical perspective, the key departure of the current ISO2394:2015 from previous versions is the introduction of a new informative Annex D on "Reliability of Geotechnical Structures". The need to achieve consistency between geotechnical and structural reliability-based design is explicitly recognized for the first time in ISO2394 with the inclusion of Annex D.

There is a gradual but perceptible shift in geotechnical design codes towards reliability-based design (RBD) in some countries such as Canada, Japan, USA, and the Netherlands. The simplified or semi-probabilistic approach is usually applied to calibrate these geotechnical RBD codes. It is useful to note that RBD can be applied in place of a full risk assessment "when the consequences of failure and damage are well understood and within normal ranges" (Clause 4.4.1, ISO2394:2015). The basic goal of RBD is to adjust a set of design parameters such that a prescribed target probability of failure is not exceeded. In the reliability literature, the term "failure" is defined in the general sense of failing to satisfy one or a set of performance requirements. RBD can be further simplified "when in addition to the consequences also the failure modes and the uncertainty representation can be categorized and standardized" (Clause 4.4.1, ISO2394:2015). This simplified RBD approach is referred to as a semi-probabilistic approach. The most popular simplified RBD format in North America is the Load and Resistance Factor Design (LRFD) format (refer to Chapter 6). The simplified (or semi-probabilistic) RBD approach may not be as widely applicable to geotechnical design as to structural design, because "standardization" is less achievable in natural geomaterials in contrast to made-to-order structural materials. The structural LRFD practice that recommends a single numerical value for a resistance factor does not allow room for the geotechnical engineer to exercise judgment in response to local site conditions and to incorporate local experience/data. *Site-specific* issues are however critical to geotechnical practice. There are merits to consider a direct probabilistic approach (refer to Chapter 7) in some situations, which may include rock engineering design.

The purpose of Chapters 3 to 7 is to explain how simplified (semi-probabilistic) and direct probabilistic approaches can be applied to geotechnical RBD in alignment with the topics covered in Annex D of ISO2394:2015 (uncertainty representation of geotechnical design parameters, statistical characterization of multivariate geotechnical data, statistical characterization of model factors, and implementation issues in geotechnical RBD). These chapters provide background information to substantiate the special considerations needed for the use of reliability concepts for geotechnical structures, as well as illustrations of approaches and procedures for uncertainty representation and the implementation of geotechnical reliability-based design. At the same time, it should be noted that Annex D is fully consistent and compliant with the general principles of reliability given by ISO2394:2015. The standard therefore provides an overall framework for the advancement of geotechnical practice that is consistent with the principles of reliability within the wider scope of buildings, infrastructure and civil engineering works. A coherent approach is provided by the standard for reliability based decision-making and design, as derived from optimized risk, expressed as performance models that accounts for the levels of knowledge and uncertainty that applies to the field of application under consideration.

Chapter 1, as an introductory chapter to this book, seeks to present a case to the geotechnical community (including the rock community) to adopt reliability principles as a basis for design and practice. It emphasizes the importance of integrating reliability principles within the prevailing body of geotechnical knowledge and experience in a judicious way to improve certain aspects of geotechnical practice, particularly those amenable to mathematical treatment and to occasions where there is considerable practical value to do so. It makes clear that RBD plays a complementary role and it does not displace or preclude well established elements of sound geotechnical practice, in soils or in rocks. The centrality of engineering judgment and its role in setting up the right lines of scientific investigation, selecting the appropriate models and parameters for calculations, and verifying the reasonableness of the results are affirmed. Practical avenues for site-specific effects to be incorporated in the RBD process are highlighted. Chapter 2 provides an outline of ISO2394:2015 as seen from the perspective of geotechnical RBD.

Editors
Kok-Kwang Phoon
Johan V. Retief

About the Editors

Kok-Kwang Phoon
kkphoon@nus.edu.sg
Kok-Kwang Phoon is Distinguished Professor in the Department of Civil and Environmental Engineering, National University of Singapore (NUS). He is a Professional Engineer in Singapore and past President of the Geotechnical Society of Singapore. His main research interests include statistical characterization of geotechnical parameters and reliability-based design in geotechnical engineering. He is the recipient of numerous research awards, including the ASCE Norman Medal in 2005 and the NUS Outstanding Researcher Award in 2010. He is the Founding Editor of Georisk and Chair of TC304 (Engineering Practice of Risk Assessment and Management) in the International Society for Soil Mechanics and Geotechnical Engineering (ISSMGE). He was former Chair of ASCE Geo-Institute Risk Assessment and Management Committee. He is Fellow of ASCE and Fellow of the Academy of Engineering Singapore.

Johan V. Retief
jvr@sun.ac.za
Johan Retief is Emeritus Professor in the Department of Civil Engineering, Stellenbosch University, South Africa. He was awarded DSc(Eng) and DEng degrees from Pretoria and Stellenbosch Universities and graduate degrees from Imperial College and Stanford University. His fields of interest are the development of risk and reliability to design standards as the basis of design, with applications to wind loading, structural concrete and geotechnical practice, amongst other related topics. As member of ISO TC98 he has been a member of the management group for ISO2394:2015 and is presently Convenor for the revision of the related ISO22111. He is a member of the SA Bureau of Standards TC98 on design standards, a fellow of SAICE and moderator of its journal and received the 2014 Jennings award for a geotechnical paper with Dr Dithinde.

Contributors

Jianye Ching
jyching@gmailcom

Jianye Ching is a Professor in the Department of Civil Engineering, National Taiwan University (NTU). His main research interests include geotechnical reliability analysis & reliability-based design, transformation uncertainties in soil properties, statistical uncertainties in site investigation, random fields & spatial variability, and geotechnical design codes. He is the secretary of TC304 (Engineering Practice of Risk Assessment and Management) in the International Society for Soil Mechanics and Geotechnical Engineering (ISSMGE) and an executive board member of Geotechnical Safety Network (GEOSNet). He is a managing editor of Georisk and an editorial board member of Canadian Geotechnical Journal. He is the recipient of the Outstanding Research Award and the Wu-Da-Yu Memorial Award from the Ministry of Science and Technology of Taiwan, Republic of China.

Mahongo Dithinde
dithinde@mopipi.ub.bw

Mahongo Dithinde is a Senior Lecturer and Head of the Department of Civil Engineering, University of Botswana. His specialisation and research interests are in the broad area of geotechnical reliability based design. He is the recipient of Jennings award 2014 from the South African Geotechnical division of the Institute of Civil Engineers for an author of a meritorious publication relevant to geotechnical engineering in South Africa. In addition to academic work, he is also a Geotechnical Partner for Mattra International where he is active in consultancy work in the field of geotechnical engineering. He is a registered Professional Engineer in the discipline of Civil Engineering in Botswana.

Wenping Gong
wenping@clemson.edu; tumugwp2007@gmail.com

Wenping Gong is a Research Assistant Professor in the Glenn Department of Civil Engineering, Clemson University. He received his PhD in civil engineering (geotechnical) from Clemson University in 2014. His main research interests include risk and reliability in geotechnical engineering, uncertainty characterization of geotechnical properties, robust geotechnical design, soil liquefaction, and shield tunnels. He is a member of ASCE Technical Committee on Risk Assessment and Management. He has received several honors including the Best Paper Award in Geo-Shanghai 2014 and the Clemson University Shrikhande Graduate Fellowship.

Dian-Qing Li
dianqing@whu.edu.cn
Dian-Qing Li is a Professor in the State Key Laboratory of Water Resources and Hydropower Engineering Science, Wuhan University. His main research interests include risk and reliability in geotechnical engineering, risk and uncertainty in dam safety, embankment dams and slopes. He is an executive board member of Geotechnical Safety Network (GEOSNet), a member of TC304 (risk) in the International Society for Soil Mechanics and Geotechnical Engineering (ISSMGE) and a member of ASCE Technical Committee on Risk Assessment and Management. Currently, he is editorial board member (EBM) of several international journals, namely Computers and Geotechnics, Soils and Foundations, Georisk, and ASCE-ASME Journal of Risk and Uncertainty in Engineering System. He is the recipient of the inaugural GEOSNet Young Researcher Award, the 2012 National Science Fund for Distinguished Young Scholars, China and the 2013 Young and Middle-Aged Leading Scientists, Engineers and Innovators, Ministry of Science and Technology, China.

Widjojo A. Prakoso
wprakoso@eng.ui.ac.id
Widjojo Prakoso is Professor and Head of the Civil Engineering Department, Universitas Indonesia. He is a licensed geotechnical engineer in Indonesia and has extensive experience in geotechnical practice in Indonesia. He is Vice President of the Indonesia Society for Geotechnical Engineering and a member of TC205 (Safety and Serviceability in Geotechnical Design) in the International Society for Soil Mechanics and Geotechnical Engineering. His research interests include geotechnical earthquake engineering, risk and reliability in geotechnical engineering, and soil-structure interaction.

Timo Schweckendiek
timo.schweckendiek@deltares.nl
Dr. Timo Schweckendiek is a specialist consultant/researcher at Deltares and research associate at Delft University of Technology, both based in Delft, Netherlands. His areas of expertise both in applied research as well as in specialist consultancy cover geotechnical reliability and risk analysis (with particular focus on reliability updating using past performance information), flood risk assessment, code calibration and risk-based site investigation. Timo is member of TC304 (risk) in the International Society for Soil Mechanics and Geotechnical Engineering (ISSMGE), the Geotechnical Safety Network (GEOSNet), the Dutch Expertise Network on Flood Protection (ENW) and editorial board member of the international journal Georisk, as well as reviewer for several international journals.

Yu Wang
yuwang@cityu.edu.hk
Yu Wang is an Associate Professor in the Department of Architecture and Civil and Architectural Engineering, City University of Hong Kong. His main research interests include geotechnical risk and reliability (e.g., probabilistic characterization of geotechnical properties, reliability-based design in geotechnical engineering, and probabilistic

slope stability analysis), seismic risk assessment of critical civil infrastructure systems (e.g., water supply systems), soil-structure interaction, and geotechnical laboratory and in situ testing. Dr Wang was the President of the American Society of Civil Engineers (ASCE) – Hong Kong Section in 2012–2013. He is also a member of several international Technical Committees (TCs), including an ASCE Geo-Institute TC on Risk and two ISSMGE TCs on Risk and In-Situ Testing, respectively. He is the recipient of the inaugural Wilson Tang Best Paper Award in 2012 and the inaugural GEOSNet Young Researcher Award in 2015.

Limin Zhang
cezhangl@ust.hk
Limin Zhang is Professor of Geotechnical Engineering and Director of Geotechnical Centrifuge Facility at the Hong Kong University of Science and Technology. His research areas include slopes and embankment dams, geotechnical risk assessment, pile foundations, multiphase flows and centrifuge modeling. He is a fellow of American Society of Civil Engineers, Past Chair of Geotechnical Safety Network (GEOSNet), Editor-in-Chief of International Journal Georisk, Associate Editor of ASCE's Journal of Geotechnical and Geoenvironmental Engineering, and an editorial board member of several other journals. He has published over 190 international journal papers.

Tengyuan Zhao
tengyzhao2-c@my.cityu.edu.hk
Tengyuan Zhao is currently a PhD student at City University of Hong Kong. His main research interests include geotechnical risk and reliability, such as characterization of geotechnical uncertainty.

Acknowledgement

This book project was initiated by the ISO2394:2015 Annex D drafting group consisting of Jianye Ching, Mahongo Dithinde, Kok-Kwang Phoon (Chair), Johan Retief, Timo Schweckendiek, Yu Wang, and Limin Zhang and Mr Janjaap Blom (Senior Publisher of CRC Press/Balkema). This book project was proposed as a follow-up to the Annex D project to provide detailed guidance on how to draft geotechnical design codes in accordance to reliability principles. The ISO2394 revision secretariat consisting of Michael Havbro Faber (Convenor), Kazuyoshi Nishijima (Secretary), and Johan Retief is credited for the initiative to draft Annex D (a new annex for geotechnical structures), from which this book project emerged from.

The Editors are grateful to Mr Janjaap Blom, Senior Publisher at CRC Press/ Balkema, who has advised and supported this project through all stages. We are also indebted to the CRC Press/Balkema production team led by Ms José van der Veer, who has ably steered and supported us in the production stage. Finally, we would also like to thank Dr Chong Tang who has organised the final proof reading stage and improved the subject index significantly. His efforts are deeply appreciated. No book project is possible without all of you.

Chapter 1

Reliability as a basis for geotechnical design

Kok-Kwang Phoon

ABSTRACT

The purpose of this introductory chapter is to present a case to the geotechnical community (including the rock community) to adopt reliability principles as a basis for design and practice. Engineers should be open to applying simplified (semi-probabilistic) or direct probabilistic approaches to reliability-based design (RBD), depending on the extent in which the design situations could be standardized. RBD refers to any design methodology that applies reliability principles, explicitly or otherwise. The intent is certainly not to advocate indiscriminate adoption of structural reliability principles, but to consider how reliability principles (which are very general) can be integrated within the larger body of geotechnical practice in a judicious way to improve certain aspects, particularly those amenable to mathematical treatment and to occasions where there is considerable practical value to do so. Aspects amenable to mathematical treatment typically fall under the category of "known unknowns" where some measured data and/or past experience exist for limited site-specific data to be supplemented by both objective regional data and subjective judgment derived from comparable sites elsewhere.

This chapter adds to the ongoing conversation on the relevance of RBD, simplified or otherwise, in geotechnical engineering. It points out that discussions on geotechnical design at times did not draw a clear distinction between performance verification strategies (examples include global factor of safety, partial factors, or RBD) and broader design considerations that affect all performance verification strategies if they were sufficiently fundamental. All performance verification strategies must operate within the prevailing norms of engineering practice and any shortcomings in these norms do not reflect shortcomings in the performance verification strategies. Confusion of this nature abound in some of the prevailing discussions pertaining to geotechnical RBD. It is primarily a performance verification methodology and one should not view it as a panacea for all afflictions affecting design calculations based on the factor of safety or geotechnical practice in general.

The key point articulated in this chapter is that useful observations raised in ongoing discussions should be viewed as providing approximate boundaries circumscribing the limits of reliability calculations *or* acting as a caution against overly simplistic reliability applications that do not respect geotechnical needs or constraints, rather than invalidating reliability principles as a whole. This chapter emphasizes the need to apply reliability principles judiciously in conjunction with other design/construction strategies. It is clear that RBD plays a complementary role. It does not displace or preclude well established elements of sound geotechnical practice, in soils or in rocks, which evolved to handle a moderate degree of "unknown unknowns". Annex D of ISO2394:2015 "Reliability of geotechnical structures" has been drafted with this central intent in mind. The remaining chapters in this book explain how simplified (semi-probabilistic) and direct probabilistic approaches can be applied to geotechnical RBD in alignment with the topics covered in Annex D of ISO2394:2015 (uncertainty representation of geotechnical design parameters, statistical characterization of multivariate geotechnical data, statistical characterization of model factors, and implementation issues in geotechnical RBD).

1.1 INTRODUCTION

There is a gradual but perceptible shift in geotechnical design codes towards reliability-based design (RBD) over the two decades, primarily in North America (Kulhawy & Phoon 2002; Phoon et al. 2003a; Scott et al. 2003; Paikowsky et al. 2009; Allen 2013; Fenton et al. 2016) and Japan (Nagao et al. 2009; Honjo et al. 2009; Honjo et al. 2010). RBD refers to any design methodology that applies reliability principles, explicitly or otherwise. The term "geotechnical design" is used in a broad sense to cover soil and rock engineering design. However, it is acknowledged that geotechnical reliability research has been focused on soils thus far. Clause 4.4.1 of ISO2394:2015 states that RBD can be applied in place of a full risk assessment *"when the consequences of failure and damage are well understood and within normal ranges"*. The basic goal of RBD is to adjust a set of design parameters such that a prescribed target probability of failure is not exceeded. For example, the depth of a bored pile or width of a footing is a practical design parameter that can be adjusted readily. The trial-and-error adjustment of a design parameter is common to RBD and the prevailing allowable stress design (ASD) method. The only difference is the design objective. The former considers a design to be satisfactory if a target probability of failure, say one in a thousand, is achieved. The latter considers a design to be satisfactory if a target global factor of safety, say three, is achieved. The advantages of using the probability of failure (or the reliability index) in place of the global factor of safety have been discussed elsewhere (Phoon et al. 2003b), but debate on the usefulness of RBD within the context of geotechnical design is still ongoing (Simpson 2011; Schuppener 2011; Vardanega & Bolton 2016). This healthy debate is on-going in part because a large part of geotechnical engineering is governed by natural geomaterials such as soils and rocks and the subsurface environment (groundwater regime is one important aspect), which can be fairly variable and complex. This challenge is further magnified by the limited availability of site information due to the volume of soils/rocks involved and possible changes in the ground conditions with time, among others. Chilès & Delfiner (1999) noted that the volume of rock sampled is a minute fraction of the total volume of a hydrocarbon reservoir in the petroleum industry. The geotechnical and the rock engineer has to live with this heightened state of uncertainty and associated risk, parts of which may not be amenable to mathematical (statistical/probabilistic) treatment. One ramification is that the design phase and the construction phase may not be as distinct as those in structural engineering as it is not uncommon for the design to be adjusted in accordance to actual ground response during construction to manage this uncertainty and associated risk in a sensible way (e.g., adjust the spacing between rock bolts during tunneling). It is crucial to appreciate "geotechnical design" in this context, because it circumscribes where reliability can be applied and it emphasizes the need to apply reliability principles judiciously in conjunction with other design/construction strategies. There is a misconception that RBD precludes or displaces existing elements of good practice and engineering judgment. Orr (2015) observed that "geotechnical designs with appropriate degrees of reliability are achieved by using calculations with partial factors ... and quality management measures related to the different stages of a geotechnical design project which are: ground investigation, design calculations, construction, and monitoring and maintenance after construction". It is useful to view design calculations, be it verified by a global factor of safety, partial factors, or RBD, as one stage of a project. In this chapter, the term "partial factors" refers to the empirical method of factoring soil parameters proposed by Hansen (1953, 1956, 1965). Partial factors are not calibrated by reliability analysis and hence, it is not a simplified RBD approach when viewed from this historical context. Although geotechnical reliability evolves from structural reliability, there are critical elements distinctive to geotechnical reliability that must be addressed for reliability principles to be integrated in a meaningful way to geotechnical design and to the broader ambit of geotechnical practice. For example, the structural Load and Resistance Factor Design (LRFD) practice that recommends a single numerical value for a resistance factor does not allow room for the geotechnical engineer to exercise judgment in response to local site conditions and to incorporate local experience/data. *Site-specific* issues are however critical to geotechnical practice. Annex D of ISO2394: 2015 "Reliability of geotechnical structures" has been drafted with this central intent in mind (Phoon et al. 2016).

Clause 4.4.1 of ISO2394:2015 also states that RBD can be further simplified *"when in addition to the consequences also the failure modes and the uncertainty representation can be categorized and standardized"*. This simplified RBD approach is referred to as a semi-probabilistic approach. It is immediately clear that the simplified RBD approach is not as widely applicable to geotechnical design as to structural design, because "standardization" is less achievable in natural geomaterials in contrast to made-to-order structural materials. There are merits to consider a direct probabilistic approach (refer to Chapter 7 Direct probability-based design methods) in some situations, which may include rock engineering design. In reference to Eurocode 7 or EC7 (EN 1997–1:2004), the Commission on Evolution of Eurocode 7 hosted by the International Society for Rock Mechanics (https://www.isrm.net/gca/index.php?id=1143) noted that "it is now widely recognised that EC7 is in many ways inappropriate – and in some circumstances inapplicable – to rock engineering." The purpose of this Commission is to "to liaise with CEN/TC250/SC7 in order to help further develop EC7 with regard to rock engineering design" during the current phase of Systematic Review of the Structural Eurocodes (2015–2018).

The most popular simplified RBD format in North America is the Load and Resistance Factor Design (LRFD) format (Allen 2013). In this chapter, the term "LRFD" refers to a design format containing load and resistance factors that are calibrated to achieve a target reliability index (Ravindra & Galambos 1978). In terms of format, LRFD may be viewed as a special case of partial factors if one were to follow the terminology in EN 1997–1:2004 (Design Approach 2). However, the partial factors "reflect more or less the different traditional practices without reference to any target safety levels." (Burlon et al. 2014). This chapter follows the North American terminology of LRFD as a design approach intrinsically based on reliability calibration, rather than simply as a format containing load and resistance factors (Paikowsky et al. 2004, 2010; Allen 2013). Chapter 6 Semi-probabilistic reliability-based design is devoted to addressing some of the challenges in geotechnical design within the context of implementing the semi-probabilistic approach. Other simplified RBD formats such as the Multiple Resistance and Load Factor Design (MRFD) (Phoon et al. 2003c), the Robust LRFD (R-LRFD) (Gong et al. 2016), and the Quantile Value Method (QVM) (Ching & Phoon 2011). In the 4th Wilson Tang Lecture, Phoon & Ching (2015) demonstrated that there are challenges in applying the simplified RBD format to even relatively common design scenarios such as deep foundations installed in layered soils. In the author's opinion, these challenges attract less attention than they deserve, because existing geotechnical LRFD or comparable simplified RBD formats focuses on standardization at the expense of dealing with site-specific considerations directly.

Simplified RBD formats in the form of LRFD and MRFD are popular because practitioners can produce designs complying with the target probability of failure (or target reliability index), albeit approximately, while retaining the simplicity of performing one *algebraic check* per trial design. No tedious Monte Carlo simulations or more sophisticated probabilistic analyses are needed. From the perspective of a practitioner, there is no difference between applying a simplified RBD format, say LRFD, and the prevailing factor of safety format, other than multiplying a set of resistance and load factors to the corresponding resistance and load components (nominal or characteristic values) mandated in such codes. The key difference is that the numerical values of these resistance and load factors are not determined purely on experience or precedents, but calibrated by the code developer using reliability analysis to achieve a desired target reliability index. Once these resistance and load factors are made available in a design code, the practitioner can use them for design without having to perform reliability analysis or to be cognizant of soil statistics other than identifying variability in broad terms, such as low, medium, or high. Given the diversity of natural geomaterials, it is immediately obvious that simplified RBD must provide a channel for the practitioner to incorporate his/her site-specific conditions. This "site-specificity" is rarely considered in structural materials, although it is clearly important for environmental loadings and structural design codes do consider this. It is possible for geotechnical RBD that ignores site-specific inputs on the material side to be viewed as insufficiently realistic or incongruent with existing sound geotechnical practice.

It is accurate to say that simplified RBD formats is the most common application of geotechnical reliability at present. A working group in the Japanese Geotechnical Society (JGS) of TC23 prepared a report to summarize "important points to note and recommendations when new design verification formulas are developed based on Level I reliability based design (RBD) format for geotechnical structures" (Honjo et al. 2009). Level I RBD format is another term for the simplified RBD format. The report further observed that "RBD seems to be the only rational tool to provide a design verification procedure that designs a structure for clearly defined limit states (i.e. performances of structures and members) and introduces sufficient safety margin. It is concluded that RBD will be used as a tool to develop design codes at least for the next several decades" (Honjo et al. 2009). Nagao et al. (2009) reported the revision of the Japanese Technical Standard for Port and Harbor Facilities to align with semi-probabilistic design. Fenton et al. (2016) highlighted that Section 6 "Foundations and Geotechnical Systems" of the most recent edition of the Canadian Highway Bridge Design Code (CAN/CSAS614:2014) incorporates significant changes with respect to reliability based geotechnical design.

The purpose of this book is to explain how simplified (semi-probabilistic) and direct probabilistic approaches can be applied to geotechnical RBD in alignment with the topics covered in Annex D of ISO2394:2015 (uncertainty representation of geotechnical design parameters, statistical characterization of multivariate geotechnical data, statistical characterization of model factors, and implementation issues in geotechnical reliability-based design). Chapter 1, as an introductory chapter to this book, seeks to present a case to the geotechnical community (including the rock community) to adopt reliability principles as a basis for design and practice. Engineers should be open to applying semi-probabilistic or direct probabilistic approaches, depending on the extent in which the design situations could be standardized. The intent is certainly not to advocate indiscriminate adoption of structural reliability principles, but to consider how reliability principles (which are very general) can be integrated within the larger body of geotechnical practice in a judicious way to improve certain aspects, particularly those amenable to mathematical treatment and to occasions where there is considerable practical value to do so. It is clear that RBD plays a complementary role and it does not displace or preclude well established elements of sound geotechnical practice, in soils or in rocks. This chapter describes the gap between geotechnical and structural design, discusses the role of engineering judgment, and contrasts the reliability and geotechnical requirements of a safety format. This "big picture" overview shows that there is ample room for RBD to play a complementary role in geotechnical design. It may help to address comments directed at some worrisome aspects of reliability calculations that in the opinion of the author, tend to miss the forest for the trees. The chapter concludes by showcasing some specific reliability applications that add value to practice to frame the technical contents presented in this book in a proper context. The focus is on the application of reliability principles to *design calculations*, particularly using the popular semi-probabilistic or simplified RBD format. It is useful to re-iterate the caveat that simplified RBD format is not applicable to all geotechnical design situations, especially when the line between design and construction is at times unclear and standardization is difficult to achieve. There are merits to considering a direct probabilistic approach for geotechnical design in these situations. Applications of reliability principles to other aspects of practice are beyond the scope of this chapter. A fuller coverage of interesting possibilities is given in ISO2394:2015.

1.2 EVOLUTION OF STRUCTURAL AND GEOTECHNICAL DESIGN

The central role of ISO2394 in providing a common basis of reliability principles for structural design standards is affirmed by the fourteen ISO Standards for which it serves as a normative reference and ten ISO member states who have adopted it as a national standard. It has been cited in national standards such as the Eurocode head standard EN1990:2002 Basis of Structural Design (Vrouwenvelder 1996), the South African standard SANS10160-1:2011 Basis of Structural Design, and the Canadian standard CSA S408 Guidelines for the Development of Limit States Design Standards, in addition to being widely used as a basis for research on the application of reliability principles. Faber (2015) explained that the key departure of the current ISO2394:2015 from previous versions is "the introduction of risk and risk-informed decision making as the fundamental basis for the regulation and standardization of safety and reliability of structures." He further elaborated that "Whereas requirements to safety and reliability in the previous edition of ISO 2394 took basis in efficiency requirements of a heuristic character, these are now based on risk considerations and socio-economic principles through utilization of the marginal life saving principle and the Life Quality Index (LQI), see Nathwani et al. (1997). This in turn facilitates a more relevant use of ISO 2394 in the context of sustainable societal developments and its adaptation for application in different nation states in accordance with prevailing economic capacity and preferences. The new revision of ISO 2394 thus facilitates regulation, verification, documentation and communication of the adequately safe and reliable performance of structures, and also to consider them in a broader sense as part of societal systems and services."

From a geotechnical perspective, the key departure of the current ISO2394:2015 from previous versions is the introduction of a new informative Annex D on "Reliability of Geotechnical Structures" (Phoon et al. 2016). The need to achieve consistency between geotechnical and structural reliability-based design is explicitly recognized for the first time in ISO2394 with the inclusion of Annex D. As highlighted previously, the emphasis in Annex D is to inject greater realism into geotechnical RBD, while respecting the principles of prevailing geotechnical practice that evolved to handle uncertainties (and risks) beyond those amenable to mathematical treatment. It is further recognized that geotechnical engineering practice is less amenable to standardization compared to structural engineering practice, because there are diverse site conditions and diverse local practices that grew and evolved over the years to suit these conditions. The gap between structural and geotechnical design at a fairly fundamental level is already evident if one were to observe that a new Annex devoted to geotechnical engineering only appears in ISO2394:2015, which is the fourth edition. The first edition of ISO2394:2015 was published in 1973, although foundations were included as structural elements.

The evolution of geotechnical design over the past six or more decades is briefly reviewed below to present a historical perspective of how structural and geotechnical design diverge due to differences in design situations and differences in emphasis. A large part of geotechnical design, particularly in the format of partial factors, was influenced by Hansen (1953, 1956, 1965). The partial factors are determined subjectively based on two guidelines: (a) a larger partial coefficient should be assigned to a more uncertain quantity, and (b) the partial coefficients should result in approximately the same design dimensions as that obtained from traditional practice (Hansen, 1965). This partial factor approach was adopted in Denmark (Ovesen 1989) and subsequently influenced geotechnical code developments in Canada (Meyerhof 1984) and Europe, notably EN 1997–1:2004 [there are 60 occurrences of the term "partial factor(s)" between Section 2 and 12 in EN 1997–1:2004]. It has been highlighted previously that the original partial factor approach suggested by Hansen (1953, 1956, 1965) differs in one aspect from the partial factor approach recommended in EN 1997–1:2004. The former approach applies the partial factor to the ground strength parameters, while the latter approach permits application of the partial factor to both ground strength parameters and ground resistances. When the partial factor in EN 1997–1:2004 is applied as a divisor to a ground resistance, it is the reciprocal of the LRFD resistance factor. The "partial factor" terminology in EN 1997–1:2004 is not adopted in this chapter. This chapter retains the original Hansen definition of a "partial factor" (which is a divisor applied to a ground strength parameter) and refers to a "resistance factor" as a multiplier to a ground resistance in the context of LRFD, where the term "resistance factor" originates from. The limitations of this approach was widely debated (e.g., Simpson et al. 1981, Baike 1985, Fleming 1989, Valsangkar & Schriver 1991), but in the opinion of the author, no satisfactory resolution emerged from these discussions. Phoon et al. (2003a) opined that: "Implementation of limit state design within a non-probabilistic framework, such as the empirical partial factors of safety method, does not appear to address adequately most of the serious drawbacks associated with the traditional factor of safety approach. For example, it is not clear how the empirical partial factors of safety method can promote communication, assist in extrapolating the experience of safe practice to new conditions, or permit full advantage to be taken of improvements in the knowledge base. The adoption of such empirical methods might pave the way for gradual rationalization of the partial factors using probabilistic means, but the desirability of trading a known system for an unknown one solely on this basis is debatable."

In a more detailed historical review on the evolution of structural and geotechnical design since the fifties, Kulhawy & Phoon (2002) noted that structural design, in the form of LRFD, "is essentially the logical end-product of a philosophical shift in mindset to probabilistic design in the first instance and a simplification of rigorous reliability-based design into a familiar 'look and feel' design format in the second". In contrast, geotechnical design predominantly involved a rearrangement of a single global factor of safety into two or more partial factors. As mentioned above, this arose in part because geotechnical engineers had to grapple with a heightened state of uncertainty and associated risk, parts of which may not be amenable to mathematical treatment. It is sensible for geotechnical engineers to seek clarity on more fundamental design considerations (e.g., what is "design"? which design situations can be standardized? is the divide between ultimate and serviceability limit states real? is it more sensible to assess the performance of a structure and foundation as a system based on ground movements? should geotechnical design migrate towards performance-based design?) and to devote less attention to performance verification which is only one step, albeit an important one, in design. It may be noted in passing that the Japanese Geo-code 21 is possibly the first performance-based foundation design code (Honjo & Kusakabe 2002, Honjo et al. 2010). In the opinion of the author, discussions on geotechnical design at times did not draw a clear distinction between performance verification strategies (examples include global factor of safety, partial factors, or RBD) and broader design considerations that affect all performance verification strategies if they were sufficiently fundamental. All performance verification strategies must operate within the prevailing norms of engineering practice and any shortcomings in these norms do not reflect shortcomings in the performance verification strategies. Confusion of this nature abound in some of the cited discussions presented in this chapter that are intended to be illustrative rather than comprehensive.

The relevance of RBD, simplified or otherwise, is still being debated in an exchange clouded occasionally by confusion due to the evolving discussion on fundamental issues at hand and their fairly tangled relationship to performance verification. The reference to "probability" or comparable terms engendered continuing controversy in the geotechnical community. For example, Schuppener (2011) shared that a similar sentiment was expressed during the National Geotechnical Conference in Germany in 1982. Schuppener (2011) summarized the following reservations raised during a discussion panel in this conference:

- The probabilistic approach does not take account of human error in design and execution although it is the main cause of damage.
- The possibilities of collecting statistical data on soil are severely limited in practice.
- The differences between geotechnical engineering and other areas of structural engineering are not only the higher coefficients of variation in the former – soil cannot be produced with clearly defined characteristics according to a set formula – but also that the geotechnical engineer only ever sees a limited part of the structure he is designing.
- Damage is usually due to risks which are connected with the soil but which go undetected.
- Distributions of geotechnical basic variables that have no upper or lower limit are unsuitable as it is not possible to measure very high and very low values, nor are such values considered likely to occur for mechanical reasons.
- Soil excavations and tests of the mechanical properties of soil never provide enough data to enable a probability calculation to be performed.

The author hastens to add that the probabilistic approach is favorably received in Canada (Fenton et al. 2016), the Netherlands (Vrouwenvelder et al. 2013), Japan (Honjo et al. 2009, 2010), and USA (Allen 2013). Chapter 7 presents new safety standards for flood defenses in the Netherlands which is the first ever national standard that adopts direct (or full) probability-based design methods (Schweckendiek et al. 2013, 2015). The author ventures to suggest that some of the reservations are based on the misconception that RBD is a panacea for all afflictions affecting design calculations based on the factor of safety or geotechnical practice in general. This aspect is clarified in the next section. Other reservations relate to the scarcity and/or incompleteness of available information. This crucial information aspect is clarified in Section 1.5. The key point here is that the reservations expressed above should be viewed as providing approximate boundaries circumscribing the limits of reliability calculations *or* acting as a caution against overly simplistic reliability applications that do not respect geotechnical needs or constraints, rather than invalidating reliability principles as a whole. For example, it is possible to incorporate upper and/or lower limits in probability distributions. It is admittedly not possible to do this using the normal or lognormal distribution, but this issue is related to over-simplification (which a part of the more theoretically oriented geotechnical reliability literature may have indulged in) rather than a fundamental limitation of reliability principles. There is no reason to retain simple reliability analysis that does not respect sound geotechnical principles. Section 1.5

provides an overview of more advanced methods that can respect prevailing geotechnical principles and add considerable value to practice. Notwithstanding the merits of using non reliability-based partial factors (all methods must have their pluses along with minuses), it suffices to note at this point that structural and geotechnical design cannot be bridged by an empirical basis. Using ISO2394:2015 as a concrete example, it is clear that there is no practical way for geotechnical design unsupported by a reliability basis to fit in. It is also difficult for our geotechnical profession to benefit from the ongoing and pervasive information technology revolution in big data and data analytics in the absence of a rational framework, be it reliability or otherwise.

1.3   ROLE OF ENGINEERING JUDGMENT

There is no doubt that a discussion on the role of reliability calculations in geotechnical design should be framed by useful caveats highlighted on a number of occasions (e.g., Simpson 2011, Schuppener 2011, Schuppener 2013, Vardanega & Bolton 2016). One noteworthy caveat is the importance of engineering judgment (e.g., Burland 2008a, Burland 2008b, Dunnicliff & Deere 1984, Focht 1994, Peck 1980, Petroski 1993, Petroski 1994). Much has been written on this topic and it will not be repeated here other than to reaffirm the centrality of engineering judgment in RBD. Certainly, it is not judicious to rely completely on calculations, regardless of their sophistication, generality, and precision. It is particularly important to repeatedly reinforce this message, given the growing power and sophistication of computing, including big data analytics and deep machine learning. Reliability analysis is merely one of the many mathematical methods routinely applied to model the complex real-world for engineering applications. It is susceptible to abuse in the absence of sound judgment in the same manner as a finite element analysis. Nonetheless, the author submits that we should take this caveat as a given and draw clearer boundaries on which aspects of practice would benefit from calculations (reliability analysis included). No one would argue that engineering practice has benefited tremendously from mathematical modelling.

Kulhawy & Phoon (1996) clarified the role of engineering judgment in RBD as follows: "The advent of powerful and inexpensive computers in the last two decades has helped to provide further impetus to the expansion and adoption of theoretical analyses in geotechnical engineering practice. The role of engineering judgment has changed as a result of these developments, but the nature of this change often has been overlooked in the enthusiastic pursuit of more sophisticated analyses ... For example, engineering judgment still is needed (and likely always will be!) in site characterization, selection of appropriate soil/rock parameters and methods of analysis, and critical evaluation of the results of analyses, measurements, and observations. The importance of engineering judgment clearly has not diminished with the growth of theory and computational tools. However, its role has become more focused on those design aspects that remain outside the scope of theoretical analyses." Examples would be provided in Section 1.5 on how reliability calculations could relieve engineering judgment from the unsuitable task of performance verification in the presence of uncertainties so that the engineer can focus on setting up the right lines of scientific investigation, selecting the appropriate models and parameters for calculations, and verifying the reasonableness of the results (Peck 1980).

The key advantage of reliability analysis is that it allows "known unknowns" to be modeled formally as random variables/fields/processes (there are merits for doing so) and to determine the uncertainty in the response (or responses) consistently from the input uncertainties. In principle, it is possible to analyse complex and large scale real-world problems. There is a practical limit to the reach of engineering judgment, particularly in the absence of precedents. In some sense, this step could be viewed as a logical progression from presumptive bearing stresses based on precedence, rules of thumb, and local experience to allowable stresses based on soil mechanics and more rational methods for analyzing stability and a factor of safety to take care of known unknowns and perhaps a moderate degree of unknown unknowns. From this perspective, reliability analysis rationalizes a part of the factor of safety, particularly characterization of the known unknowns using *objective data and to some degree, subjective experience.* As such, the probability of failure computed from a reliability analysis should be interpreted as a nominal value rather than as an actual value that matches historical rates of failure. However, the nominal probability of failure has demonstrated its value as an index that manages known unknowns in a consistent way and by doing so, allows the value of geotechnical information to be quantified in a defensible way. It is common for an engineer to grapple with the question on how many tests should be conducted, because the cost of gathering more information cannot be readily weighed against the "value" of doing so, particularly from the perspective of the client. The most extreme degree of an "unknown unknown" is referred to as a "black swan" in the catastrophic risk literature. There is no prior experience of encountering this "black swan" event that leads to disproportionate consequences. It is safe to say that reliability analysis or any computational technique is not appropriate to deal with these exceedingly rare events that may lead to catastrophic failures and in any case, no one designs for these events, be it using a global factor of safety, partial factors, or RBD. Between the known unknowns ("white swans") where some reasonable amount of data exist and extreme unknown unknowns ("black swans"), there are perhaps different shades of "grey swans" such as events reasonably foreseeable even in the absence of data (e.g., cavities in karst formations foreseeable from geologic considerations and regional experience) or events unforeseeable but do not lead to major failures (e.g., erratic soft spots below pile tips, moderate human errors, moderate accidents). The former can be dealt with by selecting an appropriate foundation system such as a raft to bridge over potential cavities. This foundation system is reasonably robust against the unknown presence of cavities, provided the cavities are reasonably smaller than the size of the raft and they are anomalous rather than prevalent subsurface features. This is where engineering judgment comes in – it is outside the scope of reliability calculations which presupposes the selection of an appropriate design, limit state, and failure mechanism to design against. The latter may be amenable to design strategies and robustness provisions given in Section F.3 of ISO2394:2015 (Table 1.1). Robust design in the sense discussed under Annex F is outside the scope of this book, although it is briefly discussed in Section 2.4. A more restricted element of robustness that accounts for the hard-to-control (i.e., cannot be easily adjusted by the practitioner) and hard-to-characterize (i.e., the uncertainty is recognized but hard to quantify due to insufficient data) noise factors, is discussed in Section 6.2 and 7.4.

One may argue that "grey swans" are considered in the global factor of safety in a broad conceptual sense. It is widely perceived that the factor of safety contains additional conservatism to take care of a moderate degree of "unknown unknowns".

Table 1.1 Classification of design methods (Source: Table F.2, ISO2394:2015. Reproduced with permission from the International Organization for Standardization (ISO). All rights reserved by ISO).

Method	Reduces	Issues to address
(a) Event control (EC)	Probability of occurrence and/or the intensity of an accidental event	– Monitoring, quality control, correction, and prevention
(b) Specific load resistance (SLR)	Probability of local damage due to an accidental event	– Strength and stiffness – Benefits of strain hardening – Ductility versus brittle failure – Post-buckling resistance – Mechanical devices
(c) Alternative load path method (ALP), including provision of ties	Probability of further damage in the case of local damage	– Multiple load path or redundancy – Progressive failure versus the zipper stopper – Second line of defence – Capacity design and the fuse element – Sacrificial and protective devices – Testing – Strength and stiffness – Continuity and ductility
(d) Reduction of consequences	Consequences of follow up damage such as progressive collapse	– Segmentation – Warnings, active intervention, and rescue – Redundancy of the services of the facilities

Meyerhof (1984) noted that partial factors were calibrated so that they result, on average, in overall factors of safety that are in agreement with existing practice. Hence, one may argue that "grey swans" are considered in empirical partial factors calibrated in this way. What about RBD? Beal (1979) voiced the concern that splitting the safety margin into components associated with loads and resistances can result in omitting some functions of the original factor of safety. Phoon et al. (2003a) proposed the following key considerations for the selection of the target reliability index for transmission line structure foundation design, taking into consideration the nominal nature of the reliability index or equivalent probability of failure:

- It should be approximately consistent with empirical rates of foundation failure, after making an appropriate adjustment for the difference between actual and calculated rates of failure.
- It should fall within the range of reliability levels implicit in existing foundation designs.
- It should be applicable for a variety of loading modes that are commonly imposed on transmission line structure foundations.
- It should exceed the target reliability index for typical transmission line structures because foundation repairs are more difficult and costly.

The second consideration is widely adopted in RBD (Ellingwood et al. 1980). Hence, it is inaccurate to say that RBD does not consider "grey swans". Kulhawy & Phoon (1996) noted that this somewhat empirical approach of calibrating the target reliability index possesses the advantage of keeping designs emerging from RBD compatible with the existing experience base. It is accurate to say, though, that the global factor of safety, partial factors, and RBD do not consider "grey swans" explicitly. The notion of a moderate degree of "unknown unknowns" is ultimately a matter of judgment. At present, judicious adjustment of the target reliability index is probably the only realistic means of incorporating less quantifiable but important considerations into RBD in a fairly consistent way. Section 7.7 presents some examples of target reliability indices. It goes without saying that existing practice does not solely rely on the global factor of safety or partial factors to handle "grey swans". Some sensible mitigating measures are presented in Table 1.1. RBD can and should be used in conjunction with these measures as well.

1.4 RELIABILITY VERSUS GEOTECHNICAL REQUIREMENTS OF A SAFETY FORMAT

This section contrasts reliability requirements and geotechnical requirements of a safety format to make clear that both sets of requirements are complementary. Clause D.5.2 of ISO2394:2015 states that the key goal in geotechnical RBD is to achieve a more uniform level of reliability than that implied in existing allowable stress design. It further clarifies that: "With regard to the semi-probabilistic approach, it is important to highlight that reliability calibration is more challenging in geotechnical engineering. One key reason is the diversity of design scenarios that shall be considered in the calibration domain, such as the range of COV resulting from different soil property estimation methodologies. Another source of diversity is the number of different soil profiles encountered even within a city size locale." COV is the abbreviation for coefficient of variation, which is defined as the ratio between the standard deviation and the mean. The COV of structural strengths is around 10%, while that of soil strengths can be considerably higher. Chapter 6 discusses how diverse design scenarios can be handled within the semi-probabilistic reliability-based design framework that requires some degree of standardization of the partial factors.

To reiterate, the goal of simplified RBD is to calibrate resistance or partial factors to achieve a target reliability index. The performance verification format only affects the ability to achieve a uniform reliability level across different design scenarios within the ambit of the design code. For a less effective verification format that cannot maintain a reasonably uniform level of reliability, an expedient solution is to partition the design space as shown in Figure 1.1. The COV partitions for undrained shear strength in Figure 1.1 are based on a reasonably practical three-tier classification scheme (Table 1.2) proposed by Phoon & Kulhawy (2008) for calibration of resistance factors in simplified RBD. This partition method has been applied in the Canadian Highway Bridge Design Code (CAN/CSA-S6-14:2014). Fenton et al. (2016) highlighted that geotechnical resistance factors for the ultimate and serviceability limit states are provided based on three levels of understanding in Table 6.2 of CAN/CSA-S6-14:2014. Tables 6.1 and 6.2 in Chapter 6 are earlier examples of an information-sensitive RBD that is more appropriate for geotechnical design. This is one of the simplest channel for site-specific conditions to be incorporated in simplified RBD. The prevailing practice of recommending one numerical value for each resistance factor does not offer such a channel. This important point is related to the value of geotechnical information and it is explained in more detail in the next section.

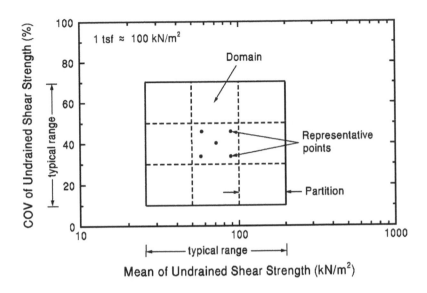

Figure 1.1 Partitioning of parameter space for calibration of resistance factors (Source: Figure D.3, ISO2394:2015. Reproduced with permission from the International Organization for Standardization (ISO). All rights reserved by ISO).

Table 1.2 Three-tier classification scheme of soil property variability for reliability calibration (Source: Table 9.7, Phoon & Kulhawy 2008).

Geotechnical parameter	Property variability	COV (%)
Undrained shear strength	Low[a]	10–30
	Medium[b]	30–50
	High[c]	50–70
Effective stress friction angle	Low[a]	5–10
	Medium[b]	10–15
	High[c]	15–20
Horizontal stress coefficient	Low[a]	30–50
	Medium[b]	50–70
	High[c]	70–90

a – typical of good quality direct lab or field measurements
b – typical of indirect correlations with good field data, except for the standard penetration test (SPT)
c – typical of indirect correlations with SPT field data and with strictly empirical correlations

From a geotechnical perspective, Simpson (2011) opined that an adequate safety format ought to include proper account of the following features:

- the designer's specific knowledge of the site, the ground conditions and their possible variability. This includes taking full account of the geology, history, geomorphology and hydrology of the site;
- an appropriate assimilation and compilation of data from all available sources, including published literature, collection of comparable case histories and test

results, often from several types of test of varying number, means of interpretation and reliability;

- a parametric study, to reveal the significance of variations of the lead variables;
- in particular, a careful assessment of the worst credible values of parameters. This will often not be obtained from a study of likely values and statistical variations around a mean;
- adequate robustness. This entails providing adequate margins for secondary actions and other variations that are not related to the primary parameters, including moderate human errors;
- adequate prescription for both ULS and SLS, noting that these may be difficult to separate.

These requirements are clearly complementary to the key RBD requirement of achieving a more uniform level of reliability. The next section argues that these geotechnical requirements can be addressed in a more advantageous way within RBD. The exception is robustness, which may be better addressed by Annex F, ISO2394:2015, rather than the semi-probabilistic approach. However, Gong et al. (2016) proposed an interesting R-LRFD approach that considers robustness in a restrictive sense. R-LRFD is a simplified version of a new design philosophy called robust geotechnical design (RGD) recently proposed by Juang et al. (2013a&b) to make the response of a geotechnical system robust against, or insensitive to, the variation of uncertain input parameters. Robust geotechnical design is an alternate approach to address the recurring comment that statistics evaluated from limited data are unreliable. It may be noted in passing that the more conventional approach is to consider "unreliable statistics" within the well-established framework of sampling errors or statistical uncertainties. Statistical uncertainties rationally assign a cost that weighs against a fairly common practice of collecting a minimum amount of geotechnical information for the sake of complying with building regulations. This is an advantage.

Reflecting on the role of reliability analyses in design calculations, Simpson (2011) observed that "Reliability analyses have the advantage that they provide a comprehensive parametric study. In the author's view, it is possible that advanced reliability analysis may be able to take account of all the aspects listed here, including consideration of extreme values. However, simple reliability analysis, such as based on a study of means and standard deviations, will not achieve this. Indeed, such an analytical approach is more likely to distract attention from the main issues relating to geology, history, geomorphology and hydrology." Indeed, the author agrees that more realism should be injected into geotechnical RBD. Chapter 7 covers direct (or full) probability-based design methods that are considerably more advanced than simpler methods relying on means and standard deviations alone.

The "simple" reliability analysis arose in part because geotechnical RBD grew from structural LRFD (e.g., Ravindra & Galambos 1978) and in the opinion of the author, insufficient attention was paid to specific needs of geotechnical practice in the past. Structural reliability has evolved rapidly beyond LRFD since the eighties (e.g., Chapter 2 describes that *risk informed decision making* is formally introduced as a recognized basis for design in ISO2394:2015), but geotechnical RBD has developed at a considerably slower pace. While it is understandable for geotechnical RBD to adopt structural LRFD concepts during its initial stage of development over the past decades, the author believes that it is timely for the geotechnical RBD community to look into how we can improve our state of practice to cater to the distinctive needs of geotechnical engineering practice. In fact, Annex D of ISO2394:2015 is an important first step to develop geotechnical RBD with practical geotechnical needs at the forefront. An excerpt of D.1 Introduction is reproduced below to illustrate this spirit.

"The emphasis in this Annex is on the identification and characterization of critical elements of the geotechnical reliability-based design process. These elements cannot be accounted for in existing deterministic geotechnical practice. The critical distinctive elements are the following:

(a) Coefficients of variation (COVs) of geotechnical design parameters can be potentially large because geomaterials are naturally occurring and in situ variability cannot be reduced (in contrast, most structural materials are manufactured with quality control).

(b) COVs for geotechnical design parameters are not unique and can vary over a wide range, depending on the procedure in which they are derived.

(c) Because geotechnical design parameter characteristics are different from one site to another, it is common to conduct a site investigation at each site. For this reason, statistical uncertainty should be handled with much care.

(d) It is common to conduct both laboratory and field tests in a site investigation. A geotechnical design parameter is typically correlated with more than one laboratory and/or field test indices. It is important to consider this multivariate correlation structure where possible because the COV of the design parameter reduces when consistent information increases.

(e) Spatial variability of geotechnical design parameters cannot be readily dismissed because the volume of geomaterial interacting with the structure is related to some multiple of the characteristic length of the structure and this characteristic length (e.g., height of slope, diameter of tunnel, depth of excavation) is typically larger than the scale of fluctuation of the design parameter, particularly in the vertical direction.

(f) There are usually many different geotechnical calculation models for the same design problem. Hence, model calibration based on local field tests and local experience is important. The proliferation of model factors, possibly site-specific, is to be expected because of the number of models and the number of calibration databases.

(g) A geotechnical system, such as a pile group and a slope is a system reliability problem containing multiple correlated failure modes. Some of these problems are further complicated by the fact that the failure surfaces are coupled to the spatial variability of the soil medium."

The elements highlighted by Simpson (2011) or in Annex D of ISO2394: 2015 are by no means comprehensive and their relative importance may be debated, but they do contribute to the ongoing conversation on how to place geotechnical RBD on a *firmer* and *more realistic* basis. Some specific applications are discussed in the following section to illustrate the key point that there is practical value to adopting reliability as a basis for geotechnical design if one were to move away from simplistic assumptions and methods that do not meet geotechnical needs and constraints.

1.5 SOME RELIABILITY APPLICATIONS

1.5.1 Multivariate soil databases

Several multivariate probability models have been constructed for a variety of clay parameters (Ching and Phoon 2012, 2013a, 2014a; Ching et al. 2014a). The details for these databases are given in Table 4.1 in Chapter 4. The constructed multivariate probability model can be used as a prior distribution to derive the multivariate distribution of design parameters based on limited but site-specific field data. Note that the entire multivariate distribution of multiple design parameters is derived, not marginal distributions or simply means and coefficients of variation (COVs) presented in Phoon & Kulhawy (1999a, 1999b) for soils and in Prakoso (2002) for rocks. Note that multiple design parameters can be updated from multiple field measurements, which is more useful than updating one design parameter using one field measurement based on current practice (for example, updating the undrained shear strength using the cone tip resistance). Details are given in Ching and Phoon (2014b) and Chapter 4.

There are clear advantages to compiling soil/rock data in a systematic way that can be applied for updating purposes. It is difficult to carry out updating in a consistent way in the presence of multiple parameters that are correlated to different degrees. The intuitive approach of taking the average of estimates from different tests does not work in all cases, because correlations between different parameters are ignored. The author recommends applying judgment as a reality check on the final outcome derived from Bayesian updating, rather than as a method of estimation. It goes without saying that an engineer would be most well placed to assess the accuracy of the input data and to remove potential outliers. Clearly, judgment is mostly effectively exercised when it is informed by all available data, knowledge, and outcomes of rationally defensible analyses. The important practical point here is that information (in which site investigation is only one source) that can be handled in a defensible way can be associated with a notion of "value of information" as elaborated below in Section 1.5.2.

Vardanega & Bolton (2016) opined that "Although reliable estimates of the mean and standard deviation are easier to ascertain than the shapes of the pdfs, there remains an unjustified tendency to rely solely on published COV values from other soil deposits. Because variability in a soil deposit is a function of the processes of geological deposition and geomorphological change that have influenced the site, intensive efforts would be necessary to draw parallels between a new site that lacks such information and previously explored sites for which COV values have been established." In the opinion of the author, it goes without saying that information gleaned from the literature should not be used indiscriminately. It has been emphasized in Section 1.3 citing Peck (1980) that selecting the appropriate parameters for calculations is within the purview of engineering judgment. Having said this, it is good practice to consider all sources of information in design, including those reported in the literature. In addition, prior information in the literature can be updated by site-specific information. This Bayesian updating approach is a powerful tool that geotechnical engineers could exploit to provide better value for the cost of site investigation. Finally, our geotechnical heritage is steeped in empiricism that includes the application of "global" correlations. Although all engineers are cognizant of the site-specific nature of geotechnical practice, one would be hard pressed to say that our current practice of applying "global" correlations that do not quite fit the specific soils encountered in a particular site is ineffective.

1.5.2 Geotechnical information: Is it an "investment" or a "cost"?

One attractive advantage of RBD is that it offers an explicit linkage between site investigation and design. Site investigation is an activity unique to geotechnical engineering practice and it is mandated in many building regulations around the world [for e.g., the number of boreholes should be the greater of (i) one borehole per 300 sq m or (ii) one borehole at every interval between 10 m to 30 m, but no less than 3 boreholes in a project site]. Site investigation is typically viewed as a cost item and it is generally an uphill task to convince clients to pay for site investigation over and above that mandated in building regulations. The practical significance of summarizing soil databases as multivariate probability models is that the COV of one soil parameter is reduced when information on a second relevant parameter (or a group of relevant parameters) is made available, particularly in the presence of some correlations between these parameters. More details on this Bayesian updating approach is given in Chapter 4. It suffices to note here that site investigation can be viewed as an investment item rather than a cost item, because reduction of uncertainties through multivariate tests can translate directly to design savings through RBD (Ching et al. 2014b). This important link between the quality/quantity of site investigation and design savings cannot be addressed systematically in our traditional factor of safety approach. In short, the advantage of RBD is that it can respond to a change of COV in a rational way explicitly related to data, while the factor of safety approach cannot. This advantage is more pronounced if the multivariate nature of geotechnical data is considered. The only requirement to realize this advantage is to adopt an information-sensitive RBD design approach. The simplest approach is to allow each resistance/partial factor to take a different numerical value depending on the level of property variability (low, medium, high) judged to be appropriate for a specific design scenario. This scheme is illustrated in Figure 1.1 (Figure D.3 of ISO2394), which illustrates a more general three-tier variability scheme shown in Table 1.2. It has been highlighted that Tables 6.1 and 6.2 in Chapter 6 are earlier examples of an information-sensitive RBD approach. It is logical to expect the resistance factor to take a higher value when property variability is low and vice-versa. This is the recommended minimum best practice to allow room for the engineer to incorporate some site-specific variability information. The existing practice of calibrating a single value for each resistance factor is inadequate for geotechnical engineering. It is possible to use a single resistance factor value if the characteristic resistance could be consistently adjusted to respond to different variability tiers, i.e. the characteristic resistance changes with COV rather than the resistance factor as discussed in Section 1.5.6. However, as pointed out in Section 1.5.7, it is virtually impossible to do this using engineering judgment alone for multiple *correlated* resistances. CAN/CSAS614:2014 has followed a similar scheme in allowing a resistance factor to take on different values depending on the degree of "understanding" (low, typical, high). The degree of understanding covers the quality of site information and the quality of performance prediction. It is possible to envisage an information-sensitive RBD that eventually considers the complete gamut of geotechnical information, which could include both pre-design information (e.g., prior experience, site investigation, small-scale model test, centrifuge test, prototype test) and post-design information (e.g., quality control, monitoring). It is safe to say that Table 6.1 in Chapter 6 or Table 6.2 in CAN/CSAS614:2014 is a step in the right direction to establish a closer and more explicit linkage between geotechnical information and design. Overall, one expects geotechnical practice to be impacted positively if information can be assigned a "value" in a defensible way.

1.5.3   Model uncertainties

The model factor for the capacity of a foundation is commonly defined as the ratio of the measured (or interpreted) capacity (Q_m) to the calculated capacity (Q_c), i.e. $M = Q_m/Q_c$. The value $M = 1$ implies that calculated capacity matches the measured capacity exactly, which is unlikely for all design scenarios. Intuition would lead us to think that M takes a different value depending on the design scenario. This intuitive observation is supported by a large number of model factor studies (see Chapter 5). It is straightforward to apply this simple definition to other responses beyond foundation capacity. Ideally, a calculation model should capture the key features of the physical system, and the remaining difference between the model and reality should be random in nature because it is caused by numerous minor factors that were left out of the model. Hence, it is reasonable to represent M as a random variable. The probability model of M (a lognormal distribution is usually adequate) is a description of these random differences resulting from model idealisations. Chapter 5 provides a useful compilation of model factor statistics for both ultimate and serviceability limit states.

This approach is entirely empirical, but it is practical way to emphasize the link between a model and reality. Engineers can easily relate to and adopt this model factor approach. It is obvious that M is a function of the prediction model (or calculation method) and to a lesser extent, the definition of the capacity on a measured load-displacement curve. Because of its empirical basis, it is potentially possible for the distribution and statistics of M to be related to the calibration load test database. It is highly recommended to validate the distribution of M or at least its statistics against an independent load test database to gauge their applicability beyond scenarios not covered in the calibration database. The model factor M is certainly *not* a function of a response such as bearing capacity or lateral capacity. Phoon & Kulhawy (2005) demonstrated using a large load test database that the mean of M for the lateral capacity of a rigid drilled shaft is function of the prediction model (five models for undrained mode, four models for drained mode) and the capacity interpretation method [lateral or moment limit (H_L), hyperbolic limit (H_h)]. Reported model statistics for a response without reference to a specific prediction model are probably ball-park figures of unknown accuracy based on experience. This is amply illustrated by the diversity of results for different models listed in Chapter 5.

Phoon (2005) demonstrated that the existing factor of safety for laterally loaded rigid drilled shafts cannot be compared when the capacity is calculated using different calculation models. Phoon (2005) illustrated this well-known limitation using a simple design example: shaft diameter = 1 m, shaft length/diameter = 5, load eccentricity = 0.5 m, a uniform undrained shear strength profile = 50 kPa, and an applied load (F) = 200 kN. The factors of safety (H_u/F), in which H_u is the lateral

Table 1.3 Values of the model factor $\gamma_{R;d}$ in the new French code for deep foundation, AFNOR (2012) (Frank 2015).

Pile type	Pressuremeter test (PMT) method		Cone penetration test (CPT) method	
	Compression	Tension	Compression	Tension
All piles, except injected piles and piles embedded in chalk	1.15	1.4	1.18	1.45
Piles embedded in chalk, except injected piles	1.4	1.7	1.45	1.75
Injected piles	2.0	2.0	2.0	2.0

capacity calculated from Reese, Broms, and Randolph & Houlsby models, are 3.1, 1.7, and 3.4, respectively. If one were to adjust the calculated capacity by the mean model factors reported in Phoon & Kulhawy (2005), the revised factors of safety (H_m/F) are 2.8, 2.6, and 2.9, respectively for $H_m = H_L$ and 4.3, 4.0, and 4.5, respectively for $H_m = H_b$. A drilled shaft designed using Broms method has to be significantly larger to achieve the same factor of safety (H_u/F), because the degree of conservatism intrinsic in the method is not included. Hence, it is useful to consider the mean model factor in design, even within our existing allowable stress design framework. BS EN 1997–1:2004, Table A.11 "Correlation factors ξ to derive characteristic values from dynamic impact tests (n – number of tested piles)" noted that the correlation factors should be multiplied by a model factor = 0.85, 1.10, and 1.20 for the following three methods of interpreting the ultimate compressive resistance from dynamic impact tests: dynamic impact tests with signal matching, a pile driving formula with measurement of the quasi-elastic pile head displacement during the impact, and a pile driving formula without measurement of the quasi-elastic pile head displacement during the impact, respectively. There are other references to model factors in BS EN 1997–1:2004, but no values are recommended in this British standard. It is useful to point out that the model factor in EN 1997–1:2004 $(\gamma_{R;d})$ and M differ in one fundamental aspect; the former is number while the latter is a random variable. From the "design point" perspective offered in Section 1.5.7 below, $\gamma_{R;d}$ can be consistently interpreted as the reciprocal of M at the design point. Frank (2015) recommended some values of $\gamma_{R;d}$ for use in the French standard in Table 1.3.

Vardanega & Bolton (2016) observed that "RBD is applied to the ultimate failure of the soil, rather than to the onset of disappointing deformations that later develop into serviceability issues, and then ultimately threaten structural collapse only if nothing has been done to interrupt the loading process or enhance the soil-foundation system. In that sense, the rigid demarcation between serviceability limit state (SLS) and ultimate limit state (ULS) failures in limit state design is unrealistic and unhelpful for a designer wishing to apply risk-based concepts." They further observed that the "challenge for geotechnical practitioners is not only to make settlement predictions, but to do so within a rigorous statistical framework." RBD can be applied to any performance function. It has understandably been applied to ULS and SLS as these limit states are commonly accepted to be important in prevailing design codes. If the geotechnical community were to consider that deformation checks are more important, it would be even more important to characterize model uncertainties associated with deformation calculations. Simpson et al. (1981) noted that structural engineers are often unsure about the confidence that geotechnical engineers actually place on their predictions of ground deformations. Vardanega & Bolton (2016) voiced a similar sentiment that deformation checks are less scrutinized and less validated than capacity checks associated with critical slip surfaces. They recommended that the mobilizable strength design (MSD) method provides engineers with a simple but realistic tool to calculate ground deformations, but a correction factor is needed to match finite element predictions. Zhang et al. (2015) extended the correction factor to a model factor in a two-step process: (a) correct MSD by finite element analysis and (b) correct finite element analysis by field measurements.

1.5.4 Scarcity of geotechnical data

This discussion is restricted to the issue of small sample size. The cost of a small sample is rationally reflected in the larger COV produced by statistical uncertainties. It is now possible to estimate statistical uncertainties even for random field parameters (Ching et al. 2016a), which is an important advance because spatial variability is a distinctive feature of geotechnical data (see Section 1.5.6 below). The impact of statistical uncertainties was found to be significant in design (Ching et al. 2016b). This is an important practical result, because it assigns a cost to the prevalent practice of keeping site investigation to a minimum. It is useful to note that EN 1997–1:2004 has considered this in the relation to "correlation factors" for pile tests. Tables A.9, A.10, and A.11 clearly increases the degree of conservatism when the number of tests decreases. The statistical basis for these correlation factors is discussed in Bauduin (2001) and Orr (2015).

It is also possible to tackle, at least partially, the difficulty of small sample size using Bayesian methods and prior knowledge (Wang and Cao 2013, Cao and Wang 2014, Wang et al. 2016, Cao et al. 2016). The prior knowledge shall include, but not limited to, engineering judgment, the designer's specific knowledge of the site, the ground conditions and their possible variability, and an appropriate assimilation and compilation of data from all available sources (including published literature, collection of comparable case histories and test results), as pointed out by Simpson (2011) and discussed in Section 1.4. The prior knowledge may be quantified in a rational and consistent manner using subjective probability assessment framework (Cao et al. 2016) and further integrated with the small-size samples from a specific site for providing the integrated knowledge on ground properties. Then, Markov chain Monte Carlo simulation may be used to transform the integrated knowledge into a large number of equivalent samples of ground properties and to bypass the difficulty in the analysis due to small sample size. Similar to traditional geotechnical practice in which the use of engineering judgment and prior knowledge is a critical element, engineering judgment and prior knowledge may play a vital role when dealing with the issue of small sample size in geotechnical RBD. Examples of using prior knowledge and Bayesian methods to deal with the issue of small sample size are given in Sections 3.9 and 7.8.

The presence of "geologic surprises" due to scarcity of data belongs to the category of "unknown unknowns". The extreme form of "geologic surprises" that may lead to disproportionately large consequences and cannot be reasonably anticipated from geologic considerations and prior experience are examples of "black swans". They cannot be captured by statistical uncertainties. It is more sensible to deal with these perhaps using a design strategy robust against such unpleasant surprises, but there are limits to robust design as well. Robust design and RBD are complementary as shown in Annex F of ISO2394:2015.

1.5.5 Probability distributions that accommodate a "worst credible" value at a prescribed quantile

Simpson (2011) urged engineers to "consciously consider the worst situations and parameter values that could be imagined on the basis of a reasonable and well informed engineering assessment." Simpson et al. (1981) termed such a value or situation the "worst credible" and suggested that "it might be assumed to have a 1 in 1000 chance of occurrence, on the basis that designers would be unlikely to be able to believe that anything more remote might happen." We defer the discussion on "worst credible situation" to Section 1.5.6. We focus on the "worst credible value" in this sub-section. For concreteness, we define a "worst credible value" as a 0.1% quantile, in other words, there is less than one in a thousand chance of finding values smaller than this "worst credible value". It is unlikely to encounter this fairly extreme value in one particular site, because the available sample size is typically not large enough. Vardanega & Bolton (2016) noted that "any required inference of extreme values, beyond the predictive limits of whatever data have been encountered on site, would have to appeal to a wider regional experience of severe deviations." The question can now be framed in a concrete way: "are there probability distributions that can accommodate a worst credible value derived from experience or judgment while respecting the objective but limited measured data at hand?"

It is easier for an engineer to understand this question from a slightly different perspective of imposing a lower limit on the probability distribution. A lower limit is a "worst value", because it is not physically or theoretically possible to produce values lower than this limit. In other words, it is a zero percent quantile, which can be very different in value from a 0.1% quantile due to strong nonlinearity at the probability tail. An obvious lower limit is zero, because many physical variables are positive valued. The lognormal distribution is commonly adopted to satisfy this limit. It is possible to use a more general shifted lognormal distribution to satisfy a lower limit larger than zero (e.g., overconsolidation ratio is larger than 1 by definition). In fact, any three-parameter distribution can be adopted to fit a prescribed lower limit in addition to the mean and COV of the data. This is a concrete example to reinforce the point that if the normal/lognormal distribution proves to be overly simplistic for the problem at hand, there are more realistic distributions available. Ching & Phoon (2015) showed that the significantly more challenging problem of fitting a multivariate probability distribution that respects both lower and upper limits can be addressed using the Johnson system of distributions.

It is also possible to use Bayesian methods to integrate the worst credible value derived from subjective experience or judgment (as prior knowledge) with the objective but limited measured data from a specific site for providing probability distributions that represent the integrated knowledge (Wang and Cao 2013, Cao and Wang 2014,

Wang et al. 2016, Wang and Aladejare 2016). Examples of generating probability distributions from Bayesian methods using both engineering experience or judgment and limited site-specific measured data are given in Sections 3.9 and 7.8. For example, a uniform probability distribution varying between the worst and best credible values may be used to quantitatively represent the prior knowledge when using Bayesian methods. When more informative prior knowledge is available from engineering experience or judgment, more sophisticated probability distributions can be obtained using subjective probability assessment framework to represent the informative prior knowledge (Cao et al. 2016).

To the knowledge of the author, there is no solution to the original question of fitting a prescribed quantile along with the mean and COV of the data in the multivariate context. The prescribed quantile or lower limit may be uncertain. This is an example of how a conversation on placing geotechnical needs at the forefront of geotechnical RBD can stimulate future research in more fruitful directions.

1.5.6 Spatial variability

Spatial variability is a common subsurface feature that can be handled systematically using random fields. The realizations emerging from random field simulations can be viewed as possible interpolations between boreholes. Extensive random finite element studies have shown that spatially variable or heterogeneous soils can produce complex failure mechanisms that are more critical than classical failure mechanisms developed in homogeneous or layered soils (Fenton & Griffiths 2008). Simpson (2011) spoke of worst credible situations and parameter values as situations/values that could be imagined on the basis of a reasonable and well informed engineering assessment. Note that it is potentially more important to "imagine" worst credible situations than worst credible parameter values. In addition, there are many worst credible situations, even if one were to focus on 1 in 1000 chance of occurrence.

The author submits that random field simulations can "imagine" these situations more systematically from borehole data than an engineer. The finite element analysis can produce more appropriate failure mechanisms consistent with mechanics and boundary conditions than an engineer. Rather than applying engineering judgment to imagining possible subsurface conditions and/or failure mechanisms, the engineer can focus on discerning the realism of these computational outputs based on his/her experience and his/her appreciation of the geological setting of the site.

EN 1997–1:2004, 2.4.5.2(2) recommends that the "characteristic value of a geotechnical parameter shall be selected as a cautious estimate of the value affecting the occurrence of the limit state." Much attention has been focused on how to obtain a "cautious estimate". For example, EN 1997–1:2004, 2.4.5.2(11) notes that "If statistical methods are used, the characteristic value should be derived such that the calculated probability of a worse value governing the occurrence of the limit state under consideration is not greater than 5%." A note to this clause clarifies that "In this respect, a cautious estimate of the mean value is a selection of the mean value of the limited set of geotechnical parameter values, with a confidence level of 95%; where local failure is concerned, a cautious estimate of the low value is a 5% fractile." There is less discussion on the "value affecting the occurrence of the limit state", particularly in the geotechnical reliability literature. One notes that the occurrence of a limit state in terms of its physical manifestation as a critical slip surface is dependent on spatial variability. The value affecting this occurrence is thus dependent on spatial variability. It is the 5% quantile of this value, rather than the 5% quantile of the borehole data (unrelated to any critical slip surface), that is relevant. The interaction between spatial variability and formation of critical slip surfaces is complex and it is not surprising to find that the 5% quantile of the "value affecting the occurrence of the limit state" is not related to the 5% quantile of the borehole data in a straightforward way (Ching & Phoon 2013b, Ching et al. 2014c, Hu & Ching 2015, Ching et al. 2016c). A realistic assessment of the characteristic value in the context of spatial variability is certainly beyond the reach of judgment. Tietje et al. (2014) quantified the characteristic shear strength along a failure surface in the spirit of EN 1997–1:2004, 2.4.5.2(11) using random field simulations.

It is useful to note that characteristic values are associated with uncertainties in the input parameters, while a reliability index is associated with uncertainties in a response (for a component) or a set of responses (for a system). The design value is defined as the ratio of a characteristic value to an appropriate material partial factor. Because the material partial factor is standardized to a single numerical value regardless of the design scenario, one suspects that the characteristic value serves as a venue for "site effects" to be included in the design. Ching & Phoon (2011) developed a Quantile Value Method (QVM) for RBD that bears some similarities to the characteristic value approach. However, Ching & Phoon (2011) demonstrated that a prescribed target reliability index cannot be achieved using a fixed quantile, such as the 5% quantile. The author hastens to add that there is no intent to achieve a prescribed target reliability index through the characteristic value and the partial factor in EN 1997–1:2004.

1.5.7 Design point from the first-order reliability method (FORM) and partial factors

The first-order reliability method or FORM is presented in standard text (Phoon 2008) and it will not be repeated here other than to draw attention to its practicality in computing the reliability index and its usefulness to complement partial factors. FORM typically requires less than ten evaluations of the performance functions. Hence, even if the performance function is defined implicitly by a finite element analysis, it is still computationally comparable to routine parametric studies undertaken in design offices. Engineers do not need to understand the mathematical details underlying FORM as it has been implemented in an EXCEL spreadsheet (Low 2008). In other words, engineers only need to acquire sufficient conceptual understanding of FORM to perform actual reliability-based design properly with the aid of software tools.

The role of the engineer should be focused on characterizing the statistical inputs, selecting an appropriate performance function, and interpreting the outputs. There are numerous outputs that are useful. FORM provides a quantitative index of the sensitivity of a response (e.g., pile head lateral deflection) to each random variable. It is possible to do this partially using conventional parametric studies, but these studies do not cover correlated variables – a concept not found in deterministic analysis but correlation is a characteristic of all geotechnical data (Section 1.5.1). It is important to appreciate sensitivity, because it may direct our data collection efforts. It may not be necessary collect more data to characterize an insensitive variable, including its statistics.

FORM produces a "design point" which is called the most probable failure point ("failure" in reliability parlance means unsatisfactory performance). For a reliability index of 3 which corresponds to a probability of failure of about 1 in 1000, the design point can be interpreted as a "worst credible" situation (or more accurately, the most probable worst credible situation). The design point consists of one design value for each random variable. For example, Low and Phoon (2015) describes a footing problem (base of a retaining wall) with width $= 4.51$ m, length $= 25$ m, depth of embedment $= 1.8$ m, resting on silty sand with a friction angle $= \phi$ and cohesion $= c$. The means of ϕ and c are 25° and 15 kN/m^2, respectively. The COVs of ϕ and c are 10% and 20%, respectively. The strength random variables, ϕ and c, are negatively correlated with a Spearman coefficient $= -0.5$. It is subjected to a horizontal load (Q_h) of mean value $= 300$ kN/m applied at a point 2.5 m above the base and a centrally applied vertical load (Q_v) of mean $= 1100$ kN/m. The COVs of Q_h and Q_v are 15% and 10%, respectively. The load random variables, Q_h and Q_v, are positively correlated with a Spearman coefficient $= 0.5$. For a reliability index $= 3$, the design point yields the following design values: 20.8° for ϕ, 15.2 kN/m^2 for c, 412.6 kN/m for Q_h, and 1184.7 kN/m for Q_v. It can be readily shown that these design values are 1.7σ below mean for ϕ, about 0.1σ *above* mean for c, 2.5σ above mean for Q_h, and 0.8σ above mean for Q_v, in which σ is the standard deviation of the respective random variables. These design values can be converted to quantile values as well to get a sense of how "extreme" these values are: 3.6% for ϕ, 56.6% for c, 98.6% for Q_h and 78.6% for Q_v. The most extreme design values occur for ϕ and Q_h. It is not possible to obtain a 3.6% quantile from say less than 10 samples, but indirect correlation with cone tip resistance can produce sufficient samples to estimate this quantile if transformation uncertainty is accounted for. A 98.6% quantile load corresponds to a 71-year return period load. The estimate of this quantile is within range of available historical load data.

Another perspective is to divide a characteristic value by the design value to produce the FORM-based partial factor for a particular variable. For ϕ, the characteristic value based on a 5% quantile $= 21.1°$. Hence, the partial factor for $\phi = 21.1/20.8 = 1.01$ [the answer is almost the same for tan(ϕ)]. This FORM-based partial factor will change according to all input statistics (not merely the COV of ϕ) and the performance function. For comparison, the partial factor for tan(ϕ) in BS EN 1997–1:2004, Table A.2, is 1.25. Orr (2015) noted that taking the 5% quantile for the characteristic value is too conservative by citing the note associated with EN 1997–1:2004, 2.4.5.2(11) relating to a cautious estimate of the mean value. If one adopts Schneider (1997)'s definition of a characteristic value being half a standard deviation below the mean, the revised characteristic value for $\phi = 23.7°$ and the revised partial factor $= 23.7/20.8 = 1.14$. Schneider & Schneider (2013) further extended the half standard deviation rule of thumb to include the reduction of COV due to spatial averaging (see Section 3.6).

One expects the "worst credible" situation to be produced by strength variables taking values below their means and load variables taking values above their means. At first glance, it is surprising to find the design value for c to be slightly above its mean value. However, a closer inspection would reveal that this bearing capacity problem is sensitive to ϕ and hence the most probable worst credible situation involves ϕ taking a value below the mean (1.7 times standard deviation below the mean). However, because ϕ and c are negatively correlated (this is created by linearization of the failure envelope that forces small ϕ to be associated with large c and vice-versa), the design value of c must be "large" when the design value of ϕ is "small". However, it is not possible to "judge" if the design value of c would appear below or above the mean without a FORM analysis. It may be possible to apply engineering judgment to assess a single worst credible value, but it will be increasingly difficult to rely on judgment alone to assess multiple worst credible values (this is a common design task) that are consistent with the correlation structure embedded in the site investigation data. There is no reason to burden engineering judgment as the design point will produce a set of worst credible values that is fully consistent with the input correlation matrix. Overall, a comparison between Eurocode partial factors and FORM-based partial factors can produce further insights into the design.

1.5.8 System reliability

System reliability needs to be addressed for problems that exhibit multiple distinct but correlated failure modes. This situation is clearly the norm rather than the exception in geotechnical systems. In fact, slope stability is an intrinsically system reliability problem. FORM identifies only the "most probable" critical slip surface in a slope and the failure probability so computed can be significantly underestimated because contributions from the second most probable critical slip surface, third most probable critical slip surface, and so on have not been included (Ching et al. 2009, Wang et al. 2011, Zhang et al. 2011).

Chapter 7 presents a gravity retaining wall example with three failure modes (i.e., sliding, overturning and bearing capacity failure). These failure modes tend to interact among each other, because loads and resistances for different failure modes are correlated. For example, self-weight of a gravity retaining wall, which is the major source of resistance against sliding and overturning failure modes, but at the same time, is also a major source of load for the bearing capacity failure mode. It is more straightforward to solve this system reliability problem using direct simulation. Section 7.6 mentions an EXCEL-based software package called UPSS (Uncertainty Propagation using Subset Simulation) that can carry out direct simulation efficiently. It may be possible to incorporate a "system" factor into a simplified RBD format for problems involving failure modes that can be standardized. However, the difficulty of retaining the simplified RBD format heightens when system reliability is coupled to spatial variability. Some system reliability methods have been developed recently to address the variation of slip surfaces, particularly when the spatial variability of soil properties is modelled in the analysis (Zhang et al. 2011, Li et al. 2013, Li et al. 2014), but they have not been simplified into the familiar LRFD or partial factor format at this point in time.

1.6 CONCLUDING THOUGHTS

Notwithstanding the unique features and conditions of geotechnical practice, the author submits that there are merits for the geotechnical community to adopt reliability as a basis for design. It is acknowledged that geotechnical engineers have to grapple with a heightened state of uncertainty and associated risk, parts of which may not be amenable to mathematical treatment, due to the complex and variable nature of natural geomaterials that cannot be readily standardized compared to made-to-order structural materials. Site-specificity is an important consideration in geotechnical practice. In this context, it is sensible for geotechnical engineers to seek clarity on more fundamental design considerations and to devote less attention to performance verification which is only one step, albeit an important one, in design.

The relevance of RBD, simplified or otherwise, is still being debated in an exchange clouded occasionally by confusion due to the evolving discussion on fundamental issues at hand and their fairly tangled relationship to performance verification. The author ventures to suggest that some of the reservations are based on the misconception that reliability is a panacea for all afflictions affecting design calculations based on the factor of safety or geotechnical practice in general. Other reservations relate to the scarcity and/or incompleteness of available information. The former includes a misconceived notion that engineering judgment is no longer necessary. This is not true. Reliability analysis is merely one of the many mathematical methods routinely applied to model the complex real-world for engineering applications. It is susceptible to abuse in the absence of sound judgment in the same manner as a finite element analysis. The importance of engineering judgment clearly has not diminished with the growth of theory and computational tools. However, its role has become more focused on those design aspects that remain outside the scope of theoretical analyses. For example, reliability calculations could relieve engineering judgment from the unsuitable task of performance verification in the presence of uncertainties so that the engineer can focus on asking the right questions, selecting the appropriate models and parameters for calculations, and performing reality check on the results.

The second reservation concerning scarce and/or incomplete information is possibly a reaction to the application of overly simplistic assumptions and methods in parts of the geotechnical reliability literature. The "simple" reliability analysis arose in part because geotechnical RBD grew from structural LRFD and more attention may have been paid in the past to reliability calculations than to developing geotechnical RBD with practical geotechnical needs at the forefront. It is timely for the geotechnical RBD community to look into how we can improve our state of practice to cater to the distinctive needs of geotechnical engineering practice. In fact, Annex D of ISO2394:2015 "Reliability of geotechnical structures" has been drafted with this central intent in mind.

The key point here is that the reservations expressed above should be viewed as providing approximate boundaries circumscribing the limits of reliability calculations *or* acting as a caution against overly simplistic reliability applications that do not respect geotechnical needs or constraints, rather than invalidating reliability principles as a whole. It is important to emphasize the need to apply reliability principles judiciously in conjunction with other design/construction strategies. In other words, RBD does not preclude or displace existing elements of good practice and engineering judgment, which evolved to handle a moderate degree of "unknown unknowns". It plays a useful complementary role. For example, RBD is very useful in handling complex real-world information (multivariate correlated data) and information imperfections (scarcity of information or incomplete information). It is also very useful in handling real-world design dimensions such as spatial variability and system reliability that cannot be easily treated using deterministic means.

Notwithstanding the merits of using non reliability-based partial factors (all methods must have their pluses along with minuses), it suffices to note at this point that structural and geotechnical design cannot be bridged by an empirical basis. Using ISO2394:2015 as a concrete example, it is clear that there is no practical way for geotechnical design unsupported by a reliability basis to fit in. It is also difficult for our geotechnical profession to benefit from the ongoing and pervasive information technology revolution in big data and data analytics in the absence of a rational framework, be it reliability or otherwise. Using Annex D as an important starting point, this book would hopefully stimulate a conversation in the wider geotechnical community on how to improve our state of practice in geotechnical reliability-based design, particularly in what ways we can revise our current RBD codes to take cognizance of the distinctive features and needs of geotechnical engineering practice.

This book elaborates the key aspects of geotechnical design that are outlined in Annex D of ISO2394:2015. It provides background information to substantiate the special considerations needed for the use of reliability concepts for geotechnical structures, as well as illustrations of approaches and procedures for uncertainty representation and the implementation of geotechnical reliability-based design. At the same time it should be noted that Annex D is fully consistent and compliant with the general principles of reliability given by ISO2394:2015. The standard therefore provides an overall framework for the advancement of geotechnical practice that is consistent with the principles of reliability within the wider scope of buildings, infrastructure and civil engineering works. A coherent approach is provided by the standard for reliability based decision-making and design, as derived from optimized risk, expressed as performance models that accounts for the levels of knowledge and uncertainty that applies to the field of application under consideration. An outline of the standard is given in Chapter 2, as seen from the perspective of geotechnical reliability based design.

ACKNOWLEDGMENTS

The authors are grateful for the valuable comments provided by Professor Jianye Ching, Johan Retief, and Yu Wang.

This chapter is also shaped in part by a discussion paper entitled "The merits of reliability calculations" that was circulated by the author before his presentation at the 13th meeting of the ISSMGE TC205 Technical committee: 'Safety and Serviceability in Geotechnical Design', 25th February 2016. The author attempted to circumscribe the applicability of RBD with reference to "black swans" and offered the suggestion that some "grey swans" could be handled using robust reliability-based design (RRBD). The author is grateful for the opportunity to share his views through this invitation extended by Dr Brian Simpson (Chair, TC205) and to learn from the distinguished practitioners in this meeting.

REFERENCES

AFNOR (2012) *NF P 94 282. Justification des ouvrages géotechniques – Normes d'application nationale de l'Eurocode 7 – Fondations profondes.* Paris, French Standard, AFNOR.

Allen, T.M. (2013) AASHTO geotechnical design specification development in the USA. In: Arnold, P., Fenton, G.A., Hicks, M.A., Schweckendiek, T. & Simpson, B. (eds.) *Modern Geotechnical Design Codes of Practice: Implementation, Application and Development.* Amsterdam, IOS Press. pp. 243–260.

Baike, L.D. (1985) Total and partial factors of safety in geotechnical engineering. *Canadian Geotechnical Journal*, 22 (4), 477–482.

Bauduin, C. (2001) Design procedure according to Eurocode 7 and analysis of test results. In: *Proceedings, Symposium on Screw Pile: Installation and Design in Stiff Clay*. Rotterdam, Balkema. pp. 275–303.

Beal, A.N. (1979) What's wrong with load factor design? *Proceedings of the Institution of Civil Engineers*, 66 (Pt 1), 595–604.

Burland, J.B. (2008a) 'Reflections on Victor de Mello, friend, engineer and philosopher.' First Victor de Mello Lecture. *Soils and Rocks*, 31 (3), 111–123.

Burland, J.B. (2008b) The founders of geotechnique. *Geotechnique*, 58 (5), 327–341.

Burlon, S., Frank, R., Baguelin, F., Harbert, J. & Legrand, S. (2014) Model factor for the bearing capacity of piles from pressuremeter test results – Eurocode 7 approach. *Geotechnique*, 64 (7), 513–525.

Cao, Z.J. & Wang, Y. (2014) Bayesian model comparison and characterization of undrained shear strength. *ASCE Journal of Geotechnical and Geoenvironmental Engineering*, 140 (6), 04014018, 1–9.

Cao, Z.J., Wang, Y. & Li, D. (2016) Quantification of prior knowledge in geotechnical site characterization. *Engineering Geology*, 203, 107–116.

CSA S408:2011. *Guidelines for the Development of Limit States Design Standards*. Mississauga, ON, Canadian Standards Organization.

CAN/CSAS614:2014. *Canadian Highway Bridge Design Code*. Mississauga, ON, Canadian Standards Organization.

Chilès, J.-P. & Delfiner, P. (1999) *Geostatistics: Modeling Spatial Uncertainty*. New York, John Wiley & Sons.

Ching, J. & Phoon, K.K. (2011) A quantile-based approach for calibrating reliability-based partial factors. *Structural Safety*, 33, 275–285.

Ching, J. & Phoon, K.K. (2012) Modeling parameters of structured clays as a multivariate normal distribution. *Canadian Geotechnical Journal*, 49 (5), 522–545.

Ching, J. & Phoon, K.K. (2013a) Multivariate distribution for undrained shear strengths under various test procedures. *Canadian Geotechnical Journal*, 50 (9), 907–923.

Ching, J. & Phoon, K.K. (2013b) Probability distribution for mobilized shear strengths of spatially variable soils under uniform stress states. *Georisk*, 7 (3), 209–224.

Ching, J. & Phoon, K.K. (2014a) Transformations and correlations among some clay parameters – The global database. *Canadian Geotechnical Journal*, 51 (6), 663–685.

Ching, J. & Phoon, K.K. (2014b) Correlations among some clay parameters – The multivariate distribution. *Canadian Geotechnical Journal*, 51 (6), 686–704.

Ching, J. & Phoon, K.K. (2015) Constructing multivariate distribution for soil parameters. In: *Risk and Reliability in Geotechnical Engineering*. Boca Raton, CRC Press, pp. 3–76.

Ching, J., Phoon, K.K. & Hu, Y.G. (2009) Efficient evaluation of reliability for slopes with circular slip surfaces using importance sampling. *Journal of Geotechnical and Geoenvironmental Engineering*, 135 (6), 768–777.

Ching, J., Phoon, K.K. & Chen, C.H. (2014a) Modeling CPTU parameters of clays as a multivariate normal distribution. *Canadian Geotechnical Journal*, 51 (1), 77–91.

Ching, J., Phoon, K.K. & Yu, J.W. (2014b) Linking site investigation efforts to final design savings with simplified reliability-based design methods. *ASCE Journal of Geotechnical and Geoenvironmental Engineering*, 140 (3), 04013032.

Ching, J., Phoon, K.K. & Kao, P.H. (2014c) Mean and variance of the mobilized shear strengths for spatially variable soils under uniform stress states. *ASCE Journal of Engineering Mechanics*, 140 (3), 487–501.

Ching, J., Wu, S.S. & Phoon, K.K. (2016a) Statistical characterization of random field parameters using frequentist and Bayesian approaches. *Canadian Geotechnical Journal*, 53 (2), 285–298.

Ching, J., Phoon, K.K. & Wu, S.H. (2016b) Impact of statistical uncertainty on geotechnical reliability estimation. *ASCE Journal of Engineering Mechanics*, 04016027.

Ching, J., Lee, S.W. & Phoon, K.K. (2016c) Undrained strength for a 3D spatially variable clay column subjected to compression or shear. *Probabilistic Engineering Mechanics*, 45, 127–139.

Dunnicliff, J. & Deere, D.U. (1984) *Judgment in Geotechnical Engineering: The Professional Legacy of Ralph B. Peck*. New York, Wiley. 332 pp.

Ellingwood, B.R., Galambos, T.V., MacGregor, J.G. & Cornell, C.A. (1980) *Development of Probability-Based Load Criterion for American National Standard A58, Special Publication 577*. Washington, National Bureau of Standards.

EN 1990:2002. *Eurocode – Basis of Structural Design*. Brussels, European Committee for Standardization (CEN).

EN 1997-1:2004. *Eurocode 7: Geotechnical Design – Part 1: General Rules*. Brussels, European Committee for Standardization (CEN).

Faber, M.H. (2015) Codified risk informed decision making for structures. In: *Symposium on Reliability of Engineering Systems (SRES2015), Hangzhou, China*.

Fenton, G.A. & Griffiths, D.V. (2008) *Risk Assessment in Geotechnical Engineering*. New York, John Wiley & Sons.

Fenton, G.A., Naghibi, F., Dundas, D., Bathurst, R.J. & Griffiths, D.V. (2016) Reliability-based geotechnical design in the 2014 Canadian Highway Bridge Design Code. *Canadian Geotechnical Journal*, 53 (2), 236–251.

Fleming, W.G.K. (1989) Limit state in soil mechanics and use of partial factors. *Ground Engineering*, 22 (7), 34–35.

Focht Jr., J.A. (1994) Lessons learned from missed predictions. *ASCE Journal of Geotechnical Engineering*, 120 (10), 1653–1683.

Frank, R. (2015) The new French standard for the application of Eurocode 7 to deep foundations. In: *Proceedings, European Conference in Geo-Environment and Construction, Tirana, Albania, 26–28 Nov 2015*. pp. 318–327.

Gong, W., Khoshnevisan, S., Juang, C.H. & Phoon, K.K. (2016) R-LRFD: Load and Resistance Factor Design considering design robustness. *Computers and Geotechnics*, 74, 74–87.

Hansen, J.B. (1953) *Earth Pressure Calculation*. Copenhagen, The Danish Technical Press.

Hansen, J.B. (1956) *Limit State and Safety Factors in Soil Mechanics*. Copenhagen, Bulletin No. 1, Danish Geotechnical Institute.

Hansen, J.B. (1965) Philosophy of foundation design: Design criteria, safety factors and settlement limits. In: *Proceedings, Symposium on Bearing Capacity and Settlement of Foundations*. Durham, Duke University. pp. 1–13.

Honjo, Y. & Kusakabe, O. (2002) Proposal of a comprehensive foundation design code: Geo-Code 21 ver. 2. In: *Proceedings, International Workshop on Foundation Design Codes and Soil Investigation in View of International Harmonization and Performance Based Design*. Lisse, Balkema. pp. 95–103.

Honjo, Y., Kieu Le, T.C., Hara, T., Shirato, M., Suzuki, M. & Kikuchi, Y. (2009) Code calibration in reliability based design level I verification format for geotechnical structures. In: *Proceedings, Second International Symposium on Geotechnical Safety & Risk*. Leiden, CRC Press/Balkema. pp. 435–452.

Honjo, Y., Kikuchi, Y. & Shirato, M. (2010) Development of the design codes grounded on the performance-based design concept in Japan. *Soils and Foundations*, 50 (6), 983–1000.

Hu, Y.G. & Ching, J. (2015) Impact of spatial variability in soil shear strength on active lateral forces, *Structural Safety*, 52, 121–131.

ISO2394:1973/1986/1998/2015. *General Principles on Reliability for Structures*. Geneva, International Organization for Standardization.

Juang, C.H., Wang, L., Liu, Z., Ravichandran, N., Huang, H. & Zhang, J. (2013a) Robust geotechnical design of drilled shafts in sand: New design perspective. *ASCE Journal of Geotechnical and Geoenvironmental Engineering*, 139 (12), 2007–2019.

Juang, C.H., Wang, L., Khoshnevisan, S. & Atamturktur, S. (2013b) Robust geotechnical design – Methodology and applications. *Journal of GeoEngineering*, 8 (3), 71–81.

Kulhawy, F.H. & Phoon, K.K. (1996) Engineering judgment in the evolution from deterministic to reliability-based foundation design. In: *Uncertainty in the Geologic Environment – From Theory to Practice (GSP 58)*. New York, ASCE. pp. 29–48.

Kulhawy, F.H. & Phoon K.K. (2002) Observations on geotechnical reliability-based design development in North America. In: *Proceedings, International Workshop on Foundation Design Codes and Soil Investigation in View of International Harmonization and Performance Based Design*. Lisse, Balkema. pp. 31–48.

Li, L., Wang, Y., Cao, Z.J. & Chu, X. (2013) Risk de-aggregation and system reliability analysis of slope stability using representative slip surfaces. *Computers and Geotechnics*, 53, 95–105.

Li, L., Wang, Y. & Cao, Z.J. (2014) Probabilistic slope stability analysis by risk aggregation. *Engineering Geology*, 176, 57–65.

Low, B.K. (2008) Practical reliability approach using spreadsheet. Chapter 3. In: Phoon, K.K. (ed.) *Reliability-Based Design in Geotechnical Engineering: Computations and Applications*. London, Taylor & Francis. pp. 134–168.

Low, B.K. & Phoon, K.K. (2015) Reliability-based design and its complementary role to Eurocode 7 design approach. *Computers and Geotechnics*, 65, 30–44.

Meyerhof, G.G. (1984) Safety factors and limit states analysis in geotechnical engineering. *Canadian Geotechnical Journal*, 21 (1), 1–7.

Nagao, T., Watabe, Y., Kikuchi, Y. & Honjo, Y. (2009) Recent revision of Japanese Technical Standard for Port and Harbor Facilities based on a performance based design concept. In: *Proceedings, Second International Symposium on Geotechnical Safety & Risk*. Leiden, CRC Press/Balkema. pp. 39–47.

Nathwani, J.S., Lind, N.C. & Pandey, M.D. (1997) *Affordable Safety by Choice: The Life Quality Method*. Waterloo, Institute for Risk Research, University of Waterloo.

Orr, T.L.L. (2015) Managing risk and achieving reliability geotechnical designs using Eurocode 7. Chapter 10. In: *Risk and Reliability in Geotechnical Engineering*. Boca Raton, CRC Press. pp. 395–433.

Ovesen, N.K. (1989) Geotechnical limit states design in Europe. In: *Proceedings, Symposium on Limit States Design in Foundation Engineering*. Toronto, Canadian Geotechnical Society. pp. 33–45.

Paikowsky, S.G., Birgisson, B., McVay, M., Nguyen, T., Kuo, C., Baecher, G.B., Ayyub, B., Stenersen, K., O'Malley, K., Chernauskas, L. & O'Neill, M. (2004) *Load and Resistance Factors Design for Deep Foundations*. NCHRP Report 507. Washington, DC, Transportation Research Board of the National Academies.

Paikowsky, S.G., Amatya, S., Lesny, K. & Kisse, A. (2009) Developing LRFD design specifications for bridge shallow foundations. In: *Proceedings, Second International Symposium on Geotechnical Safety & Risk*. Leiden, CRC Press/Balkema. pp. 97–102.

Paikowsky, S.G., Canniff, M.C., Lesney, K., Kisse, A., Amatya, S. & Muganga, R. (2010) *LRFD Design and Construction of Shallow Foundations for Highway Bridge Structures*. NCHRP Report 651. Washington, DC, Transportation Research Board of the National Academies.

Peck, R.B. (1980) 'Where has all the judgment gone?' The fifth Laurits Bjerrum memorial lecture. *Canadian Geotechnical Journal*, 17 (4), 584–590.

Petroski, H. (1993) Failure as source of engineering judgment: Case of John Roebling. *Journal of Performance of Constructed Facilities*, 7 (1), 46–58.

Petroski, H. (1994) *Design Paradigms: Case Histories of Error and Judgment in Engineering.* Cambridge, Cambridge University Press, U.K.

Phoon, K.K. (2005) Reliability-based design incorporating model uncertainties. In: *Proceedings, Third International Conference on Geotechnical Engineering Combined with Ninth Yearly Meeting of the Indonesian Society for Geotechnical Engineering, Semarang, Indonesia.* pp. 191–203.

Phoon, K.K. (2008) Numerical recipes for reliability analysis – A primer. Chapter 1. In: Phoon, K.K. (ed.) *Reliability-Based Design in Geotechnical Engineering: Computations and Applications.* London, Taylor & Francis pp. 1–75.

Phoon, K.K. & Kulhawy, F.H. (1999a) Characterization of geotechnical variability. *Canadian Geotechnical Journal,* 36 (4), 612–624.

Phoon, K.K. & Kulhawy, F.H. (1999b) Evaluation of geotechnical property variability. *Canadian Geotechnical Journal,* 36 (4), 625–639.

Phoon, K.K. & Kulhawy, F.H. (2005) Characterization of model uncertainties for laterally loaded rigid drilled shafts. *Geotechnique,* 55 (1), 45–54.

Phoon, K.K. & Kulhawy, F.H. (2008) Serviceability limit state reliability-based design. In: Phoon, K.K. (ed.) *Reliability-Based Design in Geotechnical Engineering: Computations and Applications.* London, Taylor & Francis, 344–383.

Phoon, K.K. & Ching, J. (2015) Is there anything better than LRFD for simplified geotechnical RBD? In: *Proceedings, 5th International Symposium on Geotechnical Safety and Risk (ISGSR2015), Rotterdam, Netherlands.* pp. 3–15.

Phoon, K.K., Kulhawy, F.H. & Grigoriu, M.D. (2003a) Development of a reliability-based design framework for transmission line structure foundations. *ASCE Journal of Geotechnical and Geoenvironmental Engineering,* 129 (9), 798–806.

Phoon, K.K., Becker, D.E., Kulhawy, F.H., Honjo, Y., Ovesen, N.K. & Lo, S.R. (2003b) Why consider reliability analysis in geotechnical limit state design? In: *Proceedings, International Workshop on Limit State design in Geotechnical Engineering Practice (LSD2003).* Cambridge, CDROM.

Phoon, K.K., Kulhawy, F.H. & Grigoriu, M.D. (2003c) Multiple resistance factor design (MRFD) for spread foundations. *ASCE Journal of Geotechnical and Geoenvironmental Engineering,* 129 (9), 807–818.

Phoon, K.K., Retief, J.V., Ching, J., Dithinde, M., Schweckendiek, T., Wang, Y. & Zhang, L.M. (2016). Some observations on ISO2394:2015 Annex D (Reliability of Geotechnical Structures), Structural Safety, 62, 24–33.

Prakoso, W.A. (2002) *Reliability-Based Design of Fndns. on Rock for Transmission Line & Similar Structure.* PhD Thesis. New York, Cornell University.

Ravindra, M.K. & Galambos, T.V. (1978) Load and resistance factor design for steel. *ASCE Journal of Structural Division,* 104 (ST9), 1337–1353.

SANS 10160-1:2011. *Basis of Structural Design.* Pretoria, South African National Standard, SABS.

Schneider, H.R. (1997) Definition and characterization of soil properties. In: *Proceedings, XIV ICSMGE, Hamburg.* Rotterdam, Balkema.

Schneider, H.R. & Schneider, M.A. (2013) Dealing with uncertainties in EC7 with emphasis on determination of characteristic soil properties. In: Arnold, P., Fenton, G.A., Hicks, M.A., Schweckendiek, T. & Simpson, B. (eds.) *Modern Geotechnical Design Codes of Practice: Implementation, Application and Development.* Amsterdam, IOS Press. pp. 87–101.

Schuppener, B. (2011) Reliability theory and safety in German geotechnical design. In: *Third International Symposium on Geotechnical Safety & Risk.* Germany, Federal Waterways Engineering and Research Institute. pp. 527–536.

Schuppener, B. (2013) The safety concept in German Design Codes. In: Arnold, P., Fenton, G.A., Hicks, M.A., Schweckendiek, T. & Simpson, B. (eds.) *Modern Geotechnical Design*

Codes of Practice: Implementation, Application and Development. Amsterdam, IOS Press. pp. 102–115.

Schweckendiek, T., Vrouwenvelder, T., Calle, E., Kanning, W. & Jongejan, R. (2013) Target reliabilities and partial factors for flood defenses in the Netherlands. In: Arnold, P., Fenton, G.A., Hicks, M.A., Schweckendiek, T. & Simpson, B. (eds.) *Modern Geotechnical Design Codes of Practice: Implementation, Application and Development*. Amsterdam, IOS Press. pp. 311–328.

Schweckendiek, T., Slomp, R. & Knoeff, H. (2015) New safety standards and assessment tools in the Netherlands. In: *Proceedings, Fifth Siegener Symposium "Sicherung von Dämmen, Deichen und Stauanlagen", Siegen, Germany*.

Scott, B., Kim, B.J. & Salgado, R. (2003) Assessment of current load factors for use in geotechnical load and resistance factor design. *ASCE Journal of Geotechnical and Geoenvironmental Engineering*, 129 (4), 287–295.

Simpson, B. (2011) Reliability in geotechnical design – Some fundamentals. In: *Proceedings, Third International Symposium on Geotechnical Safety & Risk*. Germany, Federal Waterways Engineering and Research Institute. pp. 393–399.

Simpson, B., Pappin, J.W. & Croft, D.D. (1981) An approach to limit state calculations in geotechnics. *Ground Engineering*, 14 (6), 21–28.

Tietje, O., Fitze, P. & Schneider, H.R. (2014) Slope stability analysis based on autocorrelated shear strength parameters. *Geotechnical and Geological Engineering*, 32 (6), 1477–1483.

Valsangkar, A.J. & Schriver, A.B. (1991) Partial and total factors of safety in anchored sheet pile design. *Canadian Geotechnical Journal*, 28 (6), 812–817.

Vardanega, P.J. & Bolton, M.D. (2016) Design of geostructural systems. *ASCE-ASME Journal of Risk and Uncertainty in Engineering Systems, Part A: Civil Engineering*, 2 (1), 04015017.

Vrouwenvelder, T. (1996) Revision of ISO 2394 General principles on reliability for structures. *IABSE Reports*, 74, 117–118.

Vrouwenvelder, T., van Seters, A. & Hannink, G. (2013) Dutch approach to geotechnical design by Eurocode 7, based on probabilistic analyses. In: Arnold, P., Fenton, G.A., Hicks, M.A., Schweckendiek, T. & Simpson, B. (eds.) *Modern Geotechnical Design Codes of Practice: Implementation, Application and Development*. Amsterdam, IOS Press. pp. 128–139.

Wang, Y. & Cao, Z.J. (2013) Probabilistic characterization of Young's modulus of soil using equivalent samples. *Engineering Geology*, 159, 106–118.

Wang, Y. & Aladejare, A.E. (2016) Bayesian characterization of correlation between uniaxial compressive strength and Young's modulus of rock. *International Journal of Rock Mechanics and Mining Sciences*, 85, 10–19.

Wang, Y., Cao, Z.J. & Au, S.K. (2011) Practical reliability analysis of slope stability by advanced Monte Carlo Simulations in spreadsheet. *Canadian Geotechnical Journal*, 48 (1), 162–172.

Wang, Y., Cao, Z.J. & Li, D. (2016) Bayesian perspective on geotechnical variability and site characterization. *Engineering Geology*, 203, 117–125.

Zhang, J., Zhang, L.M. & Tang, W.H. (2011) New methods for system reliability analysis of soil slopes. *Canadian Geotechnical Journal*, 48 (7), 1138–1148.

Zhang, D.M., Phoon, K.K., Huang, H.W. & Hu, Q.F. (2015) Characterization of model uncertainty for cantilever deflections in undrained clay. *ASCE Journal of Geotechnical and Geoenvironmental Engineering*, 141 (1), 04014088.

Chapter 2

General principles on reliability according to ISO2394

Johan V. Retief, Mahongo Dithinde, and Kok-Kwang Phoon

ABSTRACT

This chapter provides an overview of the standard ISO2394:2015 from the perspective of geotechnical engineering application. In addition to providing the context for Annex D, it summarises the overall features of the normative standard and the informative annexes, with emphasis on the advances incorporated in the 4th edition of ISO2394. As background the general role of the standard in reliability-based design is summarised. The historic development since the publication of the 1st Edition in 1973 provides a perspective on the legacy of the standard and the objectives set for the latest edition. An overview of the main components of the standard is presented next: consisting broadly of the two parts of the normative standard, firstly of basic principles and secondly the principles for applications. This is complemented by informative annexes that provide further guidance on implementation. The relevance of the standard and the balance of annexes is subsequently interpreted in general terms from the perspective of geotechnical structures; followed specifically in terms of the link to Annex D. The chapter concludes with a summary of the strengths and challenges for the advancement of reliability-based geotechnical design on the basis of the principles presented in ISO2394.

2.1 INTRODUCTION: BACKGROUND TO THE DEVELOPMENT OF ISO2394:2015

Several advances that are of specific relevance to geotechnical structures are introduced in the 4th Edition of ISO2394, published in March 2015 as ISO2394:2015. Although the latest edition represents a substantial revision from ISO2394:1998, it nevertheless builds on a legacy of the standardization of a common basis of principles for structural design. The objective of this chapter is to provide an overview of the evolution of the standard over four editions since its first publication in 1973; the background to and main features of the latest edition; the relevance of the standard to geotechnical structures, serving as a prelude to the specific link to Annex D *Reliability of Geotechnical Structures*.

This overview should serve as reference to the common principles of reliability that would ensure harmonization of geotechnical structures with related standards concerned with selected topics such as the general basis of design, stipulations on actions and their combinations, the resistance of alternative common structural materials; and finally specific classes of structures. The main objective is to demonstrate the degree to which the general principles of reliability in accordance with ISO2394:2015 can be applied compliantly with the specific characteristics of geotechnical structures.

At the most general level, ISO2394:2015 is relevant to geotechnical structures as a result of being sufficiently non-specific in the presentation and stipulation of the principles of reliability, thereby allowing sufficient freedom to accommodate the specifics of geotechnical materials and structures. At the other end of the scale the inclusion of Annex D provides an acknowledgement of the nature of uncertainties that need special consideration in geotechnical design practice.

The formulation of the standard at a high level of abstraction or generality ensures that the principles apply to the broad scope of structures for buildings and civil engineering works, including geotechnical structures. It is intended that additional information should be provided from which operational requirements and procedures can be derived using the principles of reliability presented in the standard. Such additional information should then reflect the characteristics of the intended scope of application, such as the determination of actions and their combinations or provision for specific materials. An important consideration for expressing the principles of reliability at the fundamental level is that the need for updating the standard in the future is thereby reduced.

The inclusion of Annex D on geotechnical structures can also be regarded as a specific feature of the revised standard: Whilst Annex D was a new addition to the ISO2394:2015, provision for particular areas of applications provided for in the previous edition, such as existing structures, durability or fatigue, were omitted during the revision process as being too specific. However, it should also be noted that since the publication of ISO2394:1998, new standards were published for existing structures (ISO13822:2010) and durability (ISO13823:2008). The primary motivation for the introduction of Annex D to ISO2394:2015 was to confirm that geotechnical structures do fall within the scope of the standard, the general principles presented in the standard do apply to this area of application and to enhance unification between geotechnical structures and structures constructed from concrete, steel, timber or masonry.

2.1.1   Stages of development of ISO2394

The central role of the standard *"to unify the principles governing the design calculations of constructions"* including *"all civil engineering works"* was initiated with the publication of ISO2394:1973 *General Principles for the Verification of the Safety of Structures*. The next step was the publication of a more substantial 2nd Edition ISO2394:1986 *General Principles on Reliability for Structures* followed by Addendum 1:1988 containing additional Annexes. As indicated above the present edition was preceded by ISO2394:1998, which included elaboration of both the normative requirements of the standard itself and the informative annexes, thereby almost tripling the length of the standard. A summary of the main features of each edition of the standard provided in Table 2.1 gives an indication of its evolution over the past four decades.

The overall trend of development throughout the various editions is that of an elaboration of the initial central concept provided in 1973 which contains all the main elements of reliability-based design, through a more formal and systematic treatment

Table 2.1 Evolution of the scope and contents of the series of ISO2394 editions.

Edition	Year	Scope & Contents of ISO2394
1	1973	Provides a thumb nail outline of a semi-probabilistic limit state method. – To be used by standards committees; requiring adaptation for each material; – Based on principles of probability, including optimum cost for appropriate degree of safety; – Formulating concepts such as ultimate & serviceability limit states; uncertainties of material strength, loads, models; – Design values for resistance (R^*) and load effects (S^*) based on certain coefficients (γ_m, γ_s) and characteristic values (X_k); – Verification of safety by satisfying $R^* \geq S^*$.
2	1986	Elaboration includes the stipulation of topics such as: – Fundamental requirements: Integrity to withstand local failure; use and environmental conditions; hazards from error or extreme conditions (climate, geotechnical); measures against human mistake; quality control (dedicated clause); maintenance, repair; – Principles of limit states design extensively formulated; including design situations; – Basic variables: classification of actions; materials & soils (tests, in situ observations, conversion factors & scale effects); geometrical parameters; – Analysis, calculations, model and prototype testing and their combined use; – Design format of partial coefficients: Stipulated extensively for actions & their combinations, generically for materials & soils (noting the need for different treatment of soils and existing structures, obtaining characteristic values for each case in principle by testing); geometry; model uncertainty; determination of partial factors; – Annex B provides extensive exposition of the first order probabilistic method; – Addendum 1 presents annexes on characteristic values of classes of actions.
3	1998	Whilst maintaining the same scope of application and contents of previous editions, the normative clauses are generally extended and refined, compared to the previous edition; whilst adding the following topics typically through dedicated clauses: – Formal definitions of key reliability concepts, classified under general (5 items); design (including the formal definition of reliability and related terms) (29 items); actions (20 items); structural response, resistance & material properties; geometry (6 items). – Reliability differentiation in terms of cause & mode of failure, consequences, expenses to reduce risk, societal & environmental conditions; measures related to design and quality management. – Principles of probability based design: Formal introduction of designing for reliability as the probability of failure not to exceed a specified value ($p_f \leq p_{fs}$). – Assessment of existing structures: Represented extensively in dedicated clause. – Guidance in Annexes D (experiment), E (reliability-based design) & F (action combinations) is equal in length of the normative standard.
4	2015	Objectives, approach and layout of standard revised completely in terms of optimized risk as basis for reliability-based and semi-probabilistic design: – Fundamental derivation of reliability levels from optimized risk (owner) and marginal life safety (society) using (i) risk representation; serving as input to (ii) reliability-based design for standardized consequences and (iii) semi-probabilistic design for standardized reliability classes, failure modes, material properties. – Layout structured to cover fundamental requirements (up to Clause 6); procedures for the three levels of approximation (Clauses 7–9); guidance on implementation in informative annexes (Annex A–G).

and attending to specific topics such as basic variables and characteristic values, design format and partial factors, design situations and structural integrity/robustness, together with guidance on the principles of reliability-based design. There was a subtle shift towards the treatment of actions after the 1st Edition, with some balance being restored by placing more emphasis on resistance in the 4th Edition. However, the significant change in direction can also be discerned from Table 2.1, from extension and refinement up to the 3rd Edition to the more fundamental treatment in ISO2394:2015 indicated above.

2.1.2 Status and use of ISO2394

The seminal role of ISO2394 to fulfil its intended function in providing the common basis of reliability principles for structural design standards is affirmed by the fourteen ISO Standards for which it serves as normative reference and ten ISO member states who have adopted it as a national standard. The Eurocode head standard EN1990:2002 *Basis of Structural Design* can be considered to represent an operational semi-probabilistic partial factor limit states design standard that is compliant with ISO2394 (Vrouwenvelder, 1996). In a similar fashion, the South African standard SANS10160-1:2011 *Basis of Structural Design* is compliant with SANS/ISO2394. The Canadian standard CSA S408 *Guidelines for the Development of Limit States Design Standards* is based on ISO2394 and cites it as a reference.

Even more significant than the role of ISO2394 as reference to other standards is its extensive use as a basis for research and background investigations on the application of reliability principles. In addition to direct background investigations on standardization, the literature represents an extensive body of investigations on reliability-based methodologies and fields of application which explicitly cite the standard. Reliability-based methodologies address topics such as target reliability, optimization, service life, durability, fatigue, extreme and environmental actions, fire safety, time dependent processes, sustainability, sampling, Bayesian analysis, technical diagnostics, and traffic assessment. In addition to the investigation of various aspects of conventional structural materials such as concrete, steel, composite construction, timber and masonry, various structural types are considered such as buildings, bridges, tunnels, power production and offshore structures, piles, flood protection, existing, temporary and historical structures. More than 7000 citations of ISO2394 can be identified since the year 2000; moreover the rate of citations steadily increased over this period from about 200 to 700 per year, with the related citations arising from authoritative papers.

2.1.3 Objectives and fundamental principles

A general outline of the objectives, contents and the fundamental principles for the revision of ISO2394 is provided by Faber (2015). As point of departure, the impact of reliability theory on structural engineering practice is noted: ranging from application areas such as high rise buildings, offshore structures and major infrastructure projects at the advanced level, to the formulation and calibration of safety formats for structural design standards at the operational level. The 1998 version of ISO2394 is a manifestation of the maturity reached by best practices of reliability application. Similar levels of maturity in risk informed decision making has only been achieved more recently.

The advantages that can be achieved by an integrated risk and reliability approach in terms of cost efficiency and societal risk management served as motivation and objective for the revision of the standard. Risk information could then be used rationally and transparently as basis for structural performance and associated decision making. The objective for the revision is therefore to provide a consistent best-practice risk and reliability-based decision making approach for the formulation of design standards and for application to specific projects.

Accordingly the most prominent change resulting from the revision of ISO2394 is the representation of the systematic and rational treatment of risk to implementation of reliability-based design through standards. The rational basis is derived from socio-economic principles that utilize the marginal life-saving principle to relate safety to decision making at societal level of regulation. However, in addition to the principles and requirements provided in the standard, the user is responsible for ensuring that all relevant information is available and is applied. All assumptions underlying the decisions need to be controlled and recorded or it must be ensured that the structure will perform adequately despite possible deviations from the assumptions.

The background provided by Faber (2015) furthermore provides elaboration of the fundamental concepts of risk and reliability which are used as basis for performance modelling, fully taking account of uncertainty and its modelling. The basis for the three levels of decision making and design consisting of risk informed, reliability-based and semi-probabilistic design is discussed in some detail. Special attention is given to structural robustness, its classification, measures and quantification; in addition to the consideration of life safety and optimization. These are all topics that are considered below.

2.2 OVERVIEW OF THE STANDARD ISO2394:2015

ISO2394:2015 consists of the following main parts, which are briefly defined below, together with a commentary which provides an assessment of each part:

- **Preliminary and General:** The general ISO Standards format is followed to provide introductory information (non-normative), the scope of application and the procedures, normative reference standards, terms and their definitions and the list of symbols (Introduction and Clauses 1–3).
- **Fundamental Basis and Concepts:** The conceptual basis for ensuring an adequate level of risk and reliability is expressed in fundamental terms, together with a general formulation of performance modelling in terms of basic variables and their uncertainty (Clauses 4–6).
- **Decision-Making and Design Approaches:** Three alternative design approaches are subsequently stipulated in Clauses 7–9 in a more systematic manner, as the basis for separate operational design standards. Notably *risk informed decision making* is formally introduced as a recognized basis for design (Clause 7), serving also to establish the basis for *reliability-based decision making* (Clause 8) and *semi-probabilistic methods* (Clause 9).
- **Implementation Guidance:** Informative annexes provide guidance on the way in which key concepts, expressed in general terms in the standard, should be

implemented in operational terms in design standards to ensure compliance with risk and reliability principles. In spite of their lesser status, the annexes provide essential information for the formulation of effective reliability-based standardized design procedures.

The introduction to ISO2394:2015 indicates the important addition of the methodical risk-based fundamental basis for the regulation and standardization of the safety and reliability of structures. The revised standard represents *the development of systematic and rational treatment of risk to implementation of reliability-based design through codes and standards*. However, it is confirmed that the standard is concerned with *load bearing structures relevant to the construction industry*. The scope of application is defined to cover *the majority of buildings, infrastructure and civil engineering works*. The implied limit in scope refers to the need for *adaptation and detailing in specific cases where there are potentially extreme consequences of failure*.

An important qualification in the definition of the scope of the standard is the need for *knowledge beyond what is contained* in it, with the requirement *to ensure that this knowledge is available and applied*. In addition to the obvious need for additional information for specific applications, this stipulation implies that the standard and annexes express risk and reliability-based requirements and procedures in such general terms that compliance to the standard can be achieved in principle for any specific field of application, including geotechnical design, when full account is taken of the related characteristics of that specific field. Simply stated, design verification should be based on performance modelling that fully accounts for uncertainty of basic variables applicable to the specific class of structure, actions and materials.

Interestingly, no normative standard is referenced in ISO2394:2015, similar to the practice followed by previous editions. The understandable motivation is that ISO2394 serves as the head standard. The situation has changed however with the new standard, where an appropriately formulated standard on risk should now become the starting point of the value chain. The present ISO13824:2009 *Bases for design of structures – General principles on risk assessment of systems involving structures* (ISO, 2009) does not presently provide sufficient information and procedures for risk-based optimization and the derivation of related decision criteria. The JCSS document *Risk Assessment in Engineering – Principles, System Representation and Risk Criteria* can serve as a pre-normative reference (JCSS, 2008). The concepts from the JCSS document have essentially been included in ISO2394:2015.

The primary objective of the list of terms and definitions presented in Clause 2 is simply to achieve clarity on the meaning of key concepts within the scope of the standard. This is particularly important for an international standard, including the need for translation, where ambiguity and different terms for the same or similar concepts may impact on the interpretation of the standard. A convenient way to accommodate local conflicts for example in different countries or fields of application is to introduce a dictionary of equivalent terms in a local standard, for example as given by the Canadian standard CSA S408:2011.

An important utility of the list of terms is that it provides a concise compilation of important concepts within the scope of the standard, tabulated under general (47 items); design and assessment (37 items); actions and combinations (31 items); resistance, material and geometrical properties (8 items). A pertinent example is the definition of the term *Structure – Organised combination of connected parts including geotechnical structures designed to provide resistance and rigidity against various actions*. This definition elaborates on the concept of *load bearing structures* used in the introduction and confirms the comprehensive scope of the standard.

2.3 CONCEPTUAL BASIS AND FUNDAMENTAL REQUIREMENTS

The central role of risk as the basis for structural performance decision making is stipulated in Clause 4 mainly through the formal presentation of the following concepts:

- **Risk-based decision making:** Target structural performance shall be based on optimized total risk, including loss of life and injury, damage to the environment and monetary losses (Clause 2.1.38); using a marginal life-saving risk metric as basis for safety and risk-based optimization as basis for specification by owners (Clause 4.2.2).
- **Alternative approaches:** Accordingly risk informed decision making is introduced as the overarching approach, from which reliability-based design and semi-probabilistic approaches can be derived under certain conditions; thereby formally introducing a three-level approach for risk and reliability-based decision making and design (Clause 4.4.1).

Clause 4 provides also the conceptual basis for decisions concerning structures (Clause 4.3.1); for structural performance modelling (Clause 4.3.2); and for uncertainty and the treatment of knowledge (Clause 4.3.3). Further elaboration on performance modelling is subsequently provided in Clause 5, by providing an outline of the reliability basis of limit states design. The systematic exposition of uncertainty representation and modelling is presented in Clause 6. The conceptual basis for the three alternative decision making approaches given for risk-informed decisions concerning design and assessment (Clause 4.4.2.1), reliability-based design and assessment (Clause 4.4.2.2) and semi-probabilistic approaches (Clause 4.4.3) are each stipulated by means of systematic procedures in Clauses 7, 8 and 9 respectively.

Clause 4 therefore not only introduces the concept of risk as the basis for structural performance and design, but also provides guidance on its implementation into a hierarchy of alternative design approaches. It is formulated in terms of the fundamental aims and requirements for structural performance, the conceptual basis and approaches that should be employed to ensure adequate levels of risk and reliability. Performance requirements are expressed at levels of adequate functionality; ability to withstand extreme conditions resulting from environmental exposure and use; and sufficiently robust not to result in severe damage caused by extraordinary, even unforeseen events or human error. The service life of the structure should be based on the duration of the need for the structure. The dedicated sub-clause on provisions for durability in ISO2394:1998 is replaced by reference to ISO13823 that has recently been published.

The fundamental approach to base the target performance level formally on an appropriate degree of reliability that *shall be judged with due regard to the possible consequences of failure, the associated expense and the level of efforts and procedures necessary to reduce the risk of failure and damage.* A dual process is stipulated, with safety based on societal considerations including the marginal life-saving principle which takes account of *the costs associated with saving additional lives through additional safety measures*; whilst the interests of the owner is reflected by an optimization of the cost of the construction and equivalent cost of failure. This defines performance levels in terms of risk, as opposed to previous formulations in terms of reliability.

The conceptual basis for decisions concerning structures provided in Clause 4.3 provides a fundamental formulation of the concept of structural performance modelling, which is then stipulated extensively in Clause 5. Similarly the concept of uncertainty and the treatment of knowledge are defined, with elaboration provided in Clause 6. The principal characteristics and conditions for application for alternative approaches are provided in Clause 4.4. A somewhat abridged formulation of risk informed decision making is presented in Clause 7, with formal recognition to its role in structural performance decision making. A limited formulation of the probability of failure through reliability-based decision making is provided in Clause 8, clarifying its role in structural assessment and design. The more elaborate treatment of the semi-probabilistic approach presented in Clause 9 is justified by the extensive application of this approach in the various versions of the partial factor limit states based design standards.

- **Main features of ISO2394:2015:** In addition to the pertinent features of the Standard mentioned above, a number of modifications were made during the revision.

  - The introduction of the principles of risk as the fundamental basis for establishing the performance levels of structures.
  - Accordingly, adding risk-based decision making as another level of design, complementing reliability-based and semi-probabilistic design; including a hierarchy based on the level of approximation related to the level of understanding of the consequences of failure and structural behaviour, with associated categorization of the associated risk elements.
  - Logical development of the reliability principles from fundamental concepts of structural performance and representation of uncertainty, through to alternative levels of approximation for decision making and design verification.
  - The omission of specific fields of application, such as durability and reliability of existing structures, both these cases being provided for in dedicated ISO Standards.
  - The strengthening of informative annexes providing guidance on critical aspects of structural performance such as quality management, life cycle management of integrity, design by testing, calibration for semi-probabilistic reliability parameters, robustness and risk criteria.

2.4 KEY RELIABILITY CONCEPTS

In addition to reviewing the major advances made in ISO2394:2015, it is useful to consider the key reliability concepts that serve as the building blocks for reliability-based design procedures. In most cases, these key concepts were incorporated in previous editions of the standard, but were in need for reassessment or more advanced treatment in the context of the more principled approach taken in the latest edition.

- **Decision making and service life:** The concept of design and its reliability basis is broadened to represent *decisions related to design and assessment of structures and systems involving structures over their service life* (Clause 1). This implies not only consideration of the various stages in the life cycle of the structure (Clause 4.2.1) but also that *all anticipated future consequences shall be accounted for* (Clause 4.4.2.1), with guidance given in Annex B *Lifetime management of structural integrity*. Furthermore, the *service lives for structures shall be based on the duration of the need for the structure* (Clause 4.2.1), as opposed to the previous concept that the service life is a nominally pre-selected value.

- **Quality management:** The conceptual basis of decisions concerning structures is that *quality management and quality assurance play central roles for the performance of structures and shall be completely integrated in the decision making process* (Clause 4.3.1). *In general, quality management systems for construction works shall be riskbased and according to an integral approach, encompassing human errors, design errors, and execution errors* (Clause 8.1). Whilst limited further reference is made to quality management, assurance and control in the normative standard, extensive guidance is provided in the informative Annex A with the objective of *the validation of assumptions made in the risk and reliability-based decision making process* (Clause A.1). This is a key aspect in all geotechnical construction processes. Section 4 of EN1997-1 is devoted to this aspect and *requires that all geotechnical construction processes, including the workmanship applied, must be supervised, that the performance of the structure must be monitored, both during and after construction, and that the finished structure must be adequately maintained. The nature and quality of the supervision and monitoring prescribed for a project must be commensurate with the degree of precision assumed in the design, and in the values of the engineering parameters chosen and the partial factors used in the calculations. If the reliability of the design calculations is in doubt, it may be necessary to prescribe an enhanced regime of construction supervision and monitoring.*

- **Uncertainty, knowledge and Bayesian probability:** The basis for *decisions concerning structures shall account for all uncertainties of relevance for their performance such as inherent natural variability (aleatory uncertainty) and lack of knowledge (epistemic uncertainty)* (Clause 4.3.3). It is stipulated that *the Bayesian interpretations of probability should be considered as the most adequate basis for the consistent representation of uncertainties, independent of their sources. It facilitates the joint consideration of purely subjectively assessed uncertainties, analytically assessed uncertainties and evidence as obtained through observations* (Clause 6.1.3). *In the case of relatively high uncertainties in actions, structural properties and/or models, the possibility of updating procedures shall be considered in order to accomplish a more economical design or assessment solution* (Clause 6.6). Guidance on specific application of Bayesian updating is given in Clause 4.3.3 Notes 1 & 2. Annex B.4.3 presents the general procedures for the updating of probabilistic models in order to extend the evidence in terms of observations to obtain risk and reliabilities that are gradually updated. The use of information gathered through quality control measures for updating is given in Annex A.5.5. Updating of probabilistic models based on experimental methods is considered in Annex C.5.3. *Bayesian probabilistic modelling forms the basis for semi-probabilistic design standards and regulations (see Clause 4.4.3) through calibration (Annex E).* As discussed in Chapter 3 and 4, Bayesian updating offers a natural framework to combine prior information from comparable sites with site-specific information. This updating step is crucial, because site-specific information alone is usually too limited for reliability-based design and it is consistent with existing geotechnical practice where all information is judiciously weighed, albeit using engineering judgment.

- **Robustness or damage insensitivity:** A fundamental requirement for structures is to *be robust such as not to suffer severe damage or cascading failure by extraordinary and possibly unforeseen events like natural hazards, accidents, or human errors, providing sufficient robustness* (Clause 4.2.1). Robustness is closely related to systems behaviour, as opposed to the design of individual elements (Clause 8.3). *For structures where failure and damage can imply very serious consequences, a risk-based robustness assessment shall be undertaken as part of the design and/or assessment verification* (Clause 4.4.2.1). Allowance is made for *categorization of structures in accordance with their consequences of failure* (to) *decide whether a risk-based robustness assessment is necessary or not.* For a semi-probabilistic approach *the system performance shall be ensured, depending on the consequences of system failure, either through risk-based robustness assessments or through robustness provisions. The latter includes critical member design, structural ties, and structural segmentation* (Clause 4.4.2.2). Further guidance is given in Annex F on classification of structures based on expected consequences and on appropriate measures to ensure robustness. Guidance is also given on risk-based robustness assessment. Extensive reference to robustness is also made in Annex A on quality management.

- **Analysis Models:** Although structural mechanics models for the physical behaviour of structural systems form an integral part of performance modelling (Clause 5), this topic is extensively dealt with as part of uncertainty representation and modelling under Clause 6.2 *Models for structural analysis.* The motivation for the more detailed consideration of structural models stems from the differences in nature of models for actions and environmental influences (Clause 6.2.2), and geometrical properties (Clause 6.2.3), material properties (Clause 6.2.4), structural response and resistance (Clause 6.2.5). Models for consequences now require direct consideration for the risk-based approach (Clause 6.3). As models are usually incomplete and inexact as the result of lack of knowledge or a deliberate simplification for use in operational design procedures, provision for model uncertainty must be made (Clause 6.4). In contrast to the nominal treatment of reliability modelling of resistance in previous editions of the standard, ISO2394:2015 gives increased attention to the response and resistance of the structure, on a par with the attention paid to actions on the structure. Accordingly material properties as represented as basic variables receive proper attention throughout the new standard (e.g., Clause 6.2.4). In recognition of the predominance of model uncertainty in geotechnical engineering, Annex D provides more details on this subject matter from a geotechnical perspective.

- **Design based on experimental models:** Design procedures based on calculation models can be complemented using experimental models. An important condition is that *the setup and evaluation of the tests should be performed in such a way that the structure, as designed, has at least the same reliability with respect to all relevant limit states and load conditions as structures designed on the basis of calculation models only. The test results should be evaluated on the basis of statistical methods. In principle, the tests should lead to a probability distribution for the selected unknown quantities, including the statistical uncertainties* (Clause 6.5). In Annex C, guidance is given on establishing design values or partial factors either directly or by evaluation using an evaluation model. Unlike the properties of other structural materials, soil properties are not specified but determined by testing on a site specific basis. In this regard, geotechnical design is always based on site specific tests results. A classic example is the recognition of full scale pile load tests on a given site as an accepted design method (see EN1997-1 clause 7.4.1).

- **Semi-probabilistic design approach:** The semi-probabilistic (or partial factors) method is defined as a *verification method in which allowance is made for the uncertainties and variability assigned to the basic variables by means of representative values, partial factors and, if relevant, additive quantities* (Clause 2.2.24). Despite the formal introduction of risk-based decision making in ISO2394:2015 and reliability-based design in ISO2394:1989, the semi-probabilistic design approach remains the most practical way to incorporate reliability-based design principles in operational design procedures and standards. Clause 9 *Semi-probabilistic method* therefore represents the closest format for expressing the *general principles on reliability for structures* of ISO2394:2015 in operational terms. Although this clause is completely revised from its previous version, the changes are mainly in the logical development from principles to the symbolic expressions for design verification, rather than representing any further advancement. Notably the geotechnical fraternity has embraced semi-probabilistic limit state design as the basis for the development of new generation of design codes. The commitment of the geotechnical profession to the adoption of limit state design framework is demonstrated by the establishment of a technical committee (TC 23) on Limit State Design in Geotechnical Engineering under the auspices of the International Society of Soil Mechanics and Geotechnical Engineering in 1990. TC23 was led by the late Dr. N. Krebs Ovesen of the Danish Geotechnical Society. This was two years after work began on drafting Part 1 of Eurocode 7: Geotechnical Design, General Rules. In the next seven years, the Danish Geotechnical Society remained TC23's sponsor. Not surprisingly, the emphasis was on activities in Europe and on the development of the Eurocodes in particular. The development of Eurocode 7 attracted interest outside of Europe. In recognition of this interest, and possibly of the need to shift the emphasis away from Europe, the Geotechnical Division of the South African Institution of Civil Engineers was asked to become the sponsoring Member Society of TC 23 for the period 1997 to 2001. The TC was led by Peter Day, the Chairman. The Japanese Geotechnical Society (JGS) was asked to become the sponsoring Member Society of the TC from 2001 to 2009, because the Japanese performance-based foundation design code called Geocode 21 was being developed over that period (Honjo and Kusakabe 2002). It was led by

Professor Yusuke Honjo. Since 2009, this technical committee has been led by Dr. Brian Simpson. It was renumbered as TC205 and more recently, renamed as "Safety and Serviceability in Geotechnical Design" in 2013. The committee was mandated with promoting and enhancing professional activities in the limit state design in geotechnical engineering practice. Accordingly the committee has organised several international symposia on limit state design in geotechnical engineering practice. TC 205 works closely with TC304 (formerly TC32) which focuses on engineering practice of risk assessment and management. TC32 was led by Dr. Farrokh Nadim from 2001 to 2009. It was renumbered as TC304 and led by Professor Kok-Kwang Phoon since then. Just like ISO2394, this technical committee focuses on and is tasked with promoting probabilistic site characterisation, calibration of geotechnical design codes using the semi-probabilistic approach, reliability-based design, risk-based decision analysis, and project risk management among others.

• **Risk-based conditions for semi-probabilistic design:** The alternative approaches for decision making and design are introduced as follows: *When the consequences of failure and damage are well understood and within normal ranges, reliability-based assessments can be applied instead of full risk assessments. Semi-probabilistic approaches as a further simplification are appropriate when in addition to the consequences also the failure modes and the uncertainty representation can be categorized and standardized* (Clause 4.4.1). The implications are that instead of the semi-probabilistic approach representing an improvement on experience based safety factor design, from first principles it is now necessary to fully account for risk-based acceptance criteria and for all sources of uncertainty in deriving semi-probabilistic design procedures:

  – **Conditions for target reliability:** Both reliability-based decision-making and semi-probabilistic design procedures should be derived from risk informed decision making based on categorized consequences (Clause 7). Guidance is given in Annex G on risk optimization and criteria on life safety.
  – **Additional conditions:** *For structures for which the consequences of failure and damage are well understood and the failure modes can be categorized and modelled in a standardized manner, semi-probabilistic codes are appropriate as basis for design and assessment. Standards shall serve to ensure the quality of analysis, design, materials, production, construction, operation and maintenance, and documentation, and thereby explicitly or implicitly account for the uncertainties which influence the performance of the structures. The specifications given in standards should be developed such that they quantify all known uncertainties* (Clause 4.4.3).

• **Reliability elements for semi-probabilistic design:** The following stipulation provides the basis for the reliability elements of semi-probabilistic design: *For design of structures based on codified load and resistance factor or partial safety factor design, uncertainties shall be represented through design values and characteristic values together with specified design equations, load cases, and load combination factors. The characteristic values shall, when relevant, account for available information relating, for example, to loads and material properties* (Clause 4.3.3).

The following main reliability elements for semi-probabilistic design consist of (Clause 9):

- **Safety format:** *semi-probabilistic design and assessment codes shall comprise a safety format prescribing the design equations and/or analysis procedures which shall be used for the verification of design and assessment decisions* (Clause 4.4.3). Further elaboration is given in Clause 9.4.
- **Characteristic values:** The characteristic values of basic variables should be *specified preferably on a statistical basis, so that it can be considered to have a prescribed probability of being exceeded towards unfavourable values* (Clause 2.2.30) and form part of the treatment of uncertainty and knowledge (Clause 4.3.3). *For a produced material, the characteristic value should in principle be presented as an a priori specified quantile of the statistical distribution of the material property being supplied, produced within the scope of the relevant material standard. For soils and existing structures, the values should be estimated according to the same principle and so that they are representative of the actual volume of soil or the actual part of the existing structure to be considered in the design* (Clause 9.3.2). It must be pointed out that in geotechnical engineering, there is no consensus on a clear definition of the characteristic geotechnical parameter and still there is no universally accepted methodology for its selection. Complications exist because a characteristic geotechnical parameter depends in an inter-related way on the natural (or spatial) variability and the volume of soil (or slip surface) that affects the occurrence of the limit state. Clause 2.4.5.2 of EN 1997–1:2004: "Characteristic values of geotechnical parameters" discusses some of the considerations, but leaves out a critical aspect concerning the interaction between natural variability and emergence of the slip surface. For example, a sufficiently persistent weak zone in the soil mass may force a slip surface to pass through this zone. If one were to accept that natural variability can be modelled by a random field, it would be necessary to consider that the slip surface trajectory changes from realization to realization, as the distribution of weak zones varies from realization to realization. In other words, the characteristic strength that is relevant to the critical slip surface varies (probabilistically) in a complex way. Additional observations are covered in Section 1.5.6 and Section 3.5. Details are given elsewhere (Ching & Phoon 2013; Ching et al. 2014, 2016a, 2016b).
- **Partial safety factors:** The most direct instrument available to the designer to achieve required levels of reliability is the use of partial safety factors to be applied to basic variables and design models. Guidance on the calibration of partial factors is provided in Annex E.

- **Documentation:** Recording of *decisions related to the design of structures, as well as their verification with respect to acceptance criteria . . . in a manner that is tractable and transparent* is required for the *design of individual structures, as well as the development and calibration of design codes.* The comprehensive records include *site specific data, test results, models of the performance indicators, inspection results* amongst other information (Clause 4.5). These requirements are very much in line with the current geotechnical practice whereby the site specific data should be well documented in the form of ground investigation report and the actual design in the form of a design report.

2.5 CONCLUDING SUMMARY OF ISO2394:2015

A critical assessment of ISO2394:2015 and its relevance to the reliability-based performance and design of geotechnical structures serve as summary of the standard and conclusions on its merit and utility.

- Based on four decades of capturing the essence of reliability-based performance and design of load bearing structures since 1973, the latest edition of ISO2394 represents advancement in that (i) it is based on principles of risk and reliability, (ii) as a result of this and efforts that were made to make it sufficiently general, it covers all structures for buildings and civil engineering works, and (iii) it incorporates recent advances made in the determination of performance and safety criteria based on risk optimization.
- ISO2394:2015 is the foremost platform for the advancement of structural performance based on principles of risk to determine appropriate levels of reliability and the formulation of operational design procedures that *explicitly or implicitly account for the uncertainties which influence the performance of the structures* (Clause 4.4.3).
- The principled approach represented by ISO2394:2015 makes it imperative that full account should be taken of the various uncertainties in (i) the consequences of failure, (ii) the nature of failure modes, (iii) basic variables including both actions and material properties, and (iv) the models employed for analysis.
- The fundamental approach presented requires performance levels and design levels to reflect principles of risk-based safety and provision for uncertainty. This changes the present trend of standardization where judgement based design elements are incrementally being replaced by reliability-based calibration using improved models for uncertainty. This is presented in a clearly systemized manner through the classification into the three levels of risk-based optimization, typically to obtain reliability or performance levels; reliability-based decision making or calibration to reflect uncertainties; in order to derive partial and other design factors for operational semi-probabilistic design.
- The gap between the fundamental approach where lack of knowledge is fully accounted for and traditional progress calibrated to acceptable practice, can be closed using ISO2394:2015 by expressing experience based information in terms of Bayesian probability. This would ensure that the fundamental approach does not introduce conservatism which is inconsistent with the extensive experience base for structural performance. The formal use of experience of structural performance provides a strong motivation for the harmonization of structural design, capitalising on sharing of experience internationally.
- Different selections from the standard should be followed for specific classes of design standards, such as (i) for the basis of design, to be based on target levels of reliability and consequence classes (Clause 7), limit states (Clause 5) and

robustness (Annex F) and (ii) actions and their combinations, including probability models and classification of all relevant types of actions (Clause 6), safety format (Clause 9). Similar selections can be made for the resistance of structures constructed from various structural materials or for various classes of structures such as buildings or bridges.

- The general survey of the standard reported in this chapter confirms the relevance of the risk and reliability-based approach of ISO 2394 to geotechnical structures in general, but with additional information that should be incorporated to provide for the specific nature of geotechnical structures. In fact, the standard provides a suitable platform for the advancement of the risk and reliability bases for the design and performance of geotechnical structures. A unique selection of the relevant clauses from the standard would apply to procedures applicable to geotechnical structures: In addition to the topics considered in Annex D which are mainly related to uncertainties of geotechnical materials and models, consideration should be given to risk characteristics of the performance of geotechnical structures and the consequences of their failure, the systems nature of geotechnical failures, the integral nature of geotechnical actions and resistances, special requirements for the combination of testing and experience based judgement, amongst other considerations.

- Reference to Annex D provides guidance for geotechnical structures on uncertainty representation and modelling (Clause 6), specifically considering types of uncertainty (Clause 6.1.1 Note 4). However, topics addressed in Annex D also relate to other clauses, such as reference made to reliability-based methods presented in D.5.3 (Clause 8) and the semi-probabilistic approach assessed in D.5.4 (Clause 9), specifically considering characteristic soil parameters in D.5.5 (Clauses 4.3.3 & 9.3.2) and systems reliability in D.5.7 (Clause 5.2.3).

- The standard provides a suitable platform for unification of the various fields of application that may apply to individual installations. Efforts still need to be made to ensure consistency of application for example between actions, structural and geotechnical design at the following levels:

  - Overall performance requirements for the installation under consideration, serving as input for the basis for decision making and design – this is treated comprehensively by the standard;
  - At the output stage, consistency achieved in the constructed facility such as for the design of the structure and foundation, serves as check for compliance to the standard;
  - The interface between structural and geotechnical design provides the operational basis for ensuring consistence of performance and reliability.

ACKNOWLEDGEMENTS

The authors are grateful for the valuable comments provided by Dr. Peter Day, Dr. Farrokh Nadim and Dr. Brian Simpson.

REFERENCES

Ching, J. & Phoon, K.K. (2013) Probability distribution for mobilized shear strengths of spatially variable soils under uniform stress states. *Georisk*, 7 (3), 209–224.

Ching, J., Phoon, K.K. & Kao, P.H. (2014) Mean and variance of the mobilized shear strengths for spatially variable soils under uniform stress states. *ASCE Journal of Engineering Mechanics*, 140 (3), 487–501.

Ching, J., Lee, S.W. & Phoon, K.K. (2016a) Undrained strength for a 3D spatially variable clay column subjected to compression or shear. *Probabilistic Engineering Mechanics*, 45, 127–139.

Ching, J., Hu, Y.G. & Phoon, K.K. (2016b) On characterizing spatially variable shear strength using spatial average. *Probabilistic Engineering Mechanics*, 45, 31–43.

CSA S408:2011. *Guidelines for the Development of Limit States Design Standards*. Mississauga, ON, Canadian Standards Organization.

EN 1990:2002. *Eurocode – Basis of Structural Design*. Brussels, European Committee for Standardization (CEN).

EN 1997–1:2004. *Eurocode 7: Geotechnical Design – Part 1: General Rules*. Brussels, European Committee for Standardization (CEN).

Faber, M.H. (2015) Codified risk informed decision making for structures. In: *Symposium on Reliability of Engineering Systems (SRES2015), Hangzhou, China*.

Honjo, Y. & Kusakabe, O. (2002) Proposal of a comprehensive foundation design code: Geocode 21 Ver.2. In: Honjo, Y., Kusakabe, O., Matsui, K., Kouda, M. & Pokharel, G. (eds.) *Foundation Design Codes and Soil Investigation in View of International Harmonization and Performance Based Design*. The Netherlands, A.A. Balkema Publishers. pp. 95–106.

ISO 2394:1973/1986/1998/2015. *General Principles on Reliability for Structures*. Geneva, International Organization for Standardization.

ISO 13822:2010. *Bases for Design of Structures – Assessment of Existing Structures*. Geneva, International Organization for Standardization.

ISO 13823:2008. *General Principles on the Design of Structures for Durability*. Geneva, International Organization for Standardization.

ISO 13824:2009. *Bases for Design of Structures – General Principles on Risk Assessment of Systems Involving Structures*. Geneva, International Organization for Standardization.

JCSS (June 2008) *Risk Assessment in Engineering – Principles, System Representation & Risk Criteria*. Joint Committee on Structural Safety. ISBN: 978-3-909386-78-9. Available from: http://www.jcss.byg.dtu.dk/Publications/Risk_Assessment_in_Engineering.aspx, Edited by Faber, M.H.

SANS 10160-1:2011. *Basis of Structural Design*. Pretoria, South African National Standard, SABS.

Vrouwenvelder, T. (1996) Revision of ISO 2394 General principles on reliability for structures. *IABSE Reports*, 74, 117–118.

Chapter 3

Uncertainty representation of geotechnical design parameters

Kok-Kwang Phoon, Widjojo A. Prakoso, Yu Wang, and Jianye Ching

ABSTRACT

Site investigation and the interpretation of site data are necessary aspects of sound geotechnical practice. As such, the characterization of geotechnical variability should play a central role in reliability-based design. This chapter discusses the uncertainties associated with the most basic soil/rock property evaluation task, which is to estimate a design parameter from a field test. The coefficient of variation in the estimate must be a function of the natural variability of the site, measurement error associated with the field test, and transformation uncertainty about the regression line that relates the field data to the design parameter. In addition, soil/rock properties are spatially variable. This autocorrelation (correlation between values measured at different spatial locations for the same property) effect can be quantified, for instance, by the scale of fluctuation. Useful statistical tables and guidelines for the coefficient of variation and the scale of fluctuation derived from a comprehensive survey of soil and rock databases are presented in this chapter. The cross-correlation (correlation between different properties at the same spatial location) effect is discussed in Chapter 4. Stratigraphy is also spatially variable, but this geologic uncertainty is not well studied in the literature at present.

The coefficients of variation derived from soil and rock databases may be larger than those encountered in a specific site, because they are applicable in a generic "global" sense. These generic statistics are useful as prior information in the absence of site-specific data. Measurement error is not site-specific because it is typically related to the equipment, procedure, and operator. Natural variability and transformation uncertainty are potentially site-specific. It is possible to update the statistics for natural variability and transformation uncertainty in the presence of site-specific data using Bayesian methods. However, statistical uncertainties associated with inference from spatially correlated data should be handled carefully.

3.1 INTRODUCTION

Site investigation and the interpretation of site data are necessary aspects of sound geotechnical practice. One major source of complication in the interpretation of site data is natural variability. Natural variability stems from the natural processes by which soil masses are deposited and modified over time. It may take the form of a test profile varying with depth. This profile often consists of a trend function and a fluctuating component. In the geotechnical reliability literature, this fluctuating component or natural variability is modeled, whenever possible, as a stationary (statistically homogeneous) random field. A real site generally exhibits spatial variation in both vertical and horizontal directions, most often with more pronounced variability in the vertical direction. Jaksa et al. (2003), Jaksa et al. (2005), and Goldsworthy et al. (2007) used three dimensional random fields and Monte Carlo simulation to simulate the spatially variable elastic modulus of a "virtual" site. Each spatially variable realization constitutes a plausible full information scenario. Site investigation is then carried out numerically by sampling the continuous random field at discrete locations. The site investigation data so obtained constitute the typical partial information scenario commonly encountered in practice. Site investigation data also appear in a multivariate form. For example, the Standard Penetration Test (SPT) is commonly conducted in the same borehole where undisturbed soil samples are taken. It is routine to measure properties such as unit weight, natural water content, plastic limit, liquid limit, undrained shear strength, and overconsolidation ratio from these soil samples. Data from different tests conducted at the same spatial location will be correlated, even though they are measuring different aspects of soil behavior under different boundary conditions and over different influence domains. In addition, there are different degrees of sample disturbance. In practice, a "point" refers to a sufficiently small volume of soil bounded by two adjacent boreholes/soundings at comparable depths. Ching et al. (2014a) called data simulated with the intent of replicating the multi-dimensional correlation structure underlying this basket of laboratory and field tests as "virtual site" data. Clearly, this virtual site is restricted to describing the multi-dimensional (or vector) nature of the information available within a point in the soil mass. Spatial variations between points are not considered. Finally, there is limited work on construction and characterization of non-stationary random fields, particularly in the form of one soil layer embedded in another or inclusion of pockets of different soil type within a more uniform soil mass. In existing practice, soil layers are most often assumed to be horizontal and uniform in thickness. In actuality, the depth and thickness of each layer is uncertain between boreholes. The coupled Markov chain (CMC) model has been applied to represent this type of geologic uncertainty (Li et al. 2016). It is not possible to emulate every aspect of site variability at present. The challenge lies in characterizing a full three-dimensional non-stationary vector field from borehole and/or field test data measured at only limited spatial locations [for example, FHWA (1985) stipulates one borehole every 60 m for cut slopes].

Notwithstanding the evolving research in modeling and characterizing geotechnical variability, it is useful to present statistics that: (1) are founded on actual soil databases and (2) are aligned with geotechnical engineering practice. Geotechnical engineers can relate better to reliability-based design (RBD) when they can see explicit connections to soil data and how existing practice can be enhanced in a demonstrably fruitful way. For example, geotechnical design parameters are commonly estimated from field data. The uncertainty in the design parameter (say indicated by the coefficient of variation) must be a function of the natural variability of the site, measurement error associated with the field test, and uncertainty about the regression line (called transformation uncertainty) that relates the field data to the design parameter. Phoon and Kulhawy (1999a, 1999b) characterized geotechnical variability using this simple but practical approach. Rock data have not been characterized as extensively as soil data, but some useful guidelines on the variability of rock properties are provided by Prakoso (2002) and summarized in Section 3.7. Other studies are reported by Ng et al. (2015), Kahraman (2001), Sari and Karpuz (2006), and Aladejare & Wang (2017). The coefficients of variation reported in the above references may be larger than those encountered in a specific site, because they are applicable in a generic "global" sense. These generic statistics are useful as prior information in the absence of site-specific data, although the usual caveat against indiscriminate adoption of these generic statistics without an appreciation of the underlying databases from which they are derived is applicable. Measurement error is not site-specific because it is typically related to the equipment, procedure, and operator. Natural variability and transformation uncertainty are potentially site-specific. The final two sections of this chapter deals with statistical uncertainty and Bayesian quantification of site-specific natural variability.

Another common question is how to estimate a design parameter when it is known to be correlated to different test data (say undrained shear strength can be correlated to SPT N-value, cone tip resistance, overconsolidation ratio, and others) and these tests happened to be conducted at a particular site. Chapter 4 describes a consistent approach to couple different test data and to estimate one or more design parameters from one or more test data.

3.2 SOURCES OF UNCERTAINTIES

The overall uncertainty underlying a geotechnical design parameter results from many disparate sources of uncertainties, as illustrated in Figure 3.1. There are three primary sources of geotechnical uncertainties: (1) natural (inherent) variability, (2) measurement error, and (3) transformation uncertainty. The first results primarily from the natural geologic processes that produced and continually modify the soil/rock mass in-situ. The term "natural variability" is used here to be consistent with Section 4.3.3 and Section 6.1.1 in ISO2394:2015, but this term can be used interchangeably with the term "inherent variability" or "spatial variability" (Section D.2.3). The second

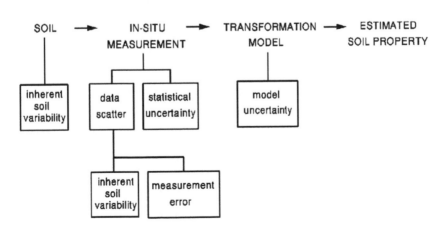

Figure 3.1 Sources of uncertainties contributing to overall uncertainty in a design soil parameter.

is caused by equipment, procedural/operator, and random testing effects. The third source of uncertainty is introduced when field or laboratory measurements are transformed into design soil/rock parameters using empirical or other correlation models.

The relative contribution of these three sources to the overall uncertainty in the design soil/rock parameter clearly depends on the site conditions, degree of equipment and procedural control, and precision of the correlation model. Therefore, statistics of soil/rock parameters that are determined from total uncertainty analyses only can be applied to the specific set of circumstances (site conditions, measurement techniques, correlation models) for which the design soil/rock parameters were derived. Geotechnical variability is now routinely discussed in texts (e.g., Chapter 10, Look 2014) and design guides (e.g., Det Norske Veritas 2010, Joint Committee on Structural Safety 2006), but the coefficients of variation are tabulated without reference to the property evaluation methodology and the soil/rock property databases. This ignores the practical reality that there are many different tests [e.g., standard penetration test (SPT), cone penetration test (CPT)] and even different transformation models linking the same pair of test and design parameters. Chapters 3 and 4 would fill this critical gap in the design guides.

3.3 NATURAL VARIABILITY

Table 3.1, 3.2, and 3.3 summarize typical natural variability of strength properties, index parameters, and field measurements, respectively. The general soil type and the approximate range of mean value for which the coefficient of variation (COV) is

Table 3.1 Summary of natural variability of strength properties (Source: Table 1, Phoon & Kulhawy 1999a).

Property[a]	Soil type	No. data groups	No. tests/group Range	No. tests/group Mean	Property value Range	Property value Mean	Property COV (%) Range	Property COV (%) Mean
s_u (UC) (kN/m^2)	Fine-grained	38	2–538	101	6–412	100	6–56	33
s_u (UU) (kN/m^2)	Clay, silt	13	14–82	33	15–363	276	11–49	22
s_u (CIUC) (kN/m^2)	Clay	10	12–86	47	130–713	405	18–42	32
s_u (kN/m^2)[b]	Clay	42	24–124	48	8–638	112	6–80	32
$\bar{\phi}$ (°)[b]	Sand	7	29–136	62	35–41	37.6	5–11	9
$\bar{\phi}$ (°)[b]	Clay, silt	12	5–51	16	9–33	15.3	10–50	21
$\bar{\phi}$ (°)[b]	–	9	–	–	17–41	33.3	4–12	9
$\tan \bar{\phi}$ (TC)	Clay, silt	4	–	–	0.24–0.69	0.509	6–46	20
$\tan \bar{\phi}$ (DS)	Clay, silt	3	–	–	–	0.615	6–46	23
$\tan \bar{\phi}$[b]	Sand	13	6–111	45	0.65–0.92	0.744	5–14	9

a – s_u = undrained shear strength; $\bar{\phi}$ = effective stress friction angle; TC = triaxial compression test; UC = unconfined compression test; UU = unconsolidated-undrained triaxial compression test; CIUC = consolidated isotropic undrained triaxial compression test; DS = direct shear test
b – laboratory test type not reported

Table 3.2 Summary of natural variability of index parameters (Source: Table 2, Phoon & Kulhawy 1999a).

Property [a]	Soil type [b]	No. data groups	No. tests/group Range	Mean	Property value Range	Mean	Property COV (%) Range	Mean
w_n (%)	Fine-grained	40	17–439	252	13–105	29	7–46	18
w_L (%)	Fine-grained	38	15–299	129	27–89	51	7–39	18
w_P (%)	Fine-grained	23	32–299	201	14–27	22	6–34	16
PI (%)	Fine-grained	33	15–299	120	12–44	25	9–57	29
LI	Clay, silt	2	32–118	75	–	0.094	60–88	74
γ (kN/m^3)	Fine-grained	6	5–3200	564	14–20	17.5	3–20	9
γ_d (kN/m^3)	Fine-grained	8	4–315	122	13–18	15.7	2–13	7
D_r (%) [c]	Sand	5	–	–	30–70	50	11–36	19
D_r (%) [d]	Sand	5	–	–	30–70	50	49–74	61

a – w_n = natural water content; w_L = liquid limit; w_P = plastic limit; PI = plasticity index; LI = liquidity index; γ = total unit weight; γ_d = dry unit weight; D_r = relative density
b – fine-grained materials derived from a variety of geologic origins, e.g., glacial deposits, tropical soils, and loess
c – total variability for direct method of determination
d – total variability for indirect determination using SPT values

Table 3.3 Summary of natural variability of field measurements (Source: Table 3, Phoon & Kulhawy 1999a).

Test type [a]	Property [a]	Soil type [b]	No. data groups	No. tests/group Range	Mean	Property value Range	Mean	Property COV (%) Range	Mean
CPT	q_c (MN/m^2)	Sand	57	10–2039	115	0.4–29.2	4.10	10–81	38
		Silty clay	12	30–53	43	0.5–2.1	1.59	5–40	27
CPT	q_T (MN/m^2)	Clay	9	–	–	0.4–2.6	1.32	2–17	8
VST	s_u(VST) (kN/m^2)	Clay	31	4–31	16	6–375	105	4–44	24
SPT	N	Sand	22	2–300	123	7–74	35	19–62	54
	N	Clay, loam	2	2–61	32	7–63	32	37–57	44
DMT	A (kN/m^2)	Sand to clayey sand	15	12–25	17	64–1335	512	20–53	33
	A (kN/m^2)	Clay	13	10–20	17	119–455	358	12–32	20
DMT	B (kN/m^2)	Sand to clayey sand	15	12–25	17	346–2435	1337	13–59	37
	B (kN/m^2)	Clay	13	10–20	17	502–876	690	12–38	20
DMT	E_D (MN/m^2)	Sand to clayey sand	15	10–25	15	9.4–46.1	25.4	9–92	50
	E_D (MN/m^2)	Sand, silt	16	–	–	10.4–53.4	21.6	7–67	36
DMT	I_D	Sand to clayey sand	15	10–25	15	0.8–8.4	2.85	16–130	53
		Sand, silt	16	–	–	2.1–5.4	3.89	8–48	30
DMT	K_D	Sand to clayey sand	15	10–25	15	1.9–28.3	15.1	20–99	44
		Sand, silt	16	–	–	1.3–9.3	4.1	17–67	38
PMT	p_L (kN/m^2)	Sand	4	–	17	1617–3566	2284	23–50	40
	p_L (kN/m^2)	Cohesive	5	10–25	–	428–2779	1084	10–32	15
PMT	E_{PMT} (MN/m^2)	Sand	4	–	–	5.2–15.6	8.97	28–68	42

a – CPT = cone penetration test; VST = vane shear test; SPT = standard penetration test; DMT = dilatometer test; PMT = pressuremeter test
b – q_c = CPT tip resistance; q_T = corrected CPT tip resistance; s_u(VST) = undrained shear strength from VST; N = SPT blow count (number of blows per foot or 305 mm); A and B = DMT A and B readings; E_D = DMT modulus; I_D = DMT material index; K_D = DMT horizontal stress in-dex; p_L = PMT limit stress; E_{PMT} = PMT modulus

applicable are also included in the tables. With respect to soil type, the COV of natural variability for sand is higher than that for clay. With respect to measurement type, the COVs of natural variability for index parameters are the lowest, with the possible exception of derived parameters such as the relative density and liquidity index. The highest COVs of natural variability seem to be associated with measurements in the horizontal direction and measurements of soil modulus. More detailed characterization at specific sites are given by Cherubini et al. (2007), Chiasson and Wang (2007), Jaksa (2007), and Uzielli et al. (2007).

3.4 MEASUREMENT ERROR

Table 3.4 and 3.5 summarize typical measurement error of laboratory tests and field tests, respectively. Statistical information on measurement error is rather limited. Based on the statistics reported by comparative testing programs, the COVs of measurement error for most laboratory strength tests are estimated to be between 5 and 15%. The COVs of measurement error for the plastic and liquid limit tests were in the range of 10–15% and 5–10%, respectively. The COV of measurement error for the natural water content was intermediate between those of the limit tests. For the plasticity index, the standard deviation of the measurement error was between 2 and 6%. The unit weight determination had the lowest COV of measurement error (~1%). As shown in Table 3.5, the measurement error for the standard penetration test is the largest among field tests, and the measurement errors for the electric cone penetration test and the

Table 3.4 Summary of total measurement error of some laboratory tests (Source: Table 5, Phoon & Kulhawy 1999a).

Property[a]	Soil type	No. data groups	No. Tests/Group Range	No. Tests/Group Mean	Property value Range	Property value Mean	Property COV (%) Range	Property COV (%) Mean
s_u(TC) (kN/m^2)	Clay, silt	11	–	13	7–407	125	8–38	19
s_u(DS) (kN/m^2)	Clay, silt	2	13–17	15	108–130	119	19–20	20
s_u(LV) (kN/m^2)	Clay	15	–	–	4–123	29	5–37	13
$\bar{\phi}$(TC) (°)	Clay, silt	4	9–13	10	2–27	19.1	7–56	24
$\bar{\phi}$(DS) (°)	Clay, silt	5	9–13	11	24–40	33.3	3–29	13
$\bar{\phi}$(DS) (°)	Sand	2	26	26	30–35	32.7	13–14	14
tan $\bar{\phi}$(TC)	Sand, silt	6	–	–	–	–	2–22	8
tan $\bar{\phi}$(DS)	Clay	2	–	–	–	–	6–22	14
w_n (%)	Fine-grained	3	82–88	85	16–21	18	6–12	8
w_L (%)	Fine-grained	26	41–89	64	17–113	36	3–11	7
w_P (%)	Fine-grained	26	41–89	62	12–35	21	7–18	10
PI (%)	Fine-grained	10	41–89	61	4–44	23	5–51	24
γ (kN/m^3)	Fine-grained	3	82–88	85	16–17	17.0	1–2	1

a – s_u = undrained shear strength; $\bar{\phi}$ = effective stress friction angle; TC = triaxial compression test; UC = unconfined compression test; DS = direct shear test; LV = laboratory vane shear test; w_n = natural water content; w_L = liquid limit; w_P = plastic limit; PI = plasticity index; γ = total unit weight

Table 3.5 Summary of measurement error of common in-situ tests (Source: Table 6, Phoon & Kulhawy 1999a).

Test	COV Equip. (%)	COV Proc. (%)	COV Random (%)	COV[a] Total (%)	COV[b] Range (%)
Standard penetration test (SPT)	5–75[c]	5–75[c]	12–15	14–100[c]	15–45
Mechanical cone penetration test (MCPT)	5	10–15[d]	10–15[d]	15–22[d]	15–25
Electrical cone penetration test (ECPT)	3	5	5–10[d]	7–12[d]	5–15
Vane shear test (VST)	5	8	10	14	10–20
Dilatometer test (DMT)	5	5	8	11	5–15
Pressuremeter test (PMT)	5	12	10	16	10–20[e]
Self-boring pressuremeter test (SBPMT)	8	15	8	19	15–25[e]

a – COV(Total) = $[COV(Equip.)^2 + COV(Proc.)^2 + COV(Random)^2]^{0.5}$
b – Because of limited data and the judgment involved in estimating COVs, ranges represent probable magnitudes of field test measurement error
c – Best to worst case scenarios, respectively, for SPT
d – Tip and side resistances, respectively, for CPT
e – It is likely that results may differ for p_o, p_f, and p_L, but the data are insufficient to clarify this issue

dilatometer test are the smallest. Because of the limited data available and the need to use judgment to estimate these errors, the last column of Table 3.5 represents the range of probable total measurement error one can expect in typical field tests.

3.5 TRANSFORMATION UNCERTAINTY

The direct measurement from a geotechnical test typically is not directly applicable to design. Instead, a transformation model is needed to relate the test measurement to an appropriate design property. Some degree of uncertainty will be introduced, because many transformation models in geotechnical engineering are obtained by empirical or semi-empirical data fitting. Transformation uncertainty would still be present even for theoretical relationships because of idealizations and simplifications in the theory. The data scatter about the transformation model can be quantified using probabilistic methods, as illustrated in Figure 3.2. In this approach, the transformation model is typically evaluated using regression analyses. More general approaches are available to quantify uncertainties beyond the pairwise correlations (refer to Chapter 4). The spread of the data about the regression curve can be modeled in many instances as an additive zero-mean random variable (ε). In this case, the standard deviation of ε (s_ε) is an indicator of the magnitude of transformation uncertainty, as shown in Figure 3.2.

It is natural to define the transformation uncertainty as the standard deviation of the regression error (ε). This definition is consistent with the regression literature. However, there are three limitations associated with this approach. One, it is difficult to compare the precision between different transformation models using a non-normalized quantity such as the standard deviation. Second, it is not applicable to transformation models that are more rules of thumb than developed from a data-driven regression analysis. These models are typically biased and this bias (over- or under-predict actual value) is important to engineers. Third, a transformation model relates a set of input variables (test measurements) to an output variable (design property).

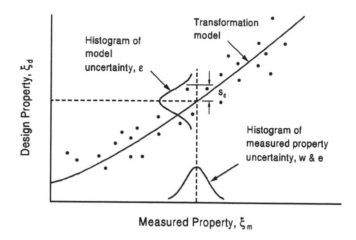

Figure 3.2 Transformation uncertainty resulting from pairwise correlation between a measured property and a desired design property.

The transformation uncertainty effectively describes the performance of this model. It is conceptually identical to the model uncertainty described in Chapter 5. The only difference is in the area of application. The transformation uncertainty is used to describe the inexact relationship between soil/rock properties. The model uncertainty is used to describe the inexact relationship between a measured and a calculated response, such as pile capacity or settlement. Ching and Phoon (2014) proposed a new definition for ε as:

$$\varepsilon = \frac{\text{actual target value}}{b \times \text{predicted target value}} \tag{3.1}$$

where the actual target value = measured value of the design property and predicted target value = estimated value of the design property from a transformation model. The product of a constant b and the predicted target value produces an unbiased prediction on the average, i.e. b = average bias. The random variable ε quantifies the deviation between the actual value and the unbiased prediction. The mean value of ε is 1 by definition. It is easy to see that the model factor defined in Eq. (5.1) is related to ε as follows:

$$M = b \times \varepsilon \tag{3.2}$$

In other words, the mean of M = b and COV of M = COV of ε. Examples of the average bias and COV of ε for clay properties and sand/gravel properties are given in Tables 4.2 and 4.3 in Chapter 4, respectively.

Transformation models are widely adopted in geotechnical engineering practice as a matter of practical expediency. Useful compilations of these models (mostly pairwise correlations) are available in the literature (e.g., Kulhawy & Mayne 1990, Mayne et al. 2001). A cursory review of these compilations would reveal a rather bewildering variety and number of models. There are many different tests (e.g., standard penetration test, cone penetration test) and even different transformation models linking the same measured parameter (e.g., cone tip resistance) and design parameter (e.g., undrained shear strength). There is a large variety of transformation models, because many were developed for a specific geomaterial type and/or a specific locale. It is not judicious to apply these models indiscriminately to other sites without a proper appreciation of the geomaterial behavior and geology. Site-specific models are generally more precise than "global" models calibrated from data covering many sites (Ching and Phoon 2012). However, site-specific models can be significantly biased when applied to another site. This "site-specific" limitation is a distinctive and fundamental feature of geotechnical engineering practice. Geotechnical RBD must take cognizance of this limitation to avoid gross oversimplification of "ground truths". Bayesian model comparison (Cao and Wang 2014a) and model selection methods (Wang and Aladejare 2015) have been developed to assist in selection of site-specific models using limited site-specific observation data and prior knowledge such as typical ranges of geotechnical properties (Wang and Cao 2013). More discussions on selection of "site-specific" transformation model are given in Section 3.10.

The transformation uncertainties associated with these models are seldom analyzed with the same degree of rigour as those presented in Ching and Phoon (2012). The majority are empirical and do not contain sufficient information for statistical characterization. A first-order estimate of the transformation uncertainty can be obtained by noting that about two thirds of the data typically fall within one standard deviation of the transformation model. Even with this simple technique, only a limited number of models could be examined, because most models have been presented without their supporting data.

Although the uncertainties in these empirical models are unknown, they are likely to be as large as those indicated in Table 4 of Ching and Phoon (2012), particularly for the case of empirical models where two (or more) parameters are being linked together that are indirectly related. A good example is the standard penetration test (SPT) N-value. The N-value is the dynamic driving resistance for a particular type of sampler, yet it has been correlated with the soil consistency, relative density, vertical and horizontal soil stress state, drained and undrained strength, modulus, and liquefaction resistance. Although these parameters undoubtedly influence N indirectly, it is too much to expect that they all (singly or collectively) can be predicted reliably without incurring significant uncertainties.

From the above observations, it is clear that the uncertainty in a design soil parameter is a function of natural soil variability, measurement error, and transformation uncertainty. These components can be combined consistently using a simple second-moment probabilistic approach described in Phoon and Kulhawy (1999b). First-order approximate guidelines for COVs of some design soil parameters as a function of the test measurement, correlation equation, and soil type is presented in Table 3.6. An illustrative COV for a spatial average over 5 m is also presented in the table to highlight the critical need to identify the characteristic design parameter governing a specific limit state. For ultimate limit state problems, this characteristic design parameter is typically a spatially averaged strength over the most critical failure path. In the presence spatial variability, the COV of this spatially averaged strength is smaller than the COV of the point strength; the degree of COV reduction is a function of the scale of fluctuation discussed in Section 3.6. It is clearly inappropriate to apply the point COV to a problem where the COV reduction is significant, say because the scale of fluctuation is short relative to some characteristic length scale of the failure path length (e.g., height of slope, diameter of tunnel, depth of excavation).

Table 3.6 Approximate guidelines for coefficients of variation of some design soil parameters (Source: Table 5, Phoon and Kulhawy 1999b).

Design Property[a]	Test[b]	Soil type	Point COV (%)	Spatial Avg. COV[c] (%)	Correlation Equation[f]
s_u(UC)	Direct (lab)	Clay	20–55	10–40	–
s_u(UU)	Direct (lab)	Clay	10–35	7–25	–
s_u(CIUC)	Direct (lab)	Clay	20–45	10–30	–
s_u(field)	VST	Clay	15–50	15–50	14
s_u(UU)	q_T	Clay	30–40[d]	30–35[d]	18
s_u(CIUC)	q_T	Clay	35–50[d]	35–40[d]	18
s_u(UU)	N	Clay	40–60	40–55	23
s_u^e	K_D	Clay	30–55	30–55	29
s_u(field)	PI	Clay	30–55[d]	–	32
ϕ	Direct (lab)	Clay, sand	7–20	6–20	–
ϕ(TC)	q_T	Sand	10–15[d]	10[d]	38
ϕ_{cv}	PI	Clay	15–20[d]	15–20[d]	43
K_o	Direct (SBPMT)	Clay	20–45	15–45	–
K_o	Direct (SBPMT)	Sand	25–55	20–55	–
K_o	K_D	Clay	35–50[d]	35–50[d]	49
K_o	N	Clay	40–75[d]	–	54
E_{PMT}	Direct (PMT)	Sand	20–70	15–70	–
E_D	Direct (DMT)	Sand	15–70	10–70	–
E_{PMT}	N	Clay	85–95	85–95	61
E_D	N	Silt	40–60	35–55	64

a – s_u = undrained shear strength; UU = unconsolidated-undrained triaxial compression test; UC = unconfined compression test; CIUC = consolidated isotropic undrained triaxial compression test; s_u(field) = corrected s_u from vane shear test; ϕ = effective stress friction angle; TC = triaxial compression; ϕ_{cv} = constant volume ϕ; K_o = in-situ horizontal stress coefficient; E_{PMT} = pressure-meter modulus; E_D = dilatometer modulus
b – VST = vane shear test; q_T = corrected cone tip resistance; N = standard penetration test blow count; K_D = dilatometer horizontal stress index; PI = plasticity index
c – averaging over 5 m
d – COV is a function of the mean; refer to COV equations in Phoon & Kulhawy (1999b) for details
e – mixture of s_u from UU, UC, and VST
f – Equation numbering in Phoon & Kulhawy (1999b)

Two observations are noteworthy here. One, it is important to distinguish between the definition of a characteristic design parameter from a physics viewpoint and the definition of a characteristic value from a statistical viewpoint. The customary definition of a lower 5% quantile as a characteristic value must be interpreted as the lower 5% quantile of a probability distribution pertaining to the appropriate spatially averaged strength or other design parameters similarly defined based on the physics of the problem. Second, the spatially averaged strength along the failure path is not identical to the spatially averaged strength along a prescribed line drawn in a spatially variable medium. The latter has been studied fairly extensively (Vanmarcke 2010), while the former is gradually being recognized (Ching and Phoon 2013; Ching et al. 2014b).

The ranges of COVs shown in Table 3.6 are based on representative statistics of natural variability and measurement error, as presented in Section 3.3 and 3.4. More accurate COVs can be calculated by substituting site-specific data on natural variability and measurement error into the closed-form COV equations given in Phoon and Kulhawy (1999b) or comparable equations based on the proposed second-moment probabilistic approach. The COVs of the undrained shear strengths determined by several different methods were found to be in the range of 10–60%. For the undrained shear strength predicted from the standard penetration N-value, higher COVs emerge when "global" relationships are used that are not calibrated to a specific geology. The probable range of COV for the undrained shear strength is estimated to be between 10 and 70%. The COV of the friction angle for sand and clay was found to be between 5 and 20%. For the in situ horizontal stress coefficient (K_o), the COV was found to be in the range of 20–80% for clay, depending on the method of evaluation. The corresponding range of COV for sand, which was found to be in the range of 25–55%, would only be applicable to the direct determination of K_o. The COV for indirect methods of evaluation could not be evaluated because the uncertainties underlying the transformation models were not available. The COV of soil modulus was found to be highest. Even for direct methods of evaluation, the COV was found to be in the range of 20–70%. Higher COVs were obtained for correlations with N, particularly if the correlation is not restricted to a specific geology. The probable range of COV for soil modulus is estimated to be on the order of 30–90%.

The upshot is that a design soil parameter and its probability distribution must depend on the site condition, the measurement method, and the transformation model. The mean or another characteristic value, say 5% quantile, is one aspect of this probability distribution that is routinely estimated in practice. The COV is merely another aspect of the probability distribution and it must also depend on the site variability, measurement precision, and transformation quality. This aspect is emphasized in Section D.2 of ISO2394:2015. It suffices to note that assigning a single COV value to a design soil parameter without reference to the property evaluation methodology is an example of gross over-simplification. For example, a COV of 30% for undrained shear strength may be appropriate for good quality laboratory measurements or direct correlations from field measurements such as the cone penetration test (CPT). It may not be appropriate for indirect correlations based on the standard penetration test (SPT). The practical impact of this observation is that it is unrealistic to calibrate a single value for each resistance/partial factor in a simplified RBD format. This practice is realistic for structural engineering, because the COV of manufactured materials can be controlled within a narrow range, say between 5 and 15%. The COV of the undrained shear strength, on the other hand, can vary between 10% and 70%. Based on reliability calibration studies for foundations (Phoon et al. 1995), Phoon and Kulhawy (2008) proposed a reasonably practical three-tier classification scheme (Table 3.7) for calibration of resistance/partial factors in simplified RBD. Based on this scheme, each resistance/partial factor can take a different numerical value depending on the level of property variability (low, medium, high) judged to be appropriate for a specific design scenario. This scheme is illustrated in Figure D.3 of ISO2394:2015. A similar approach was adopted by Paikowsky et al. (2004) in their reliability calibration of resistance factors for deep foundations. It appears that site variability is divided into low (COV < 25%), medium (25% < COV < 40%), and high (COV > 40%).

The 2014 Canadian Highway Bridge Design Code or CHBDC (CAN/CSA-S6-14:2014) also followed a similar strategy in allowing a resistance factor to take on different values depending on the degree of "understanding" (low, typical, high). The degree of understanding covers the quality of site information and the quality of performance prediction. It is possible to envisage Table 3.7 being expanded eventually

Table 3.7 Three-tier classification scheme of soil property variability for reliability calibration (Source: Table 9.7, Phoon & Kulhawy 2008).

Geotechnical parameter	Property variability	COV (%)
Undrained shear strength	Low[a]	10–30
	Medium[b]	30–50
	High[c]	50–70
Effective stress friction angle	Low[a]	5–10
	Medium[b]	10–15
	High[c]	15–20
Horizontal stress coefficient	Low[a]	30–50
	Medium[b]	50–70
	High[c]	70–90

a – typical of good quality direct lab or field measurements
b – typical of indirect correlations with good field data, except for the standard penetration test (SPT)
c – typical of indirect correlations with SPT field data and with strictly empirical correlations

to classify the complete gamut of information, which could include both pre-design information (e.g., prior experience, site investigation, prototype test) and post-design information (e.g., quality control, monitoring). It is safe to say that Table 3.7 is a step in the right direction to establish a closer linkage to good practice.

It may be noted in passing that the Load and Resistance Factor Design (LRFD) (Section 6.2) involves comparing a factored resistance with the sum of two or more factored loads. A factored resistance is the product of a resistance factor and a characteristic/nominal resistance. Ching and Phoon (2011) noted that a quantile-based characteristic resistance could be used to maintain a relatively uniform level of reliability over a wide range of COVs without applying a resistance factor. Their proposed method is called the Quantile Value Method (QVM) (Section 6.5). This alternate strategy of keeping the resistance factor constant while adjusting the characteristic resistance to handle different site conditions bears some semblance to the Eurocode 7 partial factor approach (EN 1997–1:2004). However, QVM is a form of simplified RBD while Eurocode 7 is not. The main idea here is that geotechnical design is less amenable to standardization than structural design and the engineer should be able to exercise his/her judgment to adjust the resistance factor and/or characteristic resistance to suit a particular site.

3.6 SCALE OF FLUCTUATION

Soil is a natural material that has been formed by a combination of various geologic, environmental, and physical-chemical processes. Many of these processes are continuing and can be modifying the soil in-situ. Because of these natural processes, all soil properties in-situ will vary vertically and horizontally. As illustrated in Figure 3.3, this spatial variation can be decomposed conveniently into a smoothly varying trend function [t(z)] and a fluctuating component [w(z)]. This fluctuating component represents the natural soil variability.

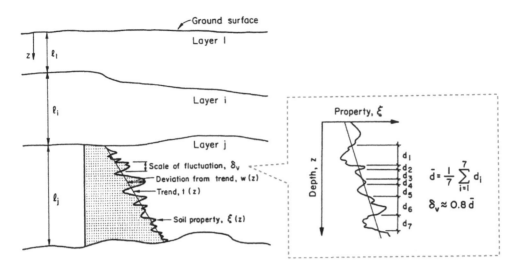

Figure 3.3 Random field model for natural soil variability (revised from Phoon and Kulhawy 1999a).

The natural variability is typically modeled as a stationary (statistically homogeneous) random field (Vanmarcke 1977). It is noteworthy that a physically homogeneous soil layer is not necessarily statistically homogeneous. Some reasonably practical methods have been proposed to identify these statistically homogeneous layers (Phoon et al. 2003; Uzielli et al. 2005). Methods are also available to identify simultaneously the statistically homogeneous layers and the associated statistical parameters in each statistically homogeneous layer (Cao and Wang 2013; Wang et al. 2013 & 2014). A critical statistical parameter that is needed to describe natural variability is the correlation distance or scale of fluctuation. The scale of fluctuation provides an indication of the distance within which the property values show relatively strong correlation. A simple but approximate method of determining the scale of fluctuation is shown as an insert in Figure 3.3. However, this estimation method is only applicable to the squared exponential autocorrelation function (Vanmarcke 1977 citing Rice 1944, 1945).

When the scale of fluctuation is very short, the property value at one point is nearly independent of the property value at another point, even if the distance apart is very short. This is manifested visually as properties varying rapidly with depth. This extreme case, called the independent case, is rare in a typical soil profile. When the scale of fluctuation is very long, the property value at one point is nearly equal to the property value at another point for a given random realization, even if the distance apart is very long. This is manifested visually as property values following a near constant trend with depth. This second extreme case, called the fully correlated (or random variable) case, is also rare in a typical soil profile. The practical importance of considering a reasonable scale of fluctuation, i.e. a reasonably realistic spatial variability, in the estimation of COV has been highlighted in Section 3.5. The assumption of independent soil parameters will produce an unconservative reduction of the point COV for the spatial average. The assumption of fully correlated soil parameters will not result in COV reduction for the spatial average, which is overly conservative. There are even more fundamental concerns beyond COV reductions related to these convenient but overly simplified assumptions. It suffices to note briefly that failure mechanisms are related to spatial variability. Examples are provided by Fenton & Griffiths (2008).

Table 3.8 provides a summary of scales of fluctuation reported in the literature. It is apparent that the amount of information on the scale of fluctuation is relatively limited in comparison to the amount of information on COV. Therefore, Table 3.8 should be viewed with caution, because there are insufficient data to establish their generality on a firm basis. However, it would appear the horizontal scale of fluctuation is about one order of magnitude larger than the vertical scale of fluctuation. Detailed studies on the scale of fluctuation are available, but rather limited in number (Jaksa 1995; Fenton 1999a; Uzielli et al. 2005).

In addition to the scale of fluctuation, a correlation function is needed in the random field modelling of natural variability, and the scale of fluctuation is generally used as an input parameter to the correlation function. Although a single exponential correlation function is frequently used in literature, several other correlation functions can also be used, such as the binary noise function and the squared exponential function (Fenton and Griffiths 2008; Cao and Wang 2014b). Based on the available site-specific observation data, the most suitable correlation function can be selected using a Bayesian model comparison method (Cao and Wang 2014b) and used in the subsequent random field modelling of natural variability.

Table 3.8 Scales of fluctuation of some geotechnical parameters (Source: Table 4, Phoon & Kulhawy 1999a).

Property[a]	Soil type	No. of studies	Scale of fluctuation (m) Range	Mean
Vertical direction				
s_u	Clay	5	0.8–6.1	2.5
q_c	Sand, clay	7	0.1–2.2	0.9
q_T	Clay	10	0.2–0.5	0.3
$s_u(VST)$	Clay	6	2.0–6.2	3.8
N	Sand	1	–	2.4
w_n	Clay, loam	3	1.6–12.7	5.7
w_L	Clay, loam	2	1.6–8.7	5.2
$\bar{\gamma}$	Clay	1	–	1.6
γ	Clay, loam	2	2.4–7.9	5.2
Horizontal direction				
q_c	Sand, clay	11	3.0–80.0	47.9
q_T	Clay	2	23.0–66.0	44.5
$s_u(VST)$	Clay	3	46.0–60.0	50.7
w_n	Clay	1	–	170.0

a – s_u = undrained shear strength from laboratory tests; $s_u(VST)$ = s_u from VST; q_c = CPT tip resistance; q_T = corrected CPT tip resistance; N = SPT blow count (number of blows per foot or 305 mm); w_n = natural water content; w_L = liquid limit; $\bar{\gamma}$ = effective unit weight; γ = total unit weight

3.7   INTACT ROCK AND ROCK MASS

3.7.1   Natural variability of intact rock

Several types of probability distribution for intact rock have been suggested in the literature. However, a type of probability distribution that is simple, but physically possible, is the lognormal probability distribution. Figure 3.4 shows an example of the application of this probability distribution on actual rock property distributions. As can be seen, the lognormal probability distribution fits well into the rock distributions, and therefore it could be used as the primary rock property probability distribution in the development of RBD procedures.

The variability of unweathered rock represented by the coefficient of variation (COV) was evaluated for several different index, strength, and stiffness properties. The index properties include the unit weight (γ and γ_d), porosity (n), Schmidt hammer hardness (R), and shore scleroscope hardness (S_h) (Kulhawy and Prakoso 2003). The strength properties include the uniaxial compressive strength (q_u), Brazilian tensile strength ($q_{t\text{-Brazilian}}$), and point load strength (I_s), while the stiffness property is the tangent Young's modulus at 50 percent of q_u ($E_{t\text{-}50}$) (Prakoso 2002). The typical COVs of inherent variability in some basic test measurements of intact rocks are summarized in Table 3.9. The mean, standard deviation (S.D.), and range of the COV are shown with the total number of data groups per test. It is noted that clastic and chemical sedimentary rocks dominate the database, followed by metamorphic non-foliated, intrusive igneous, and extrusive igneous rocks. The databases for metamorphic foliated and igneous pyroclastic rocks are small.

The mean and range of the COV of γ and γ_d are relatively small, and therefore these properties can be practically considered deterministic. The lower bound value of the COV of inherent variability of the porosity (n) remains relatively constant at about 5 to 10 percent over the range of mean values, but the upper bound on the COV appears to decrease with increasing m_n (mean of porosity). There is no apparent effect of rock type on the COV, but it can be noted that sedimentary clastic rock data dominate the higher m_n.

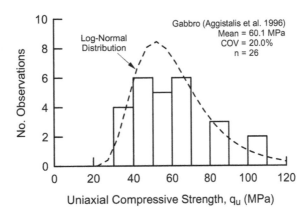

Figure 3.4 Example distributions of intact rock properties.

Table 3.9 COV of intact rock (Prakoso 2002).

Test Type	Property	Coefficient of Variation (%)			
		Number of data groups	Mean	S.D.	Range
Index	γ, γ_d	79	1.0	1.2	0.1–8.6
	n	30	24.2	18.6	3.0–71
	R	54	8.7	5.4	1.4–26
	S_h	59	11.1	8.5	1.4–38
Strength	q_u	174	14.0	11.7	0.8–61
	$q_{t\text{-Brazilian}}$	54	19.4	12.9	3.8–61
	I_s	66	20.5	14.3	2.8–59
Stiffness	$E_{t\text{-}50}$	72	20.5	16.9	1.4–69

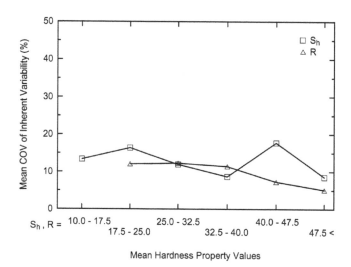

Figure 3.5 COV mean values versus mean hardness property values.

The mean values of COV inherent variability of the Schmidt hammer hardness (R) and the shore scleroscope hardness (S_h) are between 8 to 12 percent. The lower bound value of the COV of R remains relatively constant at about 5 percent over the range of mean values, while that of the COV of S_h remains relatively constant at less than 5 percent over the range of mean values. However, the upper bound of the COV for both hardness properties appears to decrease with increasing mean values. The overall trend of decreasing COV with increasing mean values for hardness properties as shown in Figure 3.5. There is no apparent effect of rock type on the COV and m_R (mean of Schmidt hammer hardness).

The strength properties considered were the uniaxial compressive strength (q_u), Brazilian tensile strength ($q_{t\text{-Brazilian}}$), and point load strength (I_s), while the stiffness property considered is the tangent Young's modulus at 50 percent of uniaxial compressive strength ($E_{t\text{-}50}$). In developing the databases, when several data groups

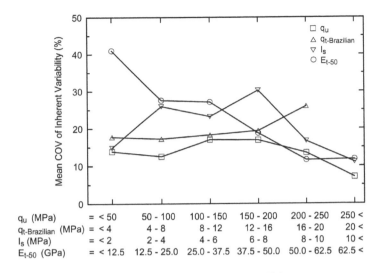

Figure 3.6 COV mean values versus mean strength and stiffness property values.

from the same site with different sample diameters (B_{sample}), moisture contents, or core orientations were available, only one data group was used. The one chosen preferably had B_{sample} closer to 50–58 mm, was saturated, and was perpendicular to the bedding or foliation.

The mean values of the COV of inherent variability of q_u, $q_{t\text{-Brazilian}}$, I_s, and $E_{t\text{-50}}$ are between 14 to 21 percent. The lower bound values of these COV ranges remain relatively constant at about 5 percent over the range of their respective mean values, m_{qu}, $m_{qt\text{-Brazilian}}$, m_{Is}, and $m_{Et\text{-50}}$. However, the upper bound values of the COV appears to decrease with increasing mean values. The overall trend of decreasing COV with increasing mean values for strength and stiffness properties as shown in Figure 3.6. There is no apparent effect of rock type on the COV and m_{qu}.

Statistical evaluation of $m_{i\text{-GSI}}$

Hoek and Brown (1980) introduced an empirical method to estimate the rock mass strength, and Hoek et al. (1995) subsequently introduced the concept of Geological Strength Index (GSI). One of the parameters required is the Hoek-Brown intact strength constant based on triaxial test results ($m_{i\text{-GSI}}$). The reader should not confuse the symbol "m" associated with the Hoek-Brown intact strength constant with the symbol "m" denoting mean in this section. Doruk (1991) conducted a comprehensive evaluation of the intact rock parameter $m_{i\text{-GSI}}$. In this study, these results were used to evaluate the uncertainty of $m_{i\text{-GSI}}$, and the statistical results are given in Table 3.10. The probability distributions of $m_{i\text{-GSI}}$ can in general be represented by a lognormal probability distribution. The mean COV of $m_{i\text{-GSI}}$ is significantly greater than the mean COV of q_u and other strength parameters previously discussed.

Table 3.10 Hoek-Brown $m_{i\text{-}GSI}$ parameters (data after Doruk, 1991).

Rock type	Hoek-Brown $m_{i\text{-}GSI}$ parameter			
	Number of data groups	Mean, $m_{m_{i\text{-}GSI}}$	Range, $r_{m_{i\text{-}GSI}}$	$COV_{m_{i\text{-}GSI}}$ (%)
Granite	18	25.3	8–43	37.7
Dolerite	4	13.2	11–15	14.7
Granodiorite	4	26.0	16–35	31.4
Sandstone	57	16.0	3–42	53.8
Mudstone	7	19.2	9–47	75.8
Shale	3	14.6	3–29	91.9
Chalk	2	7.2	–	–
Limestone	25	9.6	4–26	47.3
Dolostone	8	11.4	5–18	37.7
Carnallitite	5	20.8	3–46	94.7
Amphibolite	3	27.8	24–33	16.7
Quartzite	6	20.4	15–28	24.9
Marble	14	8.1	5–16	39.5
			Mean =	47.2
			S.D. =	27.1

COV comparison

The mean values of the COV of inherent variability of q_u, $q_{t\text{-Brazilian}}$, I_s, and $E_{t\text{-}50}$ are within a relatively narrow range of 14 to 21 percent, and the mean COV values for different tests shown in Figure 3.5 are relatively within a narrow band. This relatively minor difference in the mean COV values suggests that the effect of test type is minor.

Furthermore, q_u often is estimated from the results of the hardness tests (Kulhawy and Prakoso 2003), and $E_{t\text{-}50}$ is commonly estimated from q_u. The relationship between the COV of q_u and that of the hardness test results, and the relationship between the COV of $E_{t\text{-}50}$ and that of q_u, are evaluated. The mean and range of the COV of q_u given in Table 3.9 are greater than those of R and S_h. The individual COV values of q_u are plotted versus their corresponding COV of R and S_h in Figure 3.7, and only 55 percent of data groups falls below the 1:1 line. The mean and range of the COV of $E_{t\text{-}50}$ given given in Table 3.9 are somewhat greater than those for q_u. The individual COV values of $E_{t\text{-}50}$ are plotted versus their corresponding COV of q_u in Figure 3.7, and only 42 percent of the data groups falls below the 1:1 line shown in Figure 3.7. These results indicate that COV values of different tests are practically similar.

Effect of weathering

The variability of weathered rock is evaluated based on the results of several different testing methods. Three sets of rock data are given in Table 3.11 to illustrate the effect of different weathering conditions on the natural variability of intact rocks. The COV values of these tests tend to increase as the weathering progresses. Furthermore, Table 3.11 also indicates that the COV of properties with mixed weathering states (e.g., unweathered and slightly weathered and slightly and moderately weathered) tends to be greater than those of properties related to a single weathering state.

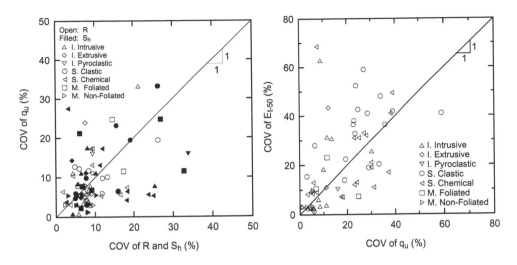

Figure 3.7 Comparison of COV values from different tests.

Table 3.11 Effect of weathering on COV (Prakoso 2002).

| Rock Type | Test Type | Coefficient of Variation (%) | | | |
		I	II	III	Combined
Gabbro	q_u	33.2	51.5	–	52.8
	I_s	46.8	47.1	–	53.9
Basalt	q_u	23.9	26.8	–	41.0
	I_s	26.1	45.7	–	39.6
Granite,	q_u	–	28.6	31.1	51.1
saturated	$q_{t-Brazilian}$	–	28.9	30.6	53.2
	I_s	–	28.5	30.2	53.4
	$q_{t-direct}$	–	27.4	36.4	56.0
	E_{t-50}	–	18.8	35.2	49.9

Note: I = fresh, II = slightly weathered, III = moderately weathered; number of samples > 10.

3.7.2 Intact rock measurement error

The intact rock measurement error and the associated transformation uncertainty have been extensively discussed by Prakoso and Kulhawy (2011). The correlations between the laboratory uniaxial compressive strength (q_u), Brazilian indirect tensile strength ($q_{t-Brazilian}$), and point load strength (I_s) and the sample diameter were developed to quantify the effect of sample diameter. The associated direct correlations among these strength test types also were developed. These correlations show relatively large uncertainties in the estimated intact rock strength. In spite of the large uncertainties associated with the use of correlations to consider the effects of sample diameter on the actual intact rock strength, the COV of these rock strength parameters with different sample diameters is not significantly affected by the sample diameter. For both q_u and $q_{t-Brazilian}$, the median value of the maximum change in the COV for individual data

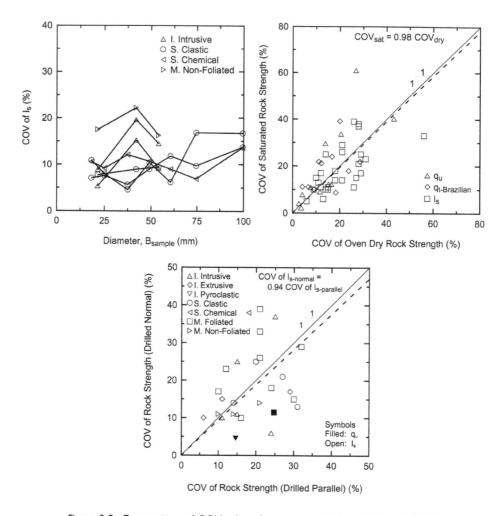

Figure 3.8 Comparison of COV values from test conditions (Prakoso 2002).

groups is about 5%. The median value of the maximum change for $q_{t\text{-Brazilian}}$ and I_s are about 8% as shown in Figure 3.8. In addition, the general rock types appear not to affect the maximum change in the COV.

The rock strength q_u, $q_{t\text{-Brazilian}}$, and I_s obtained from dry and saturated samples were also proposed. The strength decreases with increasing moisture content. In spite of the large uncertainties associated with the use of correlations to consider the effect of moisture content on the actual intact rock strength, the COV is not significantly affected by the moisture content. When the oven-dried COV data are directly compared to the saturated COV data, the data points lay about the 1:1 line as shown in Figure 3.8. For the air-dried COV data versus the saturated COV data, the data points lay about the 1:1 line, and the spread is minimal. The two observations suggest that different drying procedures most likely would not affect the level of uncertainty significantly.

The effect of the core orientation relative to loading direction on the COV of rock uniaxial compressive strength (q_u) and point load strength (I_s) was reported by Prakoso

Table 3.12 Scale of fluctuation of intact rock strength.

Rock Type	Direction	Test Type	Scale of Fluctuation, δ (m)
Shale	Vertical	$q_{t\text{-Brazilian}}$	1.22
Sandstone	Horizontal	I_s	0.61

Table 3.13 COV of rock mass Young's modulus from load tests.

		Rock Mass Modulus, E_m	
Rock Name	Number of Data	Mean, m_{Em} (MPa)	COV_{Em} (%)
Iron ore	12	406.3	66.6
"Paint" rock	4	205.5	49.9
Ash rock	4	484.4	62.0
Pomona basalt: vertical	40	9441	63.4
Pomona basalt: horizontal	36	17908	67.2
Dworshak granite	24	23265	51.1
Sandstone	5	178.6	47.2
Mudstone: S-series	17	198.4	71.9
Mudstone: M-series	21	615.3	48.2
Shale	5	354.0	21.1
Shale	3	345.0	9.9
Shale	3	3020.0	29.1
Sandstone	3	148.7	13.8
Clay-shale	3	418.3	78.6
		Mean =	47.9
		S.D =	21.8

(2002). The strength of intact rock might change with changing core orientation. The individual COV values of rocks drilled normal to the foliation or bedding are plotted versus their corresponding COV of rocks drilled parallel with foliation or bedding in Figure 3.7, and only 55 percent of data groups falls below the 1:1 line, suggesting the COV practically independent of the orientation.

3.7.3   Intact rock scale of fluctuation

The information in the literature on the scale of fluctuation of intact rock properties is very limited. Table 3.12 shows the scale of fluctuation for two sedimentary clastic rocks as reported by Prakoso (2002). For comparison, it can be seen from Table 3.8 that the horizontal scale of fluctuation for soils is typically of an order of 10 m or larger.

3.7.4   Rock mass natural variability

The rock mass natural variability herein is represented by the COV of the rock mass Young's modulus (E_m). The uncertainty of E_m was evaluated based on the back-calculated E_m from field load tests taken from 9 reports as summarized by Prakoso and Kulhawy (2004). The data base is given in Table 3.13, containing igneous intrusive, igneous extrusive, and sedimentary clastic rock types only. The mean COV of E_m is significantly greater than the mean COV of E_{t-50} previously discussed.

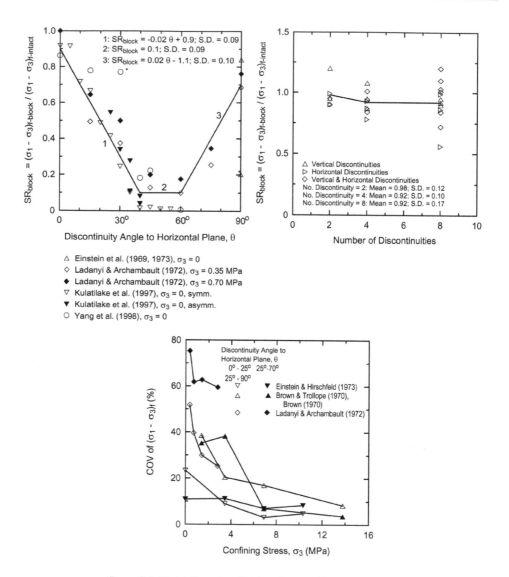

Figure 3.9 Variability of artificial rock mass (Prakoso 2002).

Furthermore, Prakoso and Kulhawy (2004) have also reported the variability of artificial rock mass made of concrete blocks, including the effects of orientation of discontinuities, number of discontinuities, and confining pressures as shown in Figure 3.9. The rock mass strength is represented by the rock mass to intact rock strength ratio (SR_{block}). The left figure indicates that the variability of rock mass strength is dependent on the rock mass predominant discontinuity angle and reaches its maximum for discontinuity angle between 40° and 60°. The right figure indicates that the variability of SR_{block} appears to increase with increasing number of discontinuities. The bottom figure indicates that the variability decreases with increasing confining pressure.

3.7.5 Rock mass transformation uncertainty

The rock mass transformation uncertainty herein is represented by the rock mass Young's modulus (E_m), which often is estimated from the intact rock Young's modulus (E_{t-50}) using the ratio α_E ($= E_m/E_{t-50}$). However, α_E is not a fundamental parameter and is a lumped parameter of the intact rock properties and the discontinuity frequency and properties. Heuze (1980) compiled and evaluated a number of physical α_E evaluations from a site, which are reproduced in Table 3.14 as a reference. The COV range is from 49 to 81 percent, relatively comparable to the COV range of direct back-calculated E_m from field load tests previously discussed.

3.8 STATISTICAL UNCERTAINTY FOR SITE-SPECIFIC NATURAL VARIABILITY

Three major variabilities are discussed so far, including natural variability (Sections 3.3 & 3.7), measurement error (Sections 3.4 & 3.7), and transformation uncertainty (Sections 3.5 & 3.7). The natural variability and transformation uncertainty are potentially site-specific. This section and the next section consider the statistical uncertainty for site-specific natural variability, whereas Section 3.10 considers the statistical uncertainty for site-specific transformation. Measurement error is not site-specific because it is typically related to the equipment, procedure, and operator.

3.8.1 Statistical uncertainty in site-specific trend

The trend for natural variability is clearly site-specific. The site-specific trend can be estimated using regression based on the site investigation data, e.g., CPT sounding. It is customary to assume that the estimated trend is the same as the actual trend. This is the underlying assumption for de-trending: data after de-trending (residuals) are treated as zero mean data (e.g., Fenton 1999a; Uzielli et al. 2005). However, the de-trended data will not have zero mean if the estimated trend is not the actual trend. Past studies have recognized that de-trending deserves more rigorous attention (e.g., Kulatilake 1991; Li 1991; Jaksa et al. 1997; Fenton 1999b).

The estimated trend is in principle not the same as the actual trend. The deviation between the estimated and actual trends is the statistical uncertainty for the trend. Honjo and Setiawan (2007) focused on the statistical uncertainty in site-specific trend. Given the site-specific investigation data, a framework is proposed to characterize the statistical uncertainty for the spatial average along a certain depth. In their framework, the standard deviation (or COV) and scale of fluctuation (SOF) for the site-specific natural variability are prescribed rather than estimated from data. The rationale for prescribing COV and SOF is based on the fact that the number of data points in site investigation is typically insufficient to estimate these second-order statistics. To circumvent this practical difficulty, Honjo and Setiawan (2007) suggested that conservative values for COV and SOF can be assumed based on past experiences.

Table 3.14 Rock mass modulus to intact rock modulus ratio (modified after Heuze 1980).

| Test Type | Number of Data | $\alpha_E = E_m/E_{t-50}$ | | |
		Mean, $m_{\alpha E}$	S.D., $s_{\alpha E}$	$COV_{\alpha E}$ (%)
Plate bearing	27	0.32	0.26	81
Full scale deformation	14	0.44	0.26	59
Flat jacks	10	0.54	0.27	50
Borehole jack or dilatometer	9	0.33	0.17	52
Pressure chamber	8	0.45	0.22	49

3.8.2 Statistical uncertainty of site-specific COV and SOF

The COV and SOF for natural variability are also site-specific, because the COV and SOF at one site are typically not the same as those at another site. The ranges summarized in previous tables represent past experiences in the literature. Although it is possible to assume conservative values for COV and SOF based on these tables, as suggested by Honjo and Setiawan (2007), there are practical difficulties for doing so. First of all, COV and SOF values in these tables vary in a wide range. For instance, the COV for the natural variability of the undrained shear strength of a clay varies from 6% to 80% (Table 3.1). Its vertical SOF varies from 0.8 to 6.2 m (Table 3.8). Its horizontal SOF is known in a very limited way [only 3 studies in Table 3.8, ranging from 46 to 60 m; another 3 studies collected by El-Ramly et al. (2003), ranging from 22 to 40 m]. If the conservative COV value is taken to be 80% (the upper bound), this would be too conservative for most sites. If 80% is excessively conservative, which COV is reasonably conservative? It is not trivial to answer this question. The same question can be asked when SOF is selected based on the ranges in Table 3.8. Moreover, SOF may depend on the problem scale (Fenton 1999a) and COV and SOF may depend on the adopted trend function and the sampling interval as well (Cafaro and Cherubini 2002). The scale considered in previous studies may not be similar to the scale applicable for the geotechnical project at hand. The trend function and sampling interval studied in the literature may not be applicable to the conditions in the project at hand.

Jaksa et al. (2005) took a different strategy: they suggested using a "worst case" SOF, which for the example of a 3-storey, nine-pad footing building examined, is equal to the spacing between footings. This "worst case" strategy circumvents the need to estimate SOF from past experiences. However, "the spacing between footing" is only applicable to the example studied in Jaksa et al. (2005). Table 3.15 shows the worse-case SOFs reported in previous studies. The worst-case SOF is typically comparable to some multiple of the characteristic length of the structure (e.g., width of footing, height of retaining wall, diameter of tunnel, depth of excavation). However, there is no universal way of determining the "worse-case" SOF.

It is more prudent to use the site investigation data to obtain the site-specific COV and SOF than to assume their values using past experiences or to assume SOF to be the "worse-case" SOF. However, the main difficulty lies in the fact that the amount of available information in a typical site investigation program is not sufficient to

Table 3.15 Worse-case SOFs reported in previous studies.

Study	Problem type	"Worse case" definition	Characteristic length	Worse-case SOF
Jaksa et al. (2005)	Settlement of a nine-pad footing system	Under-design probability is maximal	Footing spacing (S)	1 × S
Fenton and Griffiths (2003) Soubra et al. (2008)	Bearing capacity of a footing on a c-ϕ soil	Mean bearing capacity is minimal	Footing width (B)	1 × B
Fenton et al. (2005)	Active lateral force for a retaining wall	Under-design probability is maximal	Wall height (H)	0.5∼1 × H
Breysse et al. (2005)	Settlement of a footing system	Footing rotation is maximal	Footing spacing (S)	0.5 × S
		Different settlement between footings is maximal	Footing spacing (S)	f(S,B)
			Footing width (B)	(no simple equation)
Griffiths et al. (2006)	Bearing capacity of footing(s) on a ϕ = 0 soil	Mean bearing capacity is minimal	Footing width (B)	0.5∼2 × B
Ching and Phoon (2013) Ching et al. (2014b)	Overall strength of a soil column	Mean strength is minimal	Column width (W)	1 × W (compression) 0 × W (simple shear)
Hu and Ching (2015)	Active lateral force for a retaining wall	Mean active lateral force is maximal	Wall height (H)	0.2 × H

accurately determine the site-specific COV and SOF. Ching et al. (2016a) show that the vertical SOF in a soil property cannot be estimated accurately if the total depth of the investigation data is less than 20 times of the actual vertical SOF. Suppose the actual vertical SOF is 0.5 m, this means that the total depth has to be larger than 10 m. However, many soil layers have thicknesses less than 10 m. Ching et al. (2016a) also show that the COV cannot be estimated accurately if the total depth is less than 4 times of the vertical SOF. This means that the total depth has to be larger than 2 m. Besides the minimum requirement for the total depth, Ching et al. (2016a) show another requirement regarding the sampling interval. The vertical SOF cannot be estimated accurately if the sampling interval is larger than 1/2 of the vertical SOF. Suppose the actual vertical SOF is 0.5 m, this means that the sampling interval has to be less than 0.25 m. This rules out most in-situ tests, e.g., SPT test typically has sampling interval equal to 1 to 2 m. In contrast, CPT sounding has an advantage due to its small sampling interval of 2 cm.

The deviation between the estimated COV and SOF and their actual values is the statistical uncertainty for COV and SOF. As explicitly stated in Section D.1 item c ISO2394:2015, statistical uncertainty should be handled with much care because the amount of information collected in a site investigation is limited. In Eq. (26) in ISO2394:2015, it is also explicitly stated that the determination of the partial factor (resistance factor) should incorporate the statistical uncertainty in the resistance. As discussed above, significant deviation can occur if the data depth is not sufficiently large or if the sampling interval is not sufficiently small. The former issue (data depth not sufficiently large) is critical to geotechnical engineering practice, because it is quite common to have thin soil layers. Ching et al. (2016a) show that for small data depth (e.g., thin layers), there is a strong tradeoff between estimated COV and SOF: there are numerous combinations of COV and SOF that are all plausible with respect to the observed site-specific data. Moreover, these plausible combinations of COV and SOF form a manifold, a long and narrow region in the two-dimensional space. It is impossible to simultaneously estimate site-specific COV and SOF with satisfactory accuracy for a thin soil layer. However, prior information may be useful in narrowing the possible ranges for COV and SOF. The use of prior information with a Bayesian framework is discussed below.

To illustrate the statistical uncertainty in the trend, standard deviation (or COV), and SOF, consider a CPT sounding at Wufeng District in Taichung City (Taiwan). This CPT sounding was analyzed in Ching et al. (2016a). Figure 3.11a and 3.11b show the CPT data (q_t and f_s), together with the soil behaviour type index profile (I_c) (Robertson 2009) in Figure 3.11c. The vertical data interval is 0.05 m. The CPT-based stratification result based on I_c is also shown in Figure 3.11c. The statistical uncertainty for the site-specific trend, standard deviation, and SOF will be illustrated for two soil layers – one sand layer and one clay layer, shown in Figure 3.11c.

The trend, standard deviation, and SOF for the logarithm of the normalized cone resistance Q_{tn} (Robertson 2009) are of concern:

$$Q_{tn} = [(q_t - \sigma_{v0})/P_a] \times (P_a/\sigma'_{v0})^n \qquad (3.3)$$

where q_t is the (corrected) cone resistance; $P_a = 101.3 \, kN/m^2$ is one atmosphere pressure; σ'_{v0} and σ_{v0} are the effective and total overburden stresses, respectively; $n = 0.5$ for sand and $n = 1$ for clay. Figure 3.12 shows the Q_{tn} profiles within these two soil layers in the logarithmic scale. The total depths (D) for the CPT records are 0.80 m and 4.55 m for the sand and clay layers, respectively. The sand layer exemplifies a thin soil layer, whereas the clay layer exemplifies a thick soil layer. Figure 3.13 shows the multivariate probability density function (PDF) of the trend (μ), standard deviation (σ), and SOF (δ) for the sand layer (thin layer) updated by the site-specific data. Although the trend μ can be estimated with a reasonable accuracy, there is a tradeoff region (manifold) between σ and δ. It is not possible to accurately estimate σ and δ simultaneously. The statistical uncertainty is significant. Figure 3.14 shows the multivariate PDF for the clay layer (thick layer). It is clear that σ and δ can now be estimated with a reasonable accuracy. The statistical uncertainty is less significant.

Ching et al. (2016a) proposed a Bayesian framework of characterizing the statistical uncertainty for site-specific trend, COV, and SOF. The past experiences, such as the ranges summarized in Tables 3.1, 3.2, 3.3, and 3.8, are used to construct the prior probability density function (PDF) for COV and SOF, whereas a flat prior PDF is taken for the trend. Then, the site investigation data is used to update this multivariate PDF of trend, COV, and SOF. The resulting multivariate posterior PDF combines the past experiences and site investigation data. The posterior PDF is not of standard type, e.g., multivariate normal or lognormal, but its random samples can be drawn using the Markov chain Monte Carlo (MCMC) method. Ching et al.

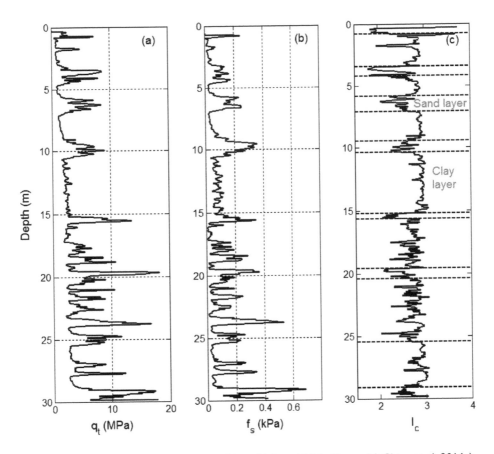

Figure 3.11 CPT data at the Wufeng site: (a) q_t; (b) f_s; and (c) I_c (Figure 14, Ching et al. 2016a).

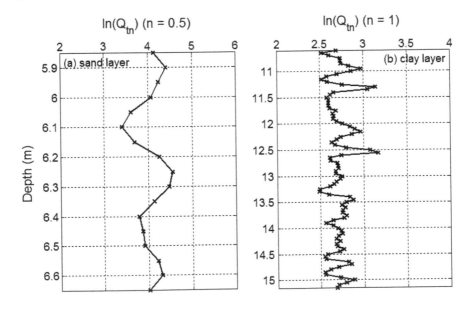

Figure 3.12 Q_{tn} profiles: (a) sand layer and (b) clay layer (Figure 15, Ching et al. 2016a).

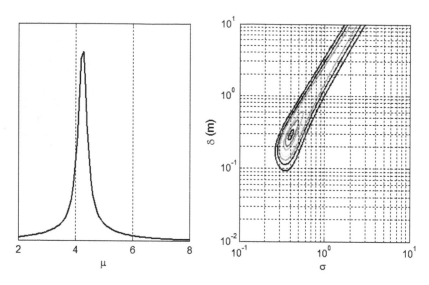

Figure 3.13 Multivariate PDF of μ, σ, and δ for the sand layer: (a) the PDF for μ; (b) the contours for the multivariate PDF of σ and δ.

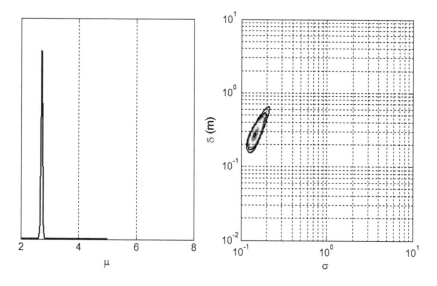

Figure 3.14 Multivariate PDF of μ, σ, and δ for the clay layer: (a) the PDF for μ; (b) the contours for the multivariate PDF of σ and δ.

(2016a) show that an equivalent multivariate normal PDF for trend, COV, and SOF can be constructed using the MCMC samples. Based on a large number of simulated examples, they further show that the 95% Bayesian confidence region (an ellipsoid) of this equivalent posterior PDF contains the actual trend, COV, and SOF with a chance that is close to 95%. This implies the equivalent posterior PDF is effective and consistent in characterizing the statistical uncertainty in trend, COV, and SOF. This conclusion holds for both thin soil layers (data depth < 20 × actual SOF) and thick soil layers (data depth ⩾ 20 × actual SOF). For thin soil layers, the equivalent posterior PDF has large uncertainty, and the confidence region is a large ellipsoid that is constrained by past experiences, and by the aforementioned tradeoff manifold as well. For thick soil layers, the equivalent posterior PDF has small uncertainty, and the confidence region is a small ellipsoid that centers at the most probable value for the trend, COV, and SOF.

Ching et al. (2016b) showed that the impact of the statistical uncertainty in the trend, COV, and SOF is to bring in extra uncertainty into the problem, hence to increase the failure probability estimate. For Load and Resistance Factor Design (LRFD), it means a smaller resistance factor is required to cater for the uncertainty in the resistance and the statistical uncertainty arising from estimation of the resistance from limited tests. They also pointed out that the common practice that only considers the point estimates for the trend, COV, and SOF is unconservative, in the sense that the failure probability can be greatly underestimated. The statistical uncertainty for site-specific natural variability deserves further research. Honjo and Setiawan (2007) and Ching et al. (2016a) have made some progress on this important subject, but both studies adopt various simplifications. Honjo and Setiawan (2007) prescribe COV and SOF values and only focus on the statistical uncertainty in the site-specific trend. Ching et al. (2016a) relax this simplification and can characterize the statistical uncertainty in the site-specific trend, COV, and SOF simultaneously. But, they assume the trend function has a fixed functional form, e.g., a linear function with unknown intercept and gradient. In reality, the functional form for the trend function is also unknown. Moreover, SOF may not even be a constant parameter, as discussed by Fenton (1999a). The SOF characterized by a small scale test may not be applicable to a large scale construction project. Finally, it is not clear how the statistical uncertainty in the site-specific natural variability can be systematically incorporated in simplified RBD, say by applying a reduction factor to the resistance factor in LRFD or to the quantile in QVM.

3.9 BAYESIAN QUANTIFICATION OF SITE-SPECIFIC NATURAL VARIABILITY

Although the natural variability, measurement errors, statistical uncertainty, and transformation uncertainty are usually lumped together as total variability and used subsequently in geotechnical reliability analyses and designs, it has been argued that only the natural variability affects the observed performance of geotechnical structures and should be differentiated from other knowledge uncertainties (e.g., measurement errors, statistical uncertainty, and transformation uncertainty) and quantified directly (Wang et al. 2016). Bayesian methods have been developed to directly quantify the site-specific natural variability, with systematic consideration of other knowledge uncertainties simultaneously (Wang and Cao 2013; Cao and Wang 2014a; Wang et al. 2015; Wang et al. 2016). The direct quantification of natural variability is formulated as a Bayesian inverse analysis problem in which the site-specific observation data is used as input to an inverse analysis model for inferring the natural variability of geo-material properties as the model output. The occurrence of natural variability, measurement errors, statistical uncertainty, and transformation uncertainty during site characterization and their propagation towards the total variability are explicitly modelled in the Bayesian inverse analysis method.

The Bayesian inverse analysis method contains three important elements: likelihood function, prior distribution, and how to solve the Bayesian equation for obtaining and expressing the posterior information in a user-friendly way. Likelihood function is the most critical one, and it should reflect, as much as possible, the physical insights into how the site-specific data are generated and observed, as well as the existing knowledge on how the observation data are expected to behave. For example, because the undrained shear strength of soil increases as its vertical effective stress increases, it is more effective to formulate the likelihood function based on the ratio of undrained shear strength over vertical effective stress than the undrained shear strength itself (Cao and Wang 2014a). In addition, complexity of the likelihood function shall be consistent with the available site-specific observation data. For example, it is extremely difficult, if not impossible, to quantify scale of fluctuation using just several discrete site-specific observation data points (e.g., SPT N values). In this case, it is probably more appropriate to model the natural variability of soil or rock properties within a statistically homogeneous layer as a random variable than a sophisticated random field. A sophisticated model with insufficient input data does not necessarily provide better results than a simple model with necessary and sound input data.

Although the exact equation of the likelihood function is problem-specific, a general process has been developed that streamlines the formulation of likelihood functions for various soil and rock properties when estimated using different field or laboratory tests (Wang et al. 2016). The streamlined process has been applied to probabilistic characterization of effective friction angle of sand using CPT data (Wang et al. 2010; Cao and Wang 2013) or SPT data (Wang et al. 2015), undrained Young's modulus (Wang and Cao 2013) and undrained shear strength (Cao and Wang 2014a) of clay, and uniaxial compressive strength of rock (Wang and Aladejare 2015), and probabilistic identification of underground soil strata using CPT (e.g., Wang et al. 2013; Cao and Wang 2013) or water content data (e.g., Wang et al. 2014).

In the Bayesian methods, prior distribution is used to quantitatively represent the prior knowledge on the site before the project (e.g., existing data in literature, engineering experience, and engineers' expertise). Two different methods have been developed to quantify the prior knowledge as prior distribution (Cao et al. 2016; Cao 2012). When there is no prevailing prior knowledge on the site, prior knowledge mainly reflects the engineering common sense and judgment, which may be quantitatively modelled by a non-informative uniform prior distribution. Only the upper and lower bounds of the uniform distribution are needed to fully specify the prior distribution. For example, the various soil and rock property statistics summarized in Sections 3.3 and 3.7 can be used to define the upper and lower bounds of the uniform distribution. Indicative prior estimates of some soil or rock properties are presented in Section 3.07: Soil Properties, JCSS Probabilistic Model Code (Joint Committee on Structural Safety 2006). As the prior knowledge improves and becomes more and more informative, a subjective probability assessment framework (SPAF) may be used to assist practitioners in quantifying their prior knowledge and engineering judgments as a proper prior distribution (Cao et al. 2016).

After the likelihood function is formulated and the prior distribution is determined, the site-specific observation data are integrated with the likelihood function and prior distribution through the Bayes' theorem to provide the posterior or updated information. When there are only sparse and limited site-specific observation data, MCMCS based Bayesian equivalent sample method may be used to numerically depict the updated natural variability through equivalent samples (Wang and Cao 2013). The Bayesian equivalent sample method combines the information from limited site-specific data with engineering experience and judgement (i.e., prior knowledge) through the Bayes' theorem and transforms the combined information into a large number of numerical samples by MCMCS. It effectively tackles the difficulty in estimating reasonable statistics of soil and rock properties from limited site-specific observation data in engineering practice. When there are extensive site-specific observation data (e.g., CPT data), Laplace asymptotic approximation based Bayesian system identification and model class selection methods may be used to estimate statistics of soil and rock properties and their scale of fluctuation (Cao and Wang 2013; Wang et al. 2013; Wang et al. 2014).

One limitation of the Bayesian methods is its mathematical and computational complexity, which may create difficulty for practitioners. To remove this mathematical hurdle, the Bayesian equivalent sample method algorithm has been implemented in a commonly available spreadsheet platform, Microsoft Excel, using its built-in programming language, Visual Basic for Applications (VBA). The program is compiled as an Excel VBA add-in, called Bayesian Equivalent Sample Toolkit (BEST), for distribution and installation. The BEST add-in can be downloaded freely from https://sites.google.com/site/yuwangcityu/best/1. The users of BEST only need to provide the site-specific observation data (e.g., several SPT N data points) and prior distribution (e.g., from the soil and rock properties statistics summarized in Sections 3.3 and 3.7). Then, the BEST will generate a large number of Bayesian equivalent samples for quantifying the site-specific natural variability of soil or rock properties.

As an illustration, the BEST is used to quantify the natural variability of undrained Young's modulus, E_u, using SPT N data obtained from the clay site of the United States National Geotechnical Experimentation Sites (NGES) at Texas A&M University (Briaud 2000). Five site-specific SPT N values are obtained within top stiff clay layer of the clay site and it is illustrated in Figure 3.15(a). Figure 3.15(b) shows the results of 42 pressuremeter tests carried out in the top clay layer at different depths (Briaud 2000) which is used for comparison purpose only. Typical ranges of undrained Young's modulus reported in the literature (e.g., Kulhawy and Mayne 1990; Phoon and Kulhawy 1999a, 1999b) are used to define the uniformly distributed prior distribution for E_u. For example, the mean of E_u is specified as uniformly distributed between 5.0 MPa and 15.0 MPa, and the standard deviation of E_u is modelled as uniformly distributed between 0.5 MPa and 13.5 MPa.

Using the 5 SPT N values and prior distribution defined above, the BEST add-in is used to generate 30000 Bayesian equivalent samples of E_u. Table 3.16 shows the statistics of the E_u samples obtained from BEST and its comparison with those measured by the pressuremeter tests. The BEST results are consistent with those from pressuremeter tests. In addition, Figure 3.16 displays the cumulative distribution functions, CDFs, of E_u estimated from the cumulative frequency diagrams of the BEST equivalent samples and the 42 pressuremeter test results which are represented by a solid line with triangle

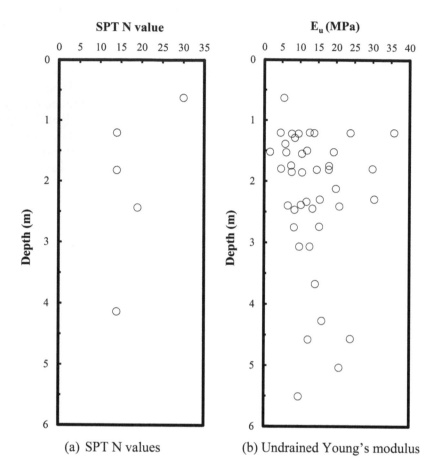

Figure 3.15 Standard penetration test (SPT) N values and undrained Young's modulus measured by pressuremeter tests, E$_u$ at the clay site of the NGES at Texas A&M University (after Briaud, 2000).

markers and open squares respectively. The open squares in Figure 3.16 plot close to the solid line, indicating that the CDF of the E$_u$ obtained from BEST compares favorably with that obtained from the 42 pressuremeter tests. The BEST add-in provides reasonable statistics for quantifying the site-specific natural variability of E$_u$.

3.10 SELECTION OF SITE-SPECIFIC TRANSFORMATION MODEL

Section 3.5 has highlighted the "site-specific" nature of transformation models as well as a wide variety and large number of models available. There are even many different transformation models relating the same measured and design parameters. This situation naturally leads to a question of how to select the "site-specific" transformation model that is most appropriate for a given project site in practice, particularly when

Table 3.16 Summary of the site-specific statistics of undrained Young's modulus.

Statistics	BEST	Pressuremeter Tests	Difference between BEST and Pressuremeter Tests
Mean (MPa)	11.46	13.50	2.04
Standard deviation (MPa)	6.00	7.50	1.50

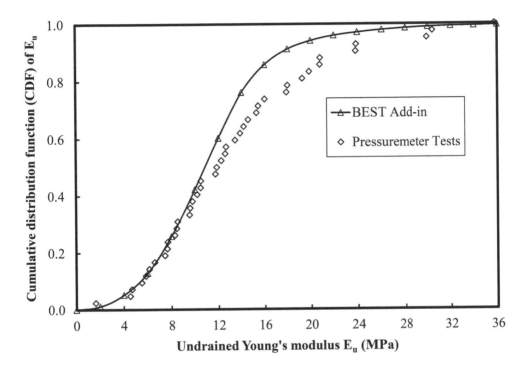

Figure 3.16 Site-specific cumulative distribution function of undrained Young's modulus E_u.

only a limited number of data on the measured parameter is obtained from a specific project and no direct measurement on the design parameter is available. This is frequently the situation that practitioners are dealing with in practice when using transformation models to estimate design parameters. Bayesian model comparison (Cao and Wang 2014a) and Bayesian model selection (Wang and Aladejare 2015) have been developed to assist engineers in selecting the most appropriate transformation models with data on the measured parameters obtained from a specific site only (i.e., no data on design parameters). The most appropriate model is the model with the highest occurrence probability for the given set of observation data on the measured parameters.

For example, Wang and Aladejare (2015) used 19 data points of point load index of granite from the Malanjkhand Copper project in India and selected the most appropriate site-specific transformation model among 4 commonly used regressions that relate the point load index to the uniaxial compressive strength, UCS, of rock. No UCS data are needed in the selection of site-specific regression model. This situation is particularly beneficial for engineering practice, because when the use or selection of regression model is needed in engineering practice, the UCS data are generally not available. The selected site-specific regression model can be further used with prior knowledge and site-specific measurement data (i.e., 19 point load index data points) for quantification of the site-specific natural variability in UCS using the Bayesian methods described in Section 3.9.

In the case where site-specific correlation data are not available, one can consider implementing generic transformation model constructed based on global correlation data. Ching and Phoon (2012) constructed such a generic transformation for the transformation between CPTU parameters and the undrained shear strength (s_u). However, one should be alerted that such generic transformation models usually have larger transformation uncertainties.

3.11   CONCLUSIONS AND FUTURE WORK

The most important element is the characterization of geotechnical variability. After all, site investigation and the interpretation of site data are necessary aspects of sound geotechnical practice. Any design methodology, be it RBD or otherwise, should place site investigation as the cornerstone of the methodology. It is not possible to emulate every aspect of site variability at present. The challenge lies in characterizing a full three-dimensional non-stationary vector field from borehole and/or field test data measured at only limited spatial locations. Notwithstanding this challenge, statistics that are founded on actual soil databases and evaluated in alignment with geotechnical engineering practice are useful. It is important to show how geotechnical RBD is related to soil data collected as part of a routine site investigation. It is also important to convince an engineer that there is value to evaluate geotechnical variability explicitly and this additional effort complements and simplifies the task of interpreting a single value (or a "cautious estimate" in Eurocode 7 parlance) from a bewildering mass of data.

This chapter discusses the uncertainties associated with the most basic soil/rock property evaluation task, which is to estimate a design parameter from a field test. The uncertainty in the estimate (say indicated by the coefficient of variation) must be a function of the natural variability of the site, measurement error associated with the field test, and transformation uncertainty about the regression line that relates the field data to the design parameter. Useful statistical tables and guidelines derived from a comprehensive survey of soil and rock databases are presented. In particular, it has been highlighted that the prevalent structural LRFD practice of assigning a single value to a resistance factor does not meet the needs of geotechnical engineering practice. Table 3.7 (or Figure D-3, ISO2394:2015) is a sensible step to close this gap and it can be expanded eventually to allow the quality of information emerging from diverse practices covering site investigation (Section D.3, ISO2394:2015), performance prediction method (Section D.4, ISO2394:2015), small-scale model test, centrifuge test, prototype test, quality control, and monitoring to be considered in a systematic way. The 2014 CHBDC (CAN/CSA-S6-14:2014) has recommended resistance factors that vary with the degree of "understanding". The degree of understanding covers the quality of site information and the quality of performance prediction. Guidelines comparable to Table 3.7 should be developed for rock properties.

It is possible to update the statistics for natural variability and transformation uncertainty in the presence of site-specific data using Bayesian methods. However, statistical uncertainties associated with inference from spatially correlated data should be handled carefully. It is also possible to update the uncertainty of a soil/rock parameter when it is correlated to data from more than one test type. It is common to conduct multiple tests (both laboratory and field tests) in a site investigation and hence, geotechnical data are typically available in a multivariate form. This aspect is discussed in Chapter 4.

Another distinctive feature of geotechnical data is that they vary spatially both in the vertical and horizontal direction. There are several practical observations associated with spatial variability. First, the spatial average along a failure surface is more relevant for a limit state, rather than the property value at a general point in the soil mass. This is obvious. Second, for limit states with failure surfaces constrained along a fixed path (e.g., side resistance is mobilized along the shaft), the COV of the spatial average is related to the COV of the property at a point and a variance reduction function (Vanmarcke 2010). This COV reduction effect increases with decreasing scale of fluctuation. Hence, the assumption of independent soil parameters will produce an excessive reduction of the point COV for the spatial average. The assumption of fully correlated soil parameters will not result in COV reduction for the spatial average, which is overly conservative. Third, Vanmarcke's classical variance reduction function does not apply to unconstrained failure surfaces that are coupled to spatial variability (e.g., slope failure surfaces). The location and shape of such a surface are different for different random realizations. In contrast, the surface where side resistance is mobilized is always constrained along the shaft regardless of the spatial variation of the soil property. The COV of the spatial average along such "unconstrained" failure surfaces is more complicated. A full solution for this more general class of failure surfaces is not available at present, although some progress has been made (Ching and Phoon 2013; Ching et al. 2014b; Hu and Ching 2015; Ching et al. 2016c). The presence of "unconstrained" failure surfaces also complicates the effort to convert the property field of a spatially variable medium into a homogeneous spatial average over a prescribed influence region, say a region below a footing equal to its width or some multiple of its width (Ching et al. 2016d). There is a practical motivation to perform this conversion, because it is obviously easier to carry out reliability-based design using a random variable (spatial average) than a random field. Fourth, there is limited work on nonstationary random fields, particularly in the form of one soil layer embedded in another or inclusion of pockets of different soil type within a more uniform soil mass (Li et al. 2016). Finally, it is significantly more difficult to characterize correlated data statistically than independent data (Phoon et al. 2003, Ching et al. 2016a). This aspect is of obvious practical significance and more research should be invested in this direction.

ACKNOWLEDGMENTS

The authors are grateful for the valuable comments provided by Dr. Marco Uzielli.

REFERENCES

Aladejare, A.E. & Wang, Y. (2017). "Evaluation of rock property variability", Georisk, in press.

Breysse, D., Niandou, H., Elachachi, S. & Houy, L. (2005) A generic approach to soil-structure interaction considering the effects of soil heterogeneity. *Geotechnique*, 55 (2), 143–150.

Briaud, J.L. (2000) The National Geotechnical Experimentation Sites at Texas A&M University: clay and sand. A Summary. National Geotechnical Experimentation Sites, Geotechnical Special Publication No. 93, 26–51.

Cafaro, F. & Cherubini, C. (2002) Large sample spacing in evaluation of vertical strength variability of clayey soil. *ASCE Journal of Geotechnical and Geoenvironmental Engineering*, 128 (7), 558–568.

CAN/CSA-S6-14:2014. *Canadian Highway Bridge Design Code*. Mississauga, ON, Canadian Standards Association.

Cao, Z.J. (2012) *Probabilistic Approaches for Geotechnical Site Characterization and Slope Stability Analysis*. PhD Thesis. Hong Kong, City University of Hong Kong.

Cao, Z.J. & Wang, Y. (2013) Bayesian approach for probabilistic site characterization using cone penetration tests. *ASCE Journal of Geotechnical and Geoenvironmental Engineering*, 139 (2), 267–276.

Cao, Z.J. & Wang, Y. (2014a) Bayesian model comparison and characterization of undrained shear strength. *ASCE Journal of Geotechnical and Geoenvironmental Engineering*, 140 (6), 04014018.

Cao, Z. & Wang, Y. (2014b) Bayesian model comparison and selection of spatial correlation functions for soil parameters. *Structural Safety*, 49, 10–17.

Cao, Z., Wang, Y. & Li, D. (2016) Quantification of prior knowledge in geotechnical site characterization. *Engineering Geology*, 203, 107–116.

Cherubini, C., Vessia, G. & Pula, W. (2007) Statistical soil characterization of Italian sites for reliability analyses. In: Tan, T.S., Phoon, K.K., Hight, D.W. & Leroueil, S. (eds.) *Characterisation and Engineering Properties of Natural Soils*. Vol. 4. Leiden, Taylor & Francis. pp. 2681–2706.

Chiasson, P. & Wang, Y.J. (2007) Spatial variability of sensitive Champlain sea clay and an application to stochastic slope stability analysis of a cut. In: Tan, T.S., Phoon, K.K., Hight, D.W. & Leroueil, S. (eds.) *Characterisation and Engineering Properties of Natural Soils*. Vol. 4. Leiden, Taylor & Francis. pp. 2707–2720.

Ching, J. & Phoon, K.K. (2011) A quantile-based approach for calibrating reliability-based partial factors. *Structural Safety*, 33, 275–285.

Ching, J. & Phoon, K.K. (2012) Establishment of generic transformations for geotechnical design parameters. *Structural Safety*, 35, 52–62.

Ching, J. & Phoon, K.K. (2013) Probability distribution for mobilized shear strengths of spatially variable soils under uniform stress states. *Georisk*, 7 (3), 209–224.

Ching, J. & Phoon, K.K. (2014) Transformations and correlations among some clay parameters – The global database. *Canadian Geotechnical Journal*, 51 (6), 663–685.

Ching, J., Phoon, K.K. & Yu, J.W. (2014a) Linking site investigation efforts to final design savings with simplified reliability-based design methods. *ASCE Journal of Geotechnical and Geoenvironmental Engineering*, 140 (3), 04013032.

Ching, J., Phoon, K.K. & Kao, P.H. (2014b) Mean and variance of the mobilized shear strengths for spatially variable soils under uniform stress states. *ASCE Journal of Engineering Mechanics*, 140 (3), 487–501.

Ching, J., Wu, S.H. & Phoon, K.K. (2016a) Statistical characterization of random field parameters using frequentist and Bayesian approaches. *Canadian Geotechnical Journal*, 53 (2), 285–298.

Ching, J., Phoon, K.K. & Wu, S.H. (2016b) Impact of statistical uncertainty on geotechnical reliability estimation. *ASCE Journal of Engineering Mechanics*, 04016027.

Ching, J., Lee, S.W. & Phoon, K.K. (2016c) Undrained strength for a 3D spatially variable clay column subjected to compression or shear. *Probabilistic Engineering Mechanics*, 45, 127–139.

Ching, J., Hu, Y.G. & Phoon, K.K. (2016d) On characterizing spatially variable shear strength using spatial average. *Probabilistic Engineering Mechanics*, 45, 31–43.

Det Norske Veritas (DNV) (2010) Recommended Practice – Statistical representation of soil data. DNV-RP-C207, Oslo, Norway.

Doruk, P. (1991) *Analysis of the Laboratory Strength Data Using the Original and Modified Hoek-Brown Criteria*. MS Thesis. Toronto, University of Toronto. 124 pp.

El-Ramly, H., Morgenstern, N.R. and Cruden, D.M. (2003) Probabilistic stability analysis of a tailings dyke on presheared clay shale. *Canadian Geotechnical Journal*, 40 (1), 192–208.

EN 1997-1:2004. *Eurocode 7: Geotechnical Design – Part 1: General Rules*. Brussels, European Committee for Standardization (CEN).

Federal Highway Administration (FHWA) (1985) *Checklist and Guidelines for Review of Geotechnical Reports and Preliminary Plans and Specifications*. Report FHWA-ED-88-053. Washington, DC.

Fenton, G.A. (1999a) Random field modeling of CPT data. *ASCE Journal of Geotechnical and Geoenvironmental Engineering*, 125 (6), 486–498.

Fenton, G.A. (1999b) Estimation for stochastic soil models. *ASCE Journal of Geotechnical and Geoenvironmental Engineering*, 125 (6), 470–485.

Fenton, G.A. & Griffiths, D.V. (2003) Bearing capacity prediction of spatially random c-ϕ soils. *Canadian Geotechnical Journal*, 40, 54–65.

Fenton, G.A. & Griffiths, D.V. (2008) *Risk Assessment in Geotechnical Engineering*. New York, John Wiley & Sons.

Fenton, G.A., Griffiths, D.V. & Williams, M.B. (2005) Reliability of traditional retaining wall design. *Geotechnique*, 55 (1), 55–62.

Goldsworthy, J.S., Jaksa, M.B., Fenton, G.A., Griffiths, D.V., Kaggwa, W.S. & Poulos, H.G. (2007) Measuring the risk of geotechnical site investigations. In: *Proc. Geo-Denver 2007, Denver*.

Griffiths, D.V., Fenton, G.A. & Manoharan, N. (2006) Undrained bearing capacity of two-strip footings on spatially random soil. *ASCE International Journal of Geomechanics*, 6 (6), 421–427.

Heuze, F.E. (1980) Scale effects in the determination of rock mass strength and deformability. *Rock Mechanics*, 12 (3–4), 167–192.

Hoek, E. & Brown, E.T. (1980) *Underground Excavations in Rock*. London, The Institution of Mining and Metallurgy. 527 pp.

Hoek, E., Kaiser, P.K. & Bawden, W.F. (1995) *Support of Underground Excavations in Hard Rock*. Rotterdam, Balkema. 215 pp.

Honjo, Y. & Setiawan, B. (2007) General and local estimation of local average and their application in geotechnical parameter estimations. *Georisk*, 1 (3), 167–176.

Hu, Y.G. & Ching, J. (2015) Impact of spatial variability in soil shear strength on active lateral forces. *Structural Safety*, 52, 121–131.

ISO2394:1973/1986/1998/2015. *General Principles on Reliability for Structures*. Geneva, International Organization for Standardization.

Jaksa, M.B. (1995) *The Influence of Spatial Variability on the Geotechnical Design Properties of a Stiff, Overconsolidated Clay*. PhD Thesis. Adelaide, University of Adelaide.

Jaksa, M.B. (2007) Modeling the natural variability of over-consolidated clay in Adelaide, South Australia. In: Tan, T.S., Phoon, K.K., Hight, D.W. & Leroueil, S. (eds.) *Characterisation and Engineering Properties of Natural Soils*. Vol. 4. Leiden, Taylor & Francis. pp. 2721–2752.

Jaksa, M.B., Brooker, P.I. & Kaggwa, W.S. (1997) Inaccuracies associated with estimating random measurement errors. *ASCE Journal of Geotechnical and Geoenvironmental Engineering*, 123 (5), 393–401.

Jaksa, M.B., Kaggwa, W.S., Fenton, G.A. & Poulos, H.G. (2003) A framework for quantifying the reliability of geotechnical investigations. In: *Applications of Statistics and Probability in Civil Engineering, ICASP9, San Francisco*. Vol. 2. Rotterdam, Mill Press. pp. 1285–1291.

Jaksa, M.B., Goldsworthy, J.S., Fenton, G.A., Kaggwa, W.S., Griffiths, D.V., Kuo, Y.L. & Poulos, H.G. (2005) Towards reliable and effective site investigations. *Geotechnique*, 55 (2), 109–121.

Joint Committee on Structural Safety (2006) Section 3.07: Soil Properties, JCSS Probabilistic Model Code, http://www.jcss.byg.dtu.dk/Publications.

Kahraman, S. (2001) Evaluation of simple methods for assessing the uniaxial compressive strength of rock. *International Journal of Rock Mechanics and Mining Sciences*, 38 (7), 981–994.

Kulatilake, P.H.S. (1991) Discussion on 'Probabilistic potentiometric surface mapping' by P.H.S. Kulatilake. *ASCE Journal of Geotechnical Engineering*, 117 (9), 1458–1459.

Kulhawy, F.H. & Mayne, P.W. (1990) *Manual on Estimating Soil Properties for Foundation Design*. Report EL-6800. Palo Alto, Electric Power Research Institute. Available online at EPRI.COM.

Kulhawy, F.H. & Prakoso, W.A. (2003) Variability of rock index properties. In: *Proceedings, Soil & Rock America, Cambridge*. pp. 2765–2770.

Li, K.S. (1991) Discussion on 'Probabilistic potentiometric surface mapping' by P.H.S. Kulatilake. *ASCE Journal of Geotechnical Engineering*, 117 (9), 1457–1458.

Li, D.Q., Qi, X.H., Cao, Z.J., Tang, X.S., Phoon, K.K. & Zhou, C.B. (2016) Evaluating slope stability uncertainty using coupled Markov chain. *Computers and Geotechnics*, 73, 72–82.

Look, B.G. (2014) Handbook of Geotechnical Investigation and Design Tables, Second Edition, CRC Press, UK.

Mayne, P.W., Christopher, B.R. & DeJong, J. (2001) *Manual on Subsurface Investigations*. National Highway Institute Publication No. FHWA NHI-01-031. Washington, DC, Federal Highway Administration.

Ng, I.T., Yuen, K.V. & Lau, C.H. (2015) Predictive model for uniaxial compressive strength for Grade III granitic rocks from Macao. *Engineering Geology*, 199, 28–37.

Paikowsky, S.G., Birgisson, B., McVay, M., Nguyen, T., Kuo, C., Baecher, G.B., Ayyub, B., Stenersen, K., O'Malley, K., Chernauskas, L. & O'Neill, M. (2004) Load and resistance factors design for deep foundations. NCHRP Report 507, Transportation Research Board of the National Academies, Washington DC.

Phoon, K.K. & Kulhawy, F.H. (1999a) Characterization of geotechnical variability. *Canadian Geotechnical Journal*, 36 (4), 612–624.

Phoon, K.K. & Kulhawy, F.H. (1999b) Evaluation of geotechnical property variability. *Canadian Geotechnical Journal*, 36 (4), 625–639.

Phoon, K.K. & Kulhawy, F.H. (2008) Serviceability limit state reliability-based design. In: Phoon, K.K. (ed.) *Reliability-Based Design in Geotechnical Engineering: Computations and Applications*. London, Taylor & Francis. pp. 344–383.

Phoon, K.K., Kulhawy, F.H. & Grigoriu, M.D. (1995) *Reliability-Based Design of Foundations for Transmission Line Structures*. Report TR-105000. Palo Alto, Electric Power Research Institute. Available online at EPRI.COM.

Phoon, K.K., Quek, S.T. & An, P. (2003) Identification of statistically homogeneous soil layers using modified Bartlett statistics. *ASCE Journal of Geotechnical and Geoenvironmental Engineering*, 129 (7), 649–659.

Prakoso, W.A. (2002) *Reliability-Based Design of Foundations on Rock for Transmission Line & Similar Structure*. PhD Thesis. New York, Cornell University.

Prakoso, W.A. & Kulhawy, F.H. (2004) Variability of rock mass engineering properties. In: Sambhandharaksa, S. et al. (ed.) *Proc. 15th SE Asian Geotech. Conf.* Bangkok. pp. 97–100.

Prakoso, W.A. & Kulhawy, F.H. (2011) Effects of testing conditions on intact rock strength and variability. *Geotechnical and Geological Engineering*, 29, 101–111.

Rice, S.O. (1944) Mathematical analysis of random noise. *Bell System Technical Journal*, 23 (3), 282–332.

Rice, S.O. (1944) Mathematical analysis of random noise. *Bell System Technical Journal*, 24 (1), 46–156.

Rehfeldt, K.R., Boggs, J.M. & Gelhar, L.W. (1992) Field study of dispersion in a heterogeneous aquifer: 3. Geostatistical analysis of hydraulic conductivity. *Water Resources Research*, 28 (12), 3309–3324.

Robertson, P.K. (2009) Interpretation of cone penetration tests – A unified approach. *Canadian Geotechnical Journal*, 46, 1337–1355.

Sari, M. & Karpuz C. (2006) Rock variability and establishing confining pressure levels for triaxial tests on rocks. *International Journal of Rock Mechanics and Mining Sciences*, 43 (2), 328–335.

Soulié, M., Montes, P. & Silvestri, V. (1990) Modelling spatial variability of soil parameters, *Canadian Geotechnical Journal*, 27 (5), 617–630.

Soubra, A.H., Massih, Y.A. & Kalfa, M. (2008) Bearing capacity of foundations resting on a spatially random soil. In: *GeoCongress 2008: Geosustainability and Geohazard Mitigation (GSP 178)*. New Orleans, LA, ASCE, pp. 66–73.

Ünlü, K., Nielsen, D.R., Biggar, J.W. & Morkoc, F. (1990) Statistical parameters characterizing the spatial variability of selected soil hydraulic properties, *Soil Science Society of America Journal*, 54, 1537–1547.

Uzielli, M., Vannucchi, G. & Phoon, K.K. (2005) Random field characterization of stress-normalized CPT Variables. *Geotechnique*, 55 (1), 3–20.

Uzielli, M., Lacasse, S., Nadim, F. & Lunne, T. (2007) Uncertainty-based characterization of Troll marine clay. In: Tan, T.S., Phoon, K.K., Hight, D.W. & Leroueil, S. (eds.) *Characterisation and Engineering Properties of Natural Soils* Vol. 4. Leiden, Taylor & Francis. pp. 2753–2782.

Vanmarcke, E.H. (1977) Probabilistic modeling of soil profiles. *ASCE Journal of Geotechnical Engineering Division*, 103 (GT11), 1227–1246.

Vanmarcke, E.H. (2010) *Random Fields: Analysis and Synthesis*. Singapore, World Scientific.

Wang, Y. & Cao, Z.J. (2013) Probabilistic characterization of Young's modulus of soil using equivalent samples. *Engineering Geology*, 159, 106–118.

Wang, Y. & Aladejare, A.E. (2015) Selection of site-specific regression model for characterization of uniaxial compressive strength of rock. *International Journal of Rock Mechanics and Mining Sciences*, 75, 73–81.

Wang, Y., Au, S.K. & Cao, Z.J. (2010) Bayesian approach for probabilistic characterization of sand friction angles. *Engineering Geology*, 114 (3–4), 354–363.

Wang, Y., Huang, K. & Cao, Z.J. (2013) Probabilistic identification of underground soil stratification using cone penetration tests. *Canadian Geotechnical Journal*, 50 (7), 766–776.

Wang, Y., Huang, K. & Cao, Z.J. (2014) Bayesian identification of soil strata in London Clay. *Geotechnique*, 64 (3), 239–246.

Wang, Y., Zhao, T. & Cao, Z. (2015) Site-specific probability distribution of geotechnical properties. *Computers and Geotechnics*, 70, 159–168.

Wang, Y., Cao, Z. & Li, D. (2016) Bayesian perspective on geotechnical variability and site characterization. *Engineering Geology*, 203, 117–125.

Chapter 4

Statistical characterization of multivariate geotechnical data

Jianye Ching, Dian-Qing Li, and Kok-Kwang Phoon

ABSTRACT

Section D.2 in ISO2394 Annex D focuses on the evaluation of the coefficient of variation (COV) for a single design soil parameter. This was the main focus for Chapter 3. However, it is more common to conduct a variety of laboratory and field tests, some of them in close proximity, in a site investigation. Data derived from two adjacent boreholes/soundings at comparable depths typically are correlated. Section D.3 in ISO2394 Annex D describes the advantage of adopting a multivariate distribution to capture these correlations in a systematic way. This is the main focus of the current chapter. The practical significance of doing this is that the COV of one soil parameter (or a group of parameters) is reduced when information on a second parameter (or a group of parameters) is made available. The reduction of the COV in the presence of multiple soil tests can translate directly to design savings through RBD. This direct link between the quality/quantity of site investigation and design savings cannot be addressed systematically in our traditional factor of safety approach.

This chapter presents several multivariate geotechnical databases available in the literature and the multivariate normal distributions constructed using these databases. Useful multivariate transformation equations will be deduced from the multivariate normal distributions. These equations not only can predict the mean value of the design soil parameter but also can predict its COV. The COV will be further reduced in the presence of multiple soil tests. As a supplement to the multivariate normal distribution, a copula-based approach for modelling the multivariate distribution of multiple soil parameters will be also introduced in this chapter. With the copula theory, it is possible to go beyond the multivariate normal distribution framework, e.g., a copula based on the multivariate t distribution can be constructed. A more robust method of estimating the correlation coefficients in the multivariate normal distribution will also be proposed.

4.1 INTRODUCTION

The emphasis in ISO2394 Annex D is to identify and characterize critical elements of the geotechnical reliability-based design (RBD) process that are distinctive from the general principles presented in the main standard. One of the critical elements is to introduce an explicit linkage between site investigation and geotechnical RBD.

Site investigation is an activity unique to geotechnical engineering practice and it is mandated in many building regulations around the world. In a site investigation program, both laboratory and field tests are commonly conducted. A geotechnical design parameter is typically correlated with more than one laboratory and/or field test indices. Chapter 3 focused on quantifying the uncertainties in design soil parameters based on univariate information from a laboratory or in-situ test. Nonetheless, multivariate information is usually available in a typical site investigation. For instance, when undisturbed samples are extracted for oedometer and triaxial tests, standard penetration test (SPT) and/or piezocone test (CPTU) may be conducted in close proximity. Moreover, data sources such as the unit weight, plastic limit (PL), liquid limit (LL), and liquidity index (LI) are commonly determined from relatively simple laboratory tests on disturbed samples. A number of these data sources could be simultaneously correlated to the design soil parameters, e.g., the undrained shear strength (s_u). It is prudent to incorporate all data sources to reduce the uncertainties in the design soil parameters. By doing so, the site investigation effort can be linked to geotechnical RBD design outcomes. This linkage is demonstrated in Ching et al. (2014a). There are two key components in this linkage: (a) a simplified RBD method that is sufficiently responsive to a wide range of geotechnical information and (b) a framework that is able to update the mean and COV of the design soil parameters based on multivariate geotechnical data. Component (a) is the subject of Chapter 6. The current chapter will address component (b).

The main framework proposed in this chapter is a multivariate probability model to couple all available sources of information together in a consistent way. The constructed multivariate probability model can be used as a prior distribution to derive the multivariate distribution of design parameters based on limited but site-specific field data. Note that the entire multivariate distribution of multiple design parameters is derived, not marginal distributions or simply means and coefficients of variation presented in Chapter 3. Multiple design parameters can be updated from multiple field measurements, which is more useful than updating one design parameter using one field measurement based on current practice (for example, updating the undrained shear strength using the cone tip resistance).

The challenge for constructing the multivariate probability model is that genuine multivariate data are rarely collected in a site investigation program, because it is not cost effective to conduct multiple tests in close proximity. There is an obvious tradeoff between conducting different tests in different locations and conducting different tests in the same location. The former strategy collects more information on the natural variability of the site. The latter strategy collects information on the correlations between all tests. It is not possible to correlate two measurements spaced more than one scale of fluctuation apart vertically and horizontally, because of natural variability. In practice, it is common to adopt an intermediate strategy involving collecting multiple sets of bivariate data in different locations, say take piezocone (CPTU) soundings next to one borehole and conduct vane shear tests (VSTs) next to a second borehole. We do not take CPTU soundings, conduct VSTs, and collect undisturbed samples for triaxial tests in three separate locations. It is not possible to produce a site-specific correlation between say the undrained shear strength from a laboratory test and the CPTU data in this case. We also do not collect CPTU, VST, and undrained shear strength data at a single location, because soil data from a single location are unlikely to be sufficiently representative of the natural variability over the entire site.

Table 4.1 Soil databases and multivariate probability models.

Database/ model	Reference	Parameters of interest	# data points	# sites/ studies	Range of properties		
					OCR	PI	S_t
CLAY/5/ 345	Ching and Phoon (2012a)	LI, s_u, s_u^{re}, σ'_p, σ'_v	345	37 sites	1–4		Sensitive to quick clays
CLAY/6/ 535	Ching et al. (2014b)	s_u/σ'_v, OCR, $(q_t - \sigma_v)/\sigma'_v$, $(q_t - u_2)/\sigma'_v$, $(u_2 - u_0)/\sigma'_v$, B_q	535	40 sites	1–6	Low to very high plasticity	Insensitive to quick clays
CLAY/7/ 6310	Ching and Phoon (2013)	s_u under 7 different test modes	6310	164 studies	1–10	Low to very high plasticity	Insensitive to quick clays
CLAY/10/ 7490	Ching and Phoon (2014a)	LL, PI, LI, σ'_v/P_a, σ'_p/P_a, s_u/σ'_v, S_t, $(q_t - \sigma_v)/\sigma'_v$, $(q_t - u_2)/\sigma'_v$, B_q	7490	251 studies	1–10	Low to very high plasticity	Insensitive to quick clays
CLAY/ 4/BN	Ching et al. (2010)	OCR, s_u, N_{60}, $(q_t - \sigma_v)/\sigma'_v$	–	–	1–50	–	–
F-CLAY/ 7/216	D'Ignazio et al. (2016)	s_u, σ'_p, σ'_v, LL, PL, w_n, S_t	216	24 sites	1–8	Low to very high plasticity	Insensitive to quick clays
SAND/ 4/BN	Ching et al. (2012)	D_r, ϕ', $(N_1)_{60}$, q_{t1}	–	–	–	–	–

LL: liquid limit; PL: plastic limit; PI: plasticity index; LI: liquidity index; w_n: natural water content; s_u: undrained shear strength; s_u^{re}: remolded s_u; σ'_p: preconsolidation stress; σ'_v: vertical effective stress; σ_v: vertical total stress; OCR: overconsolidation ratio; q_t: corrected cone tip resistance; u_2: pore pressure behind the cone; u_0: static pore pressure; B_q: CPTU pore pressure parameter; P_a: one atmosphere pressure; S_t: sensitivity; N_{60}: SPT N (corrected for energy ratio); D_r: relative density; ϕ': effective friction angle; $(N_1)_{60}$: SPT N (corrected for energy ratio & normalized by overburden stress); q_{t1}: normalized q_t (normalized by overburden stress).

This chapter will review some multivariate soil databases and the resulting multivariate probability models recently constructed in the literature. Table 4.1 shows these soil databases, labelled as (soil type)/(number of parameters of interest)/(number of data points). Comparable databases have been assembled in the literature recently (Müller et al. 2014; Liu et al. 2016). Although multivariate probability models are available in the literature, it remains an outstanding challenge to fit one with available data. The review will start from two genuine multivariate databases, CLAY/5/345 (Ching and Phoon 2012a) and CLAY/6/535 (Ching et al. 2014b), and the resulting multivariate probability models will be presented. Then, two bivariate databases CLAY/7/6310 (Ching and Phoon 2013) and CLAY/10/7490 (Ching and Phoon 2014a, 2014b) are reviewed, and the resulting multivariate probability models are presented. These four multivariate models are based on the *multivariate normal distribution*, so this special model together with the concept of correlation will be first reviewed. Then, two probability models based on the Bayesian network (Jensen 1996) will be reviewed (CLAY/4/BN and SAND/4/BN in Table 4.1). The Bayes-net model can handle situations where bivariate data points are available only for some parameter pairs but not for all pairs. For instance, OCR-s_u and s_u-CPTU data points are available, but OCR-CPTU data points are not. Under suitable conditional independence assumptions, the multivariate probability model can still be constructed. Finally, alternate multivariate probability models based on the copula theory will be demonstrated. With the copula theory, it is possible to go beyond the aforementioned multivariate normal distribution framework, e.g., a copula based on the multivariate t distribution can be constructed.

It is important to emphasize that the multivariate distributions constructed from soil databases in Table 4.1 are generic in nature, because data are drawn from many sites rather than one single site. Nonetheless, the author submits that it is reasonable to adopt these multivariate distributions as prior information for a specific site. The posterior probability distribution of a *site-specific* design property can be obtained from this prior information when it is updated by *site-specific* field data. There are occasional concerns expressed that only site-specific prior information is meaningful in this updating exercise. In other words, data gathered from the literature pertaining to comparable soils and/or sites cannot be used or more specifically, a generic multivariate distribution is not useful as prior information. This concern is understandable, but it is at odds with existing practice. The tradition of geotechnical engineering is steeped in empiricism and one notable aspect is arguably the widespread application of non-site specific generic transformation models to estimate site-specific design properties. Whether one derives a single cautious estimate or a probability distribution from a transformation model, the role of engineering judgment in selecting the appropriate transformation model and weeding out unreasonable estimates is obviously integral to this practice and needs no further emphasis.

4.2 CORRELATION

Correlation is a concept that is closely related to the transformation model discussed in Section 3.5. Transformation models are widely adopted in geotechnical engineering practice as a matter of practical expediency. Useful compilations of these models (mostly pairwise correlations) are available in the literature (e.g., Kulhawy and Mayne 1990; Mayne et al. 2001). Tables 4.2 and 4.3 list some examples. Clay and sand databases are compiled to evaluate the bias and coefficient of variation (COV) of the transformation models in the tables. The bias for a transformation model is estimated as the sample mean of the ratio (actual target value)/(predicted value), and the COV of a transformation model is estimated as the sample COV of this ratio [refer to Eqs. (3.1) and (3.2)]. For instance, for the LI-(s_u^{re}/P_a) model proposed by Locat and Demers (1988) (the second model in Table 4.2), we have $s_u^{re}/P_a \approx 0.0144 \times LI^{-2.44}$: the target value is s_u^{re}/P_a, and the predicted value is $0.0144 \times LI^{-2.44}$. For each data point with simultaneous knowledge of (LI, s_u^{re}), the ratio (actual target value)/(predicted value) $= (s_u^{re}/P_a)/(0.0144 \times LI^{-2.44})$ can be computed. There are n $= 899$ such data points, hence 899 such ratios are available. The sample mean of these ratios is equal to 1.92 (bias), whereas the sample COV of these ratios is 1.25 (COV). Tables 4.2 and 4.3 list the bias and COV for each transformation model. If the bias is close to 1, the transformation model is unbiased on the average. If the COV is small, the transformation uncertainty is small. However, Tables 4.2 and 4.3 say nothing about multivariate distributions. In particular, many transformation models in these tables only allow a single input.

To illustrate the concept of correlation, let us consider Figure 3.2. As shown in Figure 3.2, the data points (dots in the figure) inevitably scatter around the

Table 4.2 Transformation models in the literature for some clay properties (Source: Table 5, Ching and Phoon 2014a and results from D'Ignazio et al. (2016)).

Relationship	Literature	Transformation model	n	Bias	COV	Remarks
$LI-(s_u^{re}/P_a)$	Wroth and Wood (1978)	$s_u^{re}/P_a \approx 1.7 \times \exp(-4.6 \times LI)$	899	NF	NF	Based on modified Cam Clay model
	Locat and Demers (1988)	$s_u^{re}/P_a \approx 0.0144 \times LI^{-2.44}$	899	1.92	1.25	Norwegian marine clays
$LI-S_t$	Bjerrum (1954)	$S_t \approx 10^{0.8 \times LI}$	1279	2.06	1.09	Structured clays with $S_t = 2-1000$ & OCR = 1-4
	Ching and Phoon (2012a,b)	$S_t \approx 20.726 \times LI^{1.910}$	1279	0.88	1.28	
$LI-\sigma'_v/P_a-S_t$ $LI-\sigma'_p/P_a-S_t$	Mitchell (1993)	Graphical curves in page 229	–	–	–	
	NAVFAC (1982)	Graphical curves in page 7.1-142	–	–	–	
	Stas and Kulhawy (1984)	$\sigma'_p/P_a \approx 10^{1.11-1.62\times LI}$	249	2.94	1.90	Clays with $S_t < 10$
	Ching and Phoon (2012a,b)	$\sigma'_p/P_a \approx 0.235 \times LI^{-1.319} \times S_t^{0.536}$	489	1.32	0.78	Structured clays with $S_t = 2-1000$ & OCR = 1-4
$LI-s_u/\sigma'_p$	Bjerrum and Simons (1960)	Graphical curves	–	–	–	Norwegian NC clays
	Mesri (1975, 1989)	$s_u/\sigma'_p \approx 0.22$	1155	1.04	0.55	
$PI-s_u/\sigma'_p$	Jamiolkowski et al. (1985)	$s_u/\sigma'_p \approx 0.23 \times OCR^{0.8}$	1402	1.11	0.53	
$OCR-s_u/\sigma'_v$	D'Ignazio et al. (2016)	$s_u/\sigma'_v \approx 0.244 \times OCR^{0.763}$	173	0.93	0.27	Finnish soft clays; s_u = corrected field vane value
$OCR-s_u/\sigma'_v-S_t$	Ching and Phoon (2012a,b)	$s_u/\sigma'_v \approx 0.229 \times OCR^{0.823} \times S_t^{0.121}$	395	0.84	0.34	Structured clays with $S_t = 2-1000$ & OCR = 1-4
$OCR-s_u/\sigma'_v-PI$	D'Ignazio et al. (2016)	$s_u/\sigma'_v \approx 0.328 \times OCR^{0.756} \times PI^{0.165}$	173	0.95	0.29	Finnish soft clays; s_u = uncorrected field vane value
$OCR-s_u/\sigma'_v-LL$	D'Ignazio et al. (2016)	$s_u/\sigma'_v \approx 0.319 \times OCR^{0.757} \times LL^{0.333}$	173	0.94	0.26	Finnish soft clays; s_u = uncorrected field vane value
$OCR-s_u/\sigma'_v-w_n$	D'Ignazio et al. (2016)	$s_u/\sigma'_v \approx 0.296 \times OCR^{0.788} \times w_n^{0.337}$	173	0.97	0.27	Finnish soft clays; s_u = uncorrected field vane value
$OCR-s_u/\sigma'_v-LI$	D'Ignazio et al. (2016)	$s_u/\sigma'_v \approx 0.281 \times OCR^{0.77} \times LI^{-0.088}$	173	0.95	0.33	Finnish soft clays; s_u = uncorrected field vane value
$OCR-s_u/\sigma'_v-S_t$	D'Ignazio et al. (2016)	$s_u/\sigma'_v \approx 0.280 \times OCR^{0.786} \times S_t^{-0.013}$	173	0.91	0.44	Finnish soft clays; s_u = uncorrected field vane value
$LI-\sigma'_v-s_u$	Ng et al. (2015)	$s_u (kN/m^2) \approx 0.2335 \times \sigma'_v (kN/m^2) - 2.6915 \times w_n \times LI + 8.9657$	296	1.03	0.41	NC clays; s_u =field vane value
$CPTU-s_u/\sigma'_v$	Ching and Phoon (2012c)	$(q_t - \sigma_v)/s_u \approx 29.1 \times \exp(-0.513B_q)$	423	0.95	0.49	
		$(q_t - \sigma_v)/s_u \approx 34.6 \times \exp(-2.049B_q)$	428	1.11	0.57	
		$(q_t - \sigma_v)/s_u \approx 21.5 \times B_q$	423	0.94	0.49	
$CPTU-OCR$	Chen and Mayne (1996)	$OCR \approx 0.259 \times [(q_t - \sigma_v)/\sigma'_v]^{1.107}$	690	1.01	0.42	
		$OCR \approx 0.545 \times [(q_t - u_2)/\sigma'_v]^{0.969}$	542	1.06	0.57	
	Kulhawy and Mayne (1990)	$OCR \approx 1.026 \times B_q^{-1.077}$	779	1.28	0.86	
		$OCR \approx 0.32 \times (q_t - \sigma_v)/\sigma'_v$	690	1.00	0.39	
$CPTU-\sigma'_p/P_a$	Chen and Mayne (1996)	$\sigma'_p/P_a \approx 0.227 \times [(q_t - \sigma_v)/P_a]^{1.200}$	690	0.99	0.42	
		$\sigma'_p/P_a \approx 0.490 \times [(q_t - u_2)/P_a]^{1.053}$	542	1.08	0.61	
	Kulhawy and Mayne (1990)	$\sigma'_p/P_a \approx 1.274 + 0.761 \times (u_2 - u_0)/P_a$	690	NF	NF	
		$\sigma'_p \approx 0.33 \times (q_t - \sigma_v)$	690	0.97	0.39	
		$\sigma'_p \approx 0.54 \times (u_2 - u_0)$	690	1.18	0.75	

w_n: natural water content (in decimal value; e.g., if water content is 50%, $w_n = 0.5$); PI: plasticity index (in decimal value; e.g., if plasticity index is 50%, PI = 0.5); LL: liquid limit (in decimal value; e.g., if liquid limit is 50%, LL = 0.5); NF: validation data do not fit to the trend of the transformation model.

Table 4.3 Transformation models in the literature for some sand/gravel properties.

Relationship	Literature	Transformation model	n	Bias	COV	Remark
SPT – D_r	Marcuson and Bieganousky(1977)	$D_r(\%) \approx 100 \times \left\{ 12.2 + 0.75 \sqrt{222 \times N_{60} + 2311 - 711 \times OCR - 779(\sigma'_v/P_a) - 50 \times C_u^2} \right\}$	131	1.00	0.22	Sands with $N_{60} < 100$
	Kulhawy and Mayne(1990)	$D_r(\%) \approx 100 \times \sqrt{\dfrac{(N_1)_{60}}{[60 + 25 \log_{10}(D_{50})] \times OCR^{0.18}}}$	195	1.01	0.21	
CPT-D_r	Jamiolkowski et al. (1985)	$D_r(\%) \approx 68 \times [\log_{10}(q_{t1}) - 1]$	595	0.84	0.33	NC sands with $q_{t1} < 300$
	Kulhawy and Mayne(1990)	$D_r(\%) \approx 100 \times \sqrt{\dfrac{q_{t1}}{305 \times Q_C \times OCR^{0.18}}}$	823	0.97	0.34	
D_r-ϕ'	Bolton (1986)	$\phi' \approx \phi_{cv} + 3 \times (D_r[10 - \ln(p'_f)] - 1)$	431	1.02	0.047	
	Salgado et al. (2000)	$\phi' \approx \phi_{cv} + 3 \times (D_r[8.3 - \ln(p'_f)] - 0.69)$	127	1.08	0.054	Sands with 10% fines
SPT-ϕ'	Hatanaka and Uchida (1996)	$\phi' \approx \sqrt{15.4 \cdot (N_1)_{60}} + 20$	28	1.05	0.091	Sands with $(N_1)_{60} < 40$
	Chen (2004)	$\phi' \approx 27.5 + 9.2 \times \log_{10}[(N_1)_{60}]$	59	1.00	0.093	
CPT-ϕ'	Robertson and Campanella (1983)	$\phi' \approx \tan^{-1}[0.1 + 0.38 \times \log_{10}(q_t/\sigma'_v)]$	93	0.93	0.054	
	Kulhawy and Mayne (1990)	$\phi' \approx 17.6 + 11 \times \log_{10}(q_{t1})$	370	0.98	0.080	

*$q_{t1} = (q_t/P_a)/(\sigma'_v/P_a)^{0.5}$; ϕ_{cv}: critical-state friction angle (in degrees); p'_f is the mean effective stress at failure $= (\sigma'_{1f} + \sigma'_{2f} + \sigma'_{3f})/3$; $Q_C = 1.09, 1.0, 0.91$ for low, medium, high compressibility soils, respectively.

transformation model (solid line in the figure). Consider the following simple transformation between two soil parameters (Y_1, Y_2):

$$Y_1 = a + bY_2 + \varepsilon \qquad (4.1)$$

The transformation model is the functional relation $Y_1 = a + bY_2$, while ε is the zero-mean transformation uncertainty with standard deviation s_ε. The product-moment (Pearson) correlation between Y_1 and Y_2 is defined as:

$$\rho_{12} = \frac{\text{Cov}(Y_1, Y_2)}{\sqrt{\text{Var}(Y_1)}\sqrt{\text{Var}(Y_2)}} = \frac{b \times \sqrt{\text{Var}(Y_2)}}{\sqrt{b^2 \times \text{Var}(Y_2) + s_\varepsilon^2}} \qquad (4.2)$$

where $\text{Var}(Y)$ denotes the variance of Y, and $\text{Cov}(Y_1, Y_2)$ denotes the covariance between Y_1 and Y_2. It is clear that if $s_\varepsilon = 0$ (zero scattering about the transformation model), $\rho_{12} = \pm 1$ and perfect correlation exists between Y_1 and Y_2. In this case, given the information of $Y_2 = y_2$, $Y_1 = a + b \times y_2$ is deterministic, and $\text{COV} = 0$ for Y_1 (i.e., Y_1 is no longer uncertain when Y_2 is known). In contrast, if s_ε is large (large scattering about the transformation model), ρ_{12} is close to zero, and weak correlation exists between Y_1 and Y_2. In this case, given the information of $Y_2 = y_2$, $Y_1 = a + b \times y_2 + \varepsilon$ is almost the same as ε and COV is relatively larger for Y_1 (i.e., no point measuring Y_2 if the purpose is to estimate Y_1). The above simple example shows that correlation ρ_{ij} between (Y_i, Y_j) quantifies how effective one piece of information (Y_i) can be used to update a second piece of information (Y_j), and such effectiveness can be quantified by the updated COV of Y_j – the updated COV is small if ρ_{ij} is close to ± 1 and is relatively larger if ρ_{ij} is close to zero. Evans (1996) labelled $|\rho_{ij}| > 0.8$ as "very strong", $0.6 \leq |\rho_{ij}| < 0.8$ as "strong", $0.4 \leq |\rho_{ij}| < 0.6$ as "moderate", $0.2 \leq |\rho_{ij}| < 0.4$ as "weak", and $|\rho_{ij}| < 0.2$ as "very weak".

Consider another example with three soil parameters: $Y_1 = \ln(s_u/\sigma_v')$, $Y_2 = \text{LI}$, $Y_3 = \ln(\text{OCR})$ (σ_v' is the vertical effective stress; LI is the liquidity index; OCR is the overconsolidation ratio), and consider the following two transformation models:

$$\ln(s_u/\sigma_v') = -0.87 + 0.24 \times \text{LI} + \varepsilon \quad \ln(s_u/\sigma_v') = -1.47 + 0.8 \times \ln(\text{OCR}) + e \quad (4.3)$$

Note that the second equation is related to the SHANSEP concept (Ladd and Foott 1974). The question now is how to update $Y_1 = \ln(s_u/\sigma_v')$ given the bivariate information of $[\text{LI}, \ln(\text{OCR})]$? The key observation here is that the knowledge of ρ_{12} and ρ_{13} is not sufficient for the updating – we also need to know ρ_{23}. If $\rho_{23} = 1$ (this can happen if $\varepsilon = e$), one piece of the information in (LI, OCR) is redundant, and we only need the information LI (or OCR) to update $\ln(s_u/\sigma_v')$. In contrast, if ρ_{23} is relatively small (this may happen if ε and e are statistically independent), both pieces of information (LI, OCR) should be used to update $\ln(s_u/\sigma_v')$. That is to say, updating Y_1 based on multivariate information $(Y_2 = y_2, Y_3 = y_3, \ldots, Y_n = y_n)$ requires pairwise (or bivariate) correlations $(\rho_{ij}: i = 1, \ldots, n-1, j = i+1, \ldots, n)$. Note that only $n \times (n-1)/2$ correlations are needed, because $\rho_{ij} = \rho_{ji}$ by definition.

Generally speaking, for updating purposes, a multivariate probability distribution function should be estimated from multivariate information, e.g., (s_u, OCR, N) simultaneously measured at approximately the same spatial point in the soil mass.

The collection of bivariate correlations (ρ_{ij}: $i = 1, \ldots, n-1, j = i+1, \ldots, n$) is not sufficient. However, complete multivariate information is rarely available. Among multivariate probability distributions, the multivariate *normal* distribution is available analytically and can be easily constructed based on the collection of bivariate correlations alone. Because bivariate correlations between soil parameters are more commonly available, for example s_u-N, s_u-OCR, and N-OCR (Tables 4.2 and 4.3 show some examples), the multivariate normal distribution is a sensible and practical choice to capture the multivariate dependency among soil parameters in the presence of transformation uncertainties (Phoon et al. 2012). Section 4.3 presents the general framework for the multivariate normal probability distribution and how it can be exploited for Bayesian analysis.

4.3 MULTIVARIATE NORMAL PROBABILITY DISTRIBUTION FUNCTION

Many soil parameters are *not* normally distributed. Let Y denote a non-normally distributed soil parameter. One well known cumulative distribution function (CDF) transform approach can be applied to convert Y into a standard normal variable X: $X = \Phi^{-1}[F(Y)]$, where $\Phi(\cdot)$ is the CDF of the standard normal random variable, $F(\cdot)$ is the CDF of Y, and $\Phi^{-1}(\cdot)$ is the inverse function of $\Phi(\cdot)$. A set of multivariate soil parameters $\underline{Y} = (Y_1, Y_2, \ldots, Y_n)'$ can be transformed into $\underline{X} = (X_1, X_2, \ldots, X_n)'$ by mapping Y_1 to X_1, Y_2 to X_2, and so forth. By construction, X_1, X_2, $\ldots$, X_n are *individually* standard normal random variables. It is crucial to note here that *collectively* $(X_1, X_2, \ldots, X_n)'$ does not necessarily follow a multivariate normal distribution even if each component is normally distributed. Even so, recent studies by Ching et al. (2010), Ching and Phoon (2012a), Ching and Phoon (2013), Ching and Phoon (2014b), and Ching et al. (2014b) showed that the multivariate normal distribution is an acceptable approximation for selected parameters of clays, and Ching et al. (2012) arrived at the same observation for selected parameters of sands.

The multivariate (standard) normal probability density function for $\underline{X} = (X_1, X_2, \ldots, X_n)'$ can be defined uniquely by a correlation matrix:

$$f(\underline{X}) = |\mathbf{C}|^{-\frac{1}{2}} (2\pi)^{-\frac{n}{2}} \exp\left(-\frac{1}{2}\underline{X}' \cdot \mathbf{C}^{-1} \cdot \underline{X}\right) \tag{4.4}$$

where $\mathbf{C}$ is the correlation matrix. For $n = 3$, the correlation matrix is given by:

$$\mathbf{C} = \begin{bmatrix} 1 & \delta_{12} & \delta_{13} \\ \delta_{12} & 1 & \delta_{23} \\ \delta_{13} & \delta_{23} & 1 \end{bmatrix} \tag{4.5}$$

where δ_{ij} = product-moment (Pearson) correlation between X_i and X_j (not equal to the correlation ρ_{ij} between the original physical variable Y_i and Y_j). It is clear that the full multivariate dependency structure of a normal random vector only depends on a correlation matrix ($\mathbf{C}$) containing bivariate correlations between all possible pairs of components, namely X_1 and X_2, X_1 and X_3, and X_2 and X_3. One may be tempted to say that it is not necessary to measure X_1, X_2, and X_3 *simultaneously*. In other words, information on X_1 and X_2, X_1 and X_3, and X_2 and X_3 can be collected at three separate borehole locations, rather than one single borehole location (which is a more restrictive condition). However, although the former collection strategy can produce three correlation coefficients to populate C fully, it does not guarantee that C is a positive definite matrix. We discuss this abstract but important matrix property in Section 4.5.1.

The correlation coefficients δ_{ij} in the matrix C can be estimated using at least two methods (Section 4.7.4 will introduce a more robust method based on the Kendall correlations among Y data):

(a)   Full multivariate manner based on a genuine multivariate dataset $(X_1, X_2, \ldots, X_n)$:

$$C \approx \begin{bmatrix} s_1^{-1} & & \\ & \ddots & \\ & & s_n^{-1} \end{bmatrix} \times \frac{1}{N-1} \sum_{k=1}^{N} \left(\begin{bmatrix} X_1^{(k)} - m_1 \\ \vdots \\ X_n^{(k)} - m_n \end{bmatrix} \times \begin{bmatrix} X_1^{(k)} - m_1 \\ \vdots \\ X_n^{(k)} - m_n \end{bmatrix}^T \right) \times \begin{bmatrix} s_1^{-1} & & \\ & \ddots & \\ & & s_n^{-1} \end{bmatrix}$$

(4.6)

where $(X_1^{(k)}, X_2^{(k)}, \ldots, X_n^{(k)})$ is the k-th data point $(X_1, X_2, \ldots, X_n)$; N is the total number of data points; m_i and s_i are the sample mean and sample standard deviation for X_i. Note that the genuine multivariate dataset $(X_1, X_2, \ldots, X_n)$ is required for this method.

(b)   Entry-by-entry bivariate manner based on a bivariate dataset (X_i, X_j):

$$\delta_{ij} \approx \frac{\frac{1}{n_{ij}-1} \sum_{k=1}^{n_{ij}} (X_i^{(k)} - m_i) \cdot (X_j^{(k)} - m_j)}{\sqrt{\frac{1}{n_{ij}-1} \sum_{k=1}^{n_{ij}} (X_i^{(k)} - m_i)^2 \times \frac{1}{n_{ij}-1} \sum_{k=1}^{n_{ij}} (X_j^{(k)} - m_j)^2}}$$

(4.7)

where n_{ij} is the number of the bivariate (X_i, X_j) data points. The benefit of this method is that the genuine multivariate dataset $(X_1, X_2, \ldots, X_n)$ is not required. Only *all possible* bivariate datasets (X_i, X_j) are needed. We have pointed out that this method does not guarantee C to be positive definite, but it is more applicable to geotechnical data in the literature.

The framework of the multivariate normal distribution brings another technical benefit – analytical solutions for Bayesian analysis can be derived. For instance, given the information of $X_2 = x_2$, the updated mean and standard deviation of X_1 can be derived analytically:

$$E[X_1 | X_2 = x_2] = E(X_1) + Cov(X_1, X_2)Var(X_2)^{-1}x_2 = \delta_{12}x_2$$

$$Var[X_1 | X_2 = x_2] = Var(X_1) - Cov(X_1, X_2)Var(X_2)^{-1}Cov(X_2, X_1) = 1 - \delta_{12}^2$$

(4.8)

It is clear that if $\delta_{12} = 0$, no correlation between X_1 and X_2, the updated mean and standard deviation of X_1 are the same as the original values of 0 and 1, respectively. In contrast, if $\delta_{12} = 1$, perfect correlation between X_1 and X_2, the updated mean of X_1 becomes x_2 and the updated standard deviation is 0, i.e. zero uncertainty. Based on Evans (1996) classification scheme, a "very weak" correlation produces less than 5% reduction in the variance. A "weak", "moderate", or "strong" correlation produces 10%, 25%, or 50% reduction, respectively. Finally, a "very strong" correlation produces more than 2/3 reduction in the variance. The above analytical Bayesian equations can be generalized to multivariate cases: update the uncertainty in $\alpha X_i + \beta X_j + \gamma$ by multivariate data sources $(aX_m + bX_n + c, dX_p + eX_q + f, \ldots)$, where $a, b, \ldots, \beta, \gamma$ are arbitrary prescribed constants. The updated COV from multivariate information is always less than the updated COV from univariate information. This further reduction of uncertainties by multivariate information is crucial. One recurring criticism of geotechnical RBD is that there is no particular motivation to use it because it seems to produce designs comparable to existing practice. The further reduction of uncertainties by multivariate information shows that site investigation is not a *cost* item but an *investment* item, because reduction of uncertainties using multivariate tests can translate directly to design savings through RBD. There is no explicit link between how much information is collected from a site and our existing factor of safety. Hence, it is not surprising that a minimum number of boreholes is mandated in many building regulations, because it is difficult to justify site investigation costs, particularly to people outside the geotechnical profession. Simplified RBD involving partial/resistance factors that are independent of site information (e.g., resistance factor $= 0.5$ regardless of number or types of tests conducted) is effectively the same as our existing factor of safety approach insofar as linkage to site investigation is concerned. The performance of various simplified RBD methods with regards to linking site investigation efforts to final design savings is discussed in Chapter 6.

4.4 MULTIVARIATE NORMAL DISTRIBUTIONS CONSTRUCTED WITH GENUINE MULTIVARIATE DATA

4.4.1 CLAY/5/345

A multivariate database of $Y_1 = LI$ (liquidity index), $Y_2 = s_u$, $Y_3 = s_u^{re}$ (remolded undrained shear strength), $Y_4 = \sigma_p'$ (preconsolidation stress), and $Y_5 = \sigma_v'$ (effective vertical stress) is compiled in Ching and Phoon (2012a). There are 345 data points of structured clays from 37 sites worldwide, covering a wide range of sensitivity, LI, and clay types, with simultaneous knowledge of $(Y_1, Y_2, \ldots, Y_5)$. The OCR values of the data points are generally small; 97% of the values is less than 4. Fissured and organic clays are mostly left out of the database. Because s_u values depend on stress state, strain rate, stress path, etc., all s_u values are converted into mobilized s_u values following the recommendations made by Mesri and Huvaj (1997). The marginal probability density functions (PDF) for $(Y_1, Y_2, \ldots, Y_5)$ and their statistics (mean of $Y_i = \mu_i$, COV of $Y_i = V_i$, mean of $\ln(Y_i) = \lambda_i$, standard deviation of $\ln(Y_i) = \xi_i$) are summarized in Table 4.4.

Table 4.4 Distributions and statistics of $(Y_1, Y_2, \ldots, Y_5)$ for CLAY/5/345 (Source: Table 3, Ching and Phoon 2012b).

	Distribution	Mean (μ)	COV (V)	Mean of ln(Y) (λ)	Stdev of ln(Y) (ξ)
$Y_1 = LI$	Lognormal	1.251	0.487	0.122	0.459
$Y_2 = s_u$	Lognormal	31.009 kPa	0.951	3.051	0.898
$Y_3 = s_u^{re}$	Lognormal	2.514 kPa	1.516	0.226	1.191
$Y_4 = \sigma_p'$	Lognormal	105.820 kPa	0.975	4.311	0.835
$Y_5 = \sigma_v'$	Lognormal	66.631 kPa	0.803	3.891	0.823

It is useful to point out that the univariate statistics shown in Table 4.4 are not necessarily meaningful if data are collected over diverse sites and at different depths. The technical reason is that the population is not statistically homogeneous. An alternate way of saying this is that the deviation from the mean is not "random". The deviation may be explainable by, say the difference in depth or in-situ effective stress state from which the data are measured. Many geotechnical parameters are related to the effective stress state. For example, $Y_2 = s_u$ is known to be dependent on the effective vertical stress ($Y_5 = \sigma_v'$) and stress history ($Y_4 = \sigma_p'$). From this brief digression, one may say that a multivariate distribution is a more robust model than a univariate distribution. Correlations (such as between s_u, σ_v', and σ_p') are automatically captured and the "unexplainable" residuals from a multivariate model are more likely to be "random" as a result. There could be subtle correlations related to soil type or other site-specific attributes that are not readily captured because data are insufficient or incomplete (hence, "unexplainable" residuals may be explainable in the presence of new data), but this caveat applies to all real world statistical analysis. In summary, univariate statistics shown in Table 4.4 or comparable tables presented in Chapter 4 are not meant to use in a stand-alone manner without reference to the multivariate distribution (specifically, the correlation matrix). On the other hand, the univariate statistics (primarily coefficient of variation) given in Chapter 3 can be used in a stand-alone manner, because they are site-specific and evaluated at comparable depths. A discerning reader will notice that the coefficient of variation reported in Chapter 4 is larger than that in Chapter 3. This is to be expected for reasons discussed above.

For lognormal Y, the CDF transform $X = \Phi^{-1}[F(Y)]$ is simply:

$$X_i = [\ln(Y_i) - \lambda_i]/\xi_i \tag{4.9}$$

The transformed $(X_1, X_2, \ldots, X_5)$ are individually standard normal random variables. The inverse CDF transform has the following form:

$$Y_i = \exp(\lambda_i + \xi_i \cdot X_i) \tag{4.10}$$

The correlation matrix **C** for $(X_1, X_2, \ldots, X_5)$, estimated by Eq. 4.6, is shown in Table 4.5, and $(X_1, X_2, \ldots, X_5)$ is assumed to be multivariate normal with the correlation matrix listed in the table.

Table 4.5 Correlation matrix C for $(X_1, X_2, \ldots, X_5)$ in CLAY/5/345 (based on the updated database in Ching and Phoon 2012b).

		X_1	X_2	X_3	X_4	X_5
	X_1 (for LI)	1.000	$\delta_{12} = -0.128$	$\delta_{13} = -0.832$	$\delta_{14} = -0.162$	$\delta_{15} = -0.274$
	X_2 (for s_u)		1.000	$\delta_{23} = 0.272$	$\delta_{24} = 0.915$	$\delta_{25} = 0.782$
C=	X_3 (for s_u^{re})			1.000	$\delta_{34} = 0.337$	$\delta_{35} = 0.429$
	X_4 (for σ_p')		Symmetric		1.000	$\delta_{45} = 0.832$
	X_5 (for σ_v')					1.000

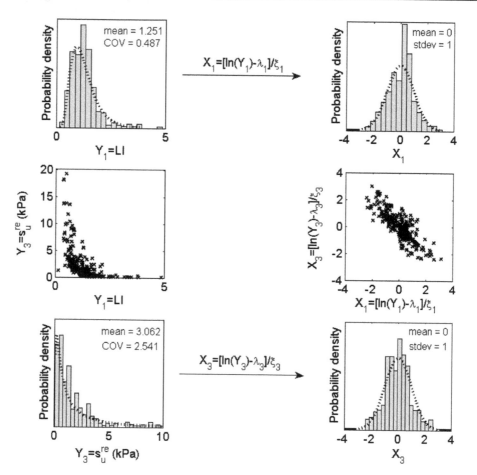

Figure 4.1 Correlations between Y_1 and Y_3 as well as between X_1 and X_3 (revised from Ching and Phoon 2012a).

The effectiveness of the CDF transform and the validity of the multivariate normal assumption are illustrated for $(Y_1 = LI, Y_3 = s_u^{re})$ in Figure 4.1. Before the CDF transform, both Y_1 and Y_3 are roughly lognormally distributed, but after the transform, both X_1 and X_3 are roughly distributed as a standard normal distribution. More

Table 4.6 Updated mean and COV of s_u/σ_v' for CLAY/5/345 under various combinations of information (Source: Table 7, Ching and Phoon 2012b).

Information	Updated mean of s_u/σ_v'	Updated COV of s_u/σ_v'
LI	$LI^{0.241} \times 0.491$	0.609
σ_p'/σ_v'	$(\sigma_p'/\sigma_v')^{0.925} \times 0.313$	0.373
LI, σ_p'/σ_v'	$LI^{0.0612} \times (\sigma_p'/\sigma_v')^{0.914} \times 0.312$	0.371
S_t, σ_p'/σ_v'	$S_t^{0.121} \times (\sigma_p'/\sigma_v')^{0.823} \times 0.229$	0.338
LI, S_t, σ_p'/σ_v'	$LI^{-0.287} \times S_t^{0.192} \times (\sigma_p'/\sigma_v')^{0.815} \times 0.194$	0.323

Table 4.7 Updated mean and COV of σ_p'/P_a for CLAY/5/345 under various combinations of information (Source: Table 8, Ching and Phoon 2012b).

Information	Updated mean of σ_p'/P_a	Updated COV of σ_p'/P_a
LI	$LI^{-0.295} \times 1.07$	0.985
LI, S_t	$LI^{-1.319} \times S_t^{0.536} \times 0.235$	0.729
LI, σ_v'/P_a	$LI^{0.130} \times (\sigma_v'/P_a)^{0.864} \times 1.507$	0.485
LI, S_t, σ_v'/P_a	$LI^{-0.398} \times S_t^{0.241} \times (\sigma_v'/P_a)^{0.727} \times 0.722$	0.433

$P_a = 101.3$ kPa is one atmosphere pressure.

interestingly, the nonlinear correlation between (Y_1, Y_3) is evident, but after the CDF transform, the correlation between (X_1, X_3) becomes fairly linear. This linear correlation is also observed for other pairs of (X_i, X_j) (Ching and Phoon 2012a). Because of the linear correlations for all pairs of (X_i, X_j), there is no strong evidence to reject the underlying multivariate normality for this particular soil database. Ching and Phoon (2015a) discussed more rigorous hypothesis testing methods for multivariate normality.

After the multivariate normal distribution of $(X_1, X_2, X_3, X_4, X_5)$ has been established, Bayesian analysis can be conducted to find useful transformation models among $(X_1, X_2, X_3, X_4, X_5)$. For instance, given $(aX_m + bX_n + c, dX_p + eX_q + f, \ldots)$, what are the analytical solutions for the updated mean and COV of $\alpha X_i + \beta X_j + \gamma$? Note that many quantities of interest can be expressed in the form of $\alpha X_i + \beta X_j + \gamma$, e.g., $\ln(s_u/\sigma_v') = \ln(Y_2) - \ln(Y_5) = \xi_2 X_2 - \xi_5 X_5 + \lambda_2 - \lambda_5$. Once the updated mean and updated standard deviation for $\ln(s_u/\sigma_v')$, respectively denoted by m and s, are derived, the updated mean and COV of s_u/σ_v' are simply $\exp(m + 0.5s^2)$ and $[\exp(s^2) - 1]^{0.5}$. Tables 4.6, 4.7, and 4.8 summarize useful transformation models developed in Ching and Phoon (2012a) using Bayesian analysis. Note that the updated probability distribution is still lognormal.

It can be seen from these three tables that the updated COV typically decreases with increasing amount of information. This demonstrates numerically that uncertainty in a design soil parameter can be reduced by multivariate correlations between the design parameter and other available pieces of site investigation information. This also illustrates the *value of information*. As stated earlier, site investigation is not a *cost* item but an *investment* item, because reduction of uncertainties using multivariate tests can translate directly to design savings through RBD. Moreover, the correlation matrix

Table 4.8 Updated mean and COV of $S_t = s_u/s_u^{re}$ for CLAY/5/345 under various combinations of information (Source: Table 9, Ching and Phoon 2012b).

Information	Updated mean of S_t	Updated COV of S_t
LI	$LI^{1.910} \times 20.726$	1.185
LI, σ_v'/P_a	$LI^{2.189} \times (\sigma_v'/P_a)^{0.597} \times 27.371$	0.982
LI, σ_p'/P_a	$LI^{2.115} \times (\sigma_p'/P_a)^{0.693} \times 21.256$	0.858
LI, σ_p'/σ_v'	$LI^{1.809} \times (\sigma_p'/\sigma_v')^{0.513} \times 16.422$	1.125
LI, s_u/σ_v'	$LI^{1.710} \times (s_u/\sigma_v')^{0.831} \times 38.262$	0.966
LI, s_u^{re}/P_a	$LI^{0.624} \times (s_u^{re}/P_a)^{-0.595} \times 1.642$	1.029
LI, $s_u^{re}/P_a, \sigma_v'/P_a$	$LI^{0.197} \times (s_u^{re}/P_a)^{-0.993} \times (\sigma_v'/P_a)^{0.880} \times 0.464$	0.599
LI, $s_u/\sigma_v', \sigma_p'/P_a$	$LI^{1.939} \times (s_u/\sigma_v')^{0.593} \times (\sigma_p'/P_a)^{0.580} \times 32.797$	0.753
LI, $s_u^{re}/P_a, \sigma_p'/P_a$	$LI^{-0.0801} \times (s_u^{re}/P_a)^{-1.058} \times (\sigma_p'/P_a)^{1.006} \times 0.237$	0.372
$s_u^{re}/P_a, \sigma_v'/P_a$	$(s_u^{re}/P_a)^{-1.081} \times (\sigma_v'/P_a)^{0.891} \times 0.359$	0.601

C in Table 4.5 contains useful information on the effectiveness of a given test type in reducing the uncertainty of another soil parameter. For instance, if the design soil parameter of interest is $s_u(X_2)$, it is clear that δ_{24} is the largest, so the oedometer test that determines the preconsolidation pressure (σ_p' or its natural logarithm transform, X_4) is the most effective test. In contrast, the knowledge of LI (or its natural logarithm transform, X_1) is not very helpful in reducing the uncertainty in s_u because δ_{12} is close to zero. The correlation matrix in Table 4.5 and the transformation models in Tables 4.6, 4.7, and 4.8 should be suitable for structured clays (sensitive or quick clays) with low OCR < 4.

4.4.2 CLAY/6/535

The CLAY/6/535 database consists of the following six dimensionless parameters simultaneously measured in close proximity (Ching et al. 2014b) in the form of $Y_1 = s_u/\sigma_v'$, $Y_2 = OCR$, $Y_3 = (q_t - \sigma_v)/\sigma_v'$, $Y_4 = (q_t - u_2)/\sigma_v'$, $Y_5 = (u_2 - u_0)/\sigma_v'$, and $Y_6 = B_q$. There are 535 genuine multivariate data points with complete (Y_1, Y_2, ..., Y_6) information from 40 sites worldwide. The clay properties cover a wide range of OCR (mostly 1~6 except for 5 sites) and wide range of plasticity index PI (10~168). Highly OC (fissured) and organic clays are mostly left out of this database. Because s_u values depend on stress state, strain rate, stress path, etc., all s_u values are converted into equivalent CIUC values. The statistics (mean and COV) and the range of each component Y_i are shown in Table 4.9. It is apparent that the range covered by each component is fairly large.

For (Y_1, Y_2, ..., Y_6), the lognormal distribution no longer provides a satisfactory fit. The marginal distributions for (Y_1, Y_2, ..., Y_6) are chosen among the Johnson system of distributions (Phoon and Ching 2013). The Johnson system of distributions contains three families of distributions (SU, SB & SL) that can be generated as a CDF transformation from a standard normal random variable X:

$$X_i = \begin{cases} b_x + a_x \cdot \sinh^{-1}[(Y_i - b_y)/a_y] & \text{for SU} \\ b_x + a_x \cdot \ln[(Y_i - b_y)/(a_y + b_y - Y_i)] & \text{for SB} \\ b_x^* + a_x \cdot \ln(Y_i - b_y) & \text{for SL} \end{cases} \quad (4.11)$$

Table 4.9 Statistics of $(Y_1, Y_2, \ldots, Y_6)$ for CLAY/6/535 (Source: Table 2, Ching et al. 2014b).

	Mean	COV	Max	Min
$Y_1 = s_u/\sigma_v'$	0.641	0.596	3.041	0.105
$Y_2 = OCR$	2.353	0.657	9.693	1.000
$Y_3 = (q_t - \sigma_v)/\sigma_v'$	9.350	0.678	58.878	2.550
$Y_4 = (q_t - u_2)/\sigma_v'$	5.280	0.885	43.694	0.605
$Y_5 = (u_2 - u_0)/\sigma_v'$	4.709	0.574	21.720	0.236
$Y_6 = B_q$	0.556	0.338	1.072	-0.093

Table 4.10 Johnson family types and parameters for $(Y_1, Y_2, \ldots, Y_6)$ in CLAY/6/535 (Source: Table 2, Ching et al. 2014b).

		Johnson parameters			
Parameter	Family type	a_x	b_x	a_y	b_y
$Y_1 = s_u/\sigma_v'$	SU	1.222	-1.742	0.141	0.250
$Y_2 = OCR$	SB	0.709	1.887	12.724	0.954
$Y_3 = (q_t - \sigma_v)/\sigma_v'$	SU	1.033	-1.438	1.723	4.157
$Y_4 = (q_t - u_2)/\sigma_v'$	SU	0.989	-1.593	0.868	1.638
$Y_5 = (u_2 - u_0)/\sigma_v'$	SU	0.971	-0.762	1.116	3.123
$Y_6 = B_q$	SU	2.961	0.049	0.544	0.570

where the inverse hyperbolic function is defined as

$$\sinh^{-1}(x) = \ln(x + \sqrt{x^2 + 1}) \tag{4.12}$$

The distribution SU is an unbounded distribution that is defined on $[-\infty, \infty]$, SB is a bounded distribution defined on $[b_y, a_y + b_y]$, and SL is a lower bounded distribution defined on $[b_y, \infty]$. The inverse CDF transforms have the following forms:

$$Y_i = \begin{cases} b_y + a_y \cdot \sinh[(X_i - b_x)/a_x] & \text{for SU} \\ (b_y + (a_y + b_y)) \cdot \exp[(X_i - b_x)/a_x])/(1 + \exp[(X_i - b_x)/a_x]) & \text{for SB} \\ b_y + \exp[(X_i - b_x^*)/a_x] & \text{for SL} \end{cases} \tag{4.13}$$

where the hyperbolic sine function is defined as

$$\sinh(x) = (e^x - e^{-x})/2 \tag{4.14}$$

Slifker and Shapiro (1980) showed that it is possible to identify the family type (SU, SB, SL) and estimate the four model parameters (a_x, b_x, a_y, b_y) based on four sample quantiles of Y (Phoon and Ching 2013; Ching et al. 2014b). Table 4.10 shows the identified Johnson family types and estimated parameters for $(Y_1, Y_2, \ldots, Y_6)$. The Johnson system can generate distributions with a wide range of mean value, COV, skewness, and kurtosis. Details for the Johnson system can be found in Phoon and Ching (2013) and Ching et al. (2014b). The chief advantage of the Johnson system is that it provides

Table 4.11 Correlation matrix **C** for $(X_1, X_2, \ldots, X_6)$ in CLAY/6/535 (Source: Table 6, Ching et al. 2014b).

		X_1	X_2	X_3	X_4	X_5	X_6
	X_1 (for s_u/σ_v')	1.00	0.62	0.67	0.61	0.49	−0.28
	X_2 (for OCR)		1.00	0.61	0.51	0.54	−0.15
C =	X_3 [for $(q_t - \sigma_v)/\sigma_v'$]			1.00	0.83	0.70	−0.45
	X_4 [for $(q_t - u_2)/\sigma_v'$]				1.00	0.31	−0.77
	X_5 [for $(u_2 - u_0)/\sigma_v'$]		Symmetric			1.00	0.28
	X_6 (for B_q)						1.00

Table 4.12 Updated Johnson distribution and parameters of s_u/σ_v' for CLAY/6/535 under various combinations of information (derived from Tables 5 and 9 in Ching et al. 2014b).

Information	Family type	a_x	b_x	a_y	b_y
OCR	SU	1.555	$(-1.742 - 0.619 \times X_2)/0.786$	0.141	0.250
$(q_t - \sigma_v)/\sigma_v'$		1.647	$(-1.742 - 0.671 \times X_3)/0.742$		
$(q_t - u_2)/\sigma_v'$		1.545	$(-1.742 - 0.612 \times X_4)/0.791$		
$(u_2 - u_0)/\sigma_v'$		1.405	$(-1.742 - 0.493 \times X_5)/0.870$		
$(q_t - \sigma_v)/\sigma_v', B_q$		1.649	$(-1.742 - 0.683 \times X_3 0.0266 \times X_6)/0.741$		
$(q_t - u_2)/\sigma_v', B_q$		1.669	$(-1.742 - 0.972 \times X_4 - 0.468 \times X_6)/0.732$		
$(u_2 - u_0)/\sigma_v', B_q$		1.619	$(-1.742 - 0.618 \times X_5 + 0.451 \times X_6)/0.755$		
$(q_t - \sigma_v)/\sigma_v', (q_t - u_2)/\sigma_v', (u_2 - u_0)/\sigma_v', B_q$		1.711	$(-1.742 - 0.443 \times X_3 - 0.609 \times X_4 + 0.124 \times X_5 - 0.422 \times X_6)/0.714$		

Note: $X_2 = 1.887 + 0.709 \times \ln[(\text{OCR} - 0.954)/(13.678 - \text{OCR})]$; $X_3 = -1.438 + 1.033 \times \sinh^{-1}\{[(q_t - \sigma_v)/\sigma_v' - 4.157]/1.723\}$; $X_4 = -1.593 + 0.989 \times \sinh^{-1}\{[(q_t - u_2)/\sigma_v' - 1.638]/0.868\}$; $X_5 = -0.762 + 0.971 \times \sinh^{-1}\{[(u_2 - u_0)/\sigma_v' - 3.123]/1.116\}$; $X_6 = 0.049 + 2.961 \times \sinh^{-1}[(B_q 0.570)/0.544]$.

Table 4.13 Updated Johnson distribution and parameters of OCR for CLAY/6/535 under various combinations of information (derived from Tables 5 and 10 in Ching et al. 2014b)

Information	Family type	a_x	b_x	a_y	b_y
$(q_t - \sigma_v)/\sigma_v'$	SB	0.894	$(0.709 - 0.610 \times X_3)/0.793$	12.724	0.954
$(q_t - u_2)/\sigma_v'$		0.826	$(0.709 - 0.514 \times X_4)/0.858$		
$(u_2 - u_0)/\sigma_v'$		0.844	$(0.709 - 0.542 \times X_5)/0.840$		
B_q		0.717	$(0.709 + 0.148 \times X_6)/0.989$		
$(q_t - \sigma_v)/\sigma_v', B_q$		0.909	$(0.709 - 0.681 \times X_3 - 0.158 \times X_6)/0.780$		
$(q_t - u_2)/\sigma_v', B_q$		0.927	$(0.709 - 0.982 \times X_4 - 0.608 \times X_6)/0.765$		
$(q_t - \sigma_v)/\sigma_v', (q_t - u_2)/\sigma_v', (u_2 - u_0)/\sigma_v', B_q$		0.944	$(0.709 - 0.257 \times X_3 - 0.602 \times X_4 - 0.0589 \times X_5 - 0.415 \times X_6)/0.751$		

Note: X_3, X_4, X_5, and X_6 are defined in Table 4.10.

more flexibility in distribution fitting while retaining the closed-form simplicity of the CDF transform $X = \Phi^{-1}[F(Y)]$.

$(Y_1, Y_2, \ldots, Y_6)$ are transformed into $(X_1, X_2, \ldots, X_6)$ by using the CDF transforms given in Eq. 4.11. The resulting $(X_1, X_2, \ldots, X_6)$ are roughly standard normal.

The correlation matrix C for $(X_1, X_2, \ldots, X_6)$ can be readily estimated based on Eq. 4.6, as shown in Table 4.11. Tables 4.12 and 4.13 summarize useful transformation models developed in Ching et al. (2014b) using Bayesian analysis. Note that the updated probability distribution is still Johnson (another practical advantage of the Johnson system). The correlation matrix C in Table 4.11 and the transformation models in Tables 4.12 and 4.13 should be suitable for normally to medium consolidated clays with OCR $\leqslant 6$.

4.5 MULTIVARIATE NORMAL DISTRIBUTIONS CONSTRUCTED WITH *BIVARIATE* DATA

4.5.1 CLAY/7/6310

The CLAY/7/6310 database (Ching and Phoon 2013) consists of a large number of s_u data points obtained from different test procedures from 164 studies worldwide. The clay properties cover a wide range of OCR (mostly 1–10, 92% of the studies are associated with OCR <10, but 99.5% of the studies are associated with OCR <50) and a wide range of sensitivity S_t (sites with $S_t = 1 \sim$ tens or hundreds are fairly typical). The seven clay parameters are normalized s_u values under various test modes: $Y_1 = (\bar{s}_u/\sigma'_v)_{CIUC}$, $Y_2 = (\bar{s}_u/\sigma'_v)_{CK_0UC}$, $Y_3 = (\bar{s}_u/\sigma'_v)_{CK_0UE}$, $Y_4 = (\bar{s}_u/\sigma'_v)_{DSS}$, $Y_5 = (\bar{s}_u/\sigma'_v)_{FV}$, $Y_6 = (\bar{s}_u/\sigma'_v)_{UU}$, and $Y_7 = (\bar{s}_u/\sigma'_v)_{UC}$, where CIUC = isotropically consolidated undrained compression test; $CK_0UC = K_0$-consolidated undrained compression test; $CK_0UE = K_0$-consolidated undrained extension test; DSS = direct simple shear test; FV = field vane test; UU = unconsolidated undrained compression test; UC = unconfined compression test. $\bar{s}_u/\sigma'_v$ is the normalized undrained shear strength of a normally consolidated clay with PI = 20 subjected to a 1%/hr strain rate – three factors, namely PI, OCR, and strain rate, are standardized, but the test mode is not standardized although the effects of test mode on a_{OCR}, a_{rate}, and a_{PI} are considered:

$$\bar{s}_u/\sigma'_v = (s_u/\sigma'_v)/(a_{OCR} \times a_{rate} \times a_{PI}) \tag{4.15}$$

The a_{OCR}, a_{rate}, and a_{PI} factors are given in Table 4.14. Table 4.15 shows the statistics of Y_i.

The Y data points for each test mode are roughly lognormally distributed, i.e., $X_i = [\ln(Y_i) - \lambda_i]/\xi_i$ is roughly standard normal [λ_i is the sample mean of $\ln(Y_i)$, and ξ_i is the sample standard deviation], as noted in Ching et al. (2013). The lognormal parameters λ and ξ are shown in Table 4.15. Table 4.16 lists the number of pairwise data where (X_i, X_j) values under two test modes are simultaneously known. The diagonals are the numbers of available data points for individual s_u test modes (same as the third column for parameter 'n' in Table 4.15). The correlation matrix C cannot be estimated using Eq. 4.6 because this equation requires genuine multivariate data. Instead, each correlation coefficient δ_{ij} can be estimated with an *entry-by-entry* bivariate manner based on a bivariate dataset (X_i, X_j) using Eq. 4.7. Difficulties arise when estimating δ_{ij} for pairs with limited data points, including those involve X_5 (FV) (solid box in Table 4.16) and the following pairs: (X_2, X_6), (X_2, X_7), (X_3, X_6), (X_3, X_7), (X_4, X_6), and (X_4, X_7) (dashed box in Table 4.16). The correlation coefficient δ_{ij} estimated with limited (X_i, X_j) data can be misleading. Reasonably practical procedures for dealing with this sample size difficulty are proposed in Ching and Phoon (2013) to

Table 4.14 a_{OCR}, a_{rate}, and a_{PI} factors (Source: Table 3 in Ching and Phoon 2013).

Factor	Test type	Formula
$a_{OCR} = OCR^{\Lambda}$	CIUC	$OCR^{0.602}$
	CK_0UC	$OCR^{0.681}$
	CK_0UE	$OCR^{0.898}$
	DSS	$OCR^{0.749}$
	FV	$OCR^{0.902}$
	UU	$OCR^{0.800}$
	UC	$OCR^{0.932}$
a_{rate}		$1.0 + 0.1 \times log_{10}(hourly\ strain\ rate/1\%)$
$a_{PI} = (PI/20)^{\beta}$	CIUC	$(PI/20)^0 = 1$
	CK_0UC	$(PI/20)^0 = 1$
	CK_0UE	$(PI/20)^{0.178}$
	DSS	$(PI/20)^{0.0655}$
	FV	$(PI/20)^{0.124}$
	UU	$(PI/20)^0 = 1$
	UC	$(PI/20)^0 = 1$

Table 4.15 Statistics of $(Y_1, Y_2, \ldots, Y_7)$ for CLAY/7/6310 (Source: Table 4, Ching and Phoon 2013).

Variable		n	Mean	COV	Min	Max	$E[ln(Y)]$ (λ)	$\sigma[ln(Y)]$ (ξ)
Y_1	$(\bar{s}_u/\sigma'_v)_{CIUC}$	637	0.404	0.316	0.12	0.82	−0.955	0.315
Y_2	$(\bar{s}_u/\sigma'_v)_{CK_0UC}$	555	0.350	0.318	0.063	1.72	−1.090	0.280
Y_3	$(\bar{s}_u/\sigma'_v)_{CK_0UE}$	224	0.184	0.324	0.055	0.45	−1.748	0.355
Y_4	$(\bar{s}_u/\sigma'_v)_{DSS}$	573	0.241	0.399	0.081	1.83	−1.468	0.277
Y_5	$(\bar{s}_u/\sigma'_v)_{FV}$	1057	0.275	0.416	0.068	1.25	−1.363	0.372
Y_6	$(\bar{s}_u/\sigma'_v)_{UU}$	435	0.243	0.504	0.067	1.44	−1.523	0.463
Y_7	$(\bar{s}_u/\sigma'_v)_{UC}$	387	0.223	0.611	0.039	1.01	−1.640	0.523

Table 4.16 Numbers of available (X_i, X_j) data pairs (Source: Table 5 in Ching and Phoon 2013).

	X_1 (CIUC)	X_2 (CK_0UC)	X_3 (CK_0UE)	X_4 (DSS)	X_5 (FV)	X_6 (UU)	X_7 (UC)
X_1 (CIUC)	637	129	30	24	20	84	38
X_2 (CK_0UC)		555	69	135	79	13	14
X_3 (CK_0UE)			224	66	43	7	14
X_4 (DSS)				573	58	18	14
X_5 (FV)					1057	123	140
X_6 (UU)		Symmetric				435	53
X_7 (UC)							387

estimate those δ_{ij} values. The resulting correlation matrix C for $(X_1, X_2, \ldots, X_7)$ is shown in Table 4.17. We note in passing that limited data is the norm in geotechnical engineering and the development of statistical methods that are robust under this practical constraint remains one major challenge in geotechnical reliability.

Table 4.17 Correlation matrix **C** for $(X_1, X_2, ..., X_7)$ in CLAY/7/6310 (Source: Table 7, Ching and Phoon 2013).

	X_1 (CIUC)	X_2 (CK$_0$UC)	X_3 (CK$_0$UE)	X_4 (DSS)	X_5 (FV)	X_6 (UU)	X_7 (UC)
X_1 (CIUC)	1.00	0.84	0.47	0.72	0.63	0.88	0.85
X_2 (CK$_0$UC)		1.00	0.39	0.78	0.35	0.7	0.6
X_3 (CK$_0$UE)			1.00	0.45	0.41	0.4	0.3
C = X_4 (DSS)				1.00	0.73	0.6	0.5
X_5 (FV)					1.00	0.64	0.46
X_6 (UU)		Symmetric				1.00	0.68
X_7 (UC)							1.00

Based on this correlation matrix, Ching and Phoon (2015b) further derived the equations to calculate the updated mean (m) and standard deviation (s) for the natural logarithm of the mobilized undrained shear strength, denoted by ln[s_u(mob)], given the s_u values from other test modes. s_u(mob) is the undrained shear strength mobilized in a full-scale undrained failure in the field (Mesri and Huvaj 2007). The equations have the following generic form:

$$m = \text{updated mean of } \ln[s_u(\text{mob})/\sigma_v']$$
$$= a_0 + a_1 \ln[s_u(\text{CIUC})/\sigma_v'] + a_2 \ln[s_u(\text{CK}_0\text{UC})/\sigma_v'] + a_3 \ln[s_u(\text{CK}_0\text{UE})/\sigma_v']$$
$$+ a_4 \ln[s_u(\text{DSS})/\sigma_v'] + a_5 \ln[s_u(\text{FV})/\sigma_v'] + a_6 \ln[s_u(\text{UU})/\sigma_v'] + a_7 \ln[s_u(\text{UC})/\sigma_v']$$
$$+ \Lambda' \ln(\text{OCR}) + \beta' \ln(\text{PI}/20) + \ln(a_{\text{rate}}) \tag{4.16}$$

$$s^2 = \text{updated variance of } \ln[s_u(\text{mob})\sigma_v']$$

where s_u(test mode) is the undrained shear strength obtained under a certain test mode, e.g., CIUC; a_{rate} is the strain rate correction factor (see Table 4.14). Ching and Phoon (2015b) derived the equations for three types of s_u(mob): (a) the s_u(mob) for an embankment failure; (b) the s_u(mob) for the active state failure; and (c) the s_u(mob) for the passive state failure. Table 4.18 lists the values of (a_0, a_1, a_2, a_3, a_4, a_5, a_6, a_7, Λ', β') and s for various combinations of s_u information for the s_u(mob) for an embankment failure. The tables for the other two s_u(mob) can be found in Ching and Phoon (2015b).

The usefulness of Table 4.18 is illustrated using the following example. At the depth of 11 m of a clay site, s_u(FV) = 33.8 kN/m² and s_u(UC) = 25 kN/m² are known. The vertical effective stress (σ_v') at this depth is 98.4 kN/m², OCR = 2.06, and PI = 10. Here we demonstrate the estimation of the mean and standard deviation of ln[s_u(mob)/σ_v'] (embankment failure) using s_u(FV)/σ_v' and s_u(UC)/σ_v' [s_u(FV)/σ_v' = 33.8/94.8 = 0.356; s_u(UC)/σ_v' = 25/94.8 = 0.264]. According to Table 4.18, when information from s_u(FV)/σ_v' and s_u(FV)/σ_v' is available, m and s are

$$m = a_0 + a_5 \ln\left[\frac{s_u(\text{FV})}{\sigma_v'}\right] + a_7 \ln\left[\frac{s_u(\text{UC})}{\sigma_v'}\right] + \Lambda' \ln(\text{OCR}) + \beta' \ln(\text{PI}/20) + \ln(a_{\text{rate}})$$
$$s = 0.187 \tag{4.17}$$

Table 4.18 Coefficients (a_0, a_1, a_2, a_3, a_4, a_5, a_6, a_7, Λ', β') and s for the s_u(mob) of an embankment failure (Source: Table 5, Ching and Phoon 2015b, with permission from ASCE).

Information	a_0	a_1	a_2	a_3	a_4	a_5	a_6	a_7	Λ'	β'	s
None (prior)	−1.435								0.776	0.081	0.251
A single test mode											
CIUC	−0.827	0.636							0.393	0.081	0.151
CK$_0$UC	−0.612		0.755						0.262	0.081	0.136
CK$_0$UE	−0.469			0.553					0.279	−0.017	0.157
DSS	−0.278				0.788				0.185	0.030	0.124
FV	−0.956					0.399			0.416	0.032	0.203
UU	−0.942						0.363		0.486	0.081	0.187
UC	−1.047							0.263	0.531	0.081	0.210
Two test modes											
CK$_0$UC & CK$_0$UE	−0.155		0.568	0.378					0.049	0.014	0.056
CIUC & FV	−0.779	0.563				0.099			0.348	0.069	0.149
CIUC & UU	−0.837	0.740					−0.080		0.395	0.081	0.150
CIUC & UC	−0.856	0.955						−0.226	0.411	0.081	0.138
CK$_0$UC & FV	−0.454		0.649			0.228			0.128	0.053	0.110
CK$_0$UC & UU	−0.604		0.657				0.085		0.261	0.081	0.133
CK$_0$UC & UC	−0.604		0.719					0.032	0.256	0.081	0.135
DSS & FV	−0.263				0.850	−0.062			0.196	0.033	0.123
DSS & UU	−0.292				0.663		0.125		0.179	0.038	0.115
DSS & UC	−0.271				0.720			0.073	0.169	0.034	0.120
FV & UU	−0.848					0.187	0.267		0.394	0.058	0.179
FV & UC	−0.838					0.291		0.168	0.357	0.045	0.187
UU & UC	−0.906						0.299	0.083	0.459	0.081	0.184
Three test modes											
CIUC, FV & UU	−0.783	0.698				0.120	−0.117		0.341	0.066	0.147
CIUC, FV & UC	−0.825	0.892				0.062		−0.214	0.382	0.074	0.137
CIUC, UU & UC	−0.882	1.226					−0.169	−0.262	0.418	0.081	0.133
CK$_0$UC, FV & UU	−0.428		0.728			0.278	−0.088		0.100	0.047	0.107
CK$_0$UC, FV & UC	−0.453		0.682			0.243		−0.035	0.126	0.051	0.109
CK$_0$UC, UU & UC	−0.603		0.656				0.084	0.002	0.260	0.081	0.133
DSS, FV & UU	−0.260				0.772	−0.150	0.163		0.203	0.049	0.110
DSS, FV & UC	−0.250				0.796	−0.085		0.080	0.182	0.040	0.118
DSS, UU & UC	−0.288				0.657		0.109	0.024	0.175	0.038	0.115
FV, UU & UC	−0.817					0.182	0.210	0.077	0.372	0.059	0.177

where $a_0 = -0.838$, $a_5 = 0.291$, $a_7 = 0.168$, $\Lambda' = 0.357$, and $\beta' = 0.045$. The constants $a_1 = a_2 = a_3 = a_4 = a_6 = 0$ because CIUC, CK$_0$UC, CK$_0$UE, DSS, and UU are not known. It is then clear that $m = -1.135 + \ln(a_{\text{rate}})$ and $s = 0.187$. These are the updated mean and standard deviation for $\ln[s_u(\text{mob})/\sigma_v']$. Because $s_u(\text{mob})/\sigma_v'$ is lognormal, the mean of $s_u(\text{mob})/\sigma_v'$ is $\exp(m + s^2/2) = 0.327 \times a_{\text{rate}}$ and its COV is $[\exp(s^2) - 1]^{0.5} = 0.189$.

4.5.2   CLAY/10/7490

The CLAY/10/7490 database (Ching and Phoon 2014a, 2014b) compiles data from 251 studies worldwide. The number of data points associated with each study varies

Table 4.19 Statistics of the tenclay parameters in CLAY/10/7490 (Source: Table 3, Ching and Phoon 2014a).

Parameter	n	Mean	COV	Min	Max
LL	3822	67.7	0.80	18.1	515
PI	4265	39.7	1.08	1.9	363
LI	3661	1.01	0.78	−0.75	6.45
σ_v'/P_a	3370	1.80	1.47	4.13E−3	38.74
σ_p'/P_a	2028	4.37	2.31	0.094	193.30
s_u/σ_v'	3538	0.51	1.25	3.7E−3	7.78
S_t	1589	35.0	2.88	1.0	1467
B_q	1016	0.58	0.35	0.01	1.17
$(q_t - \sigma_v)/\sigma_v'$	862	8.90	1.17	0.48	95.98
$(q_t - u_2)/\sigma_v'$	668	5.34	1.37	0.61	108.20
OCR	3531	3.85	1.56	1.0	60.23

Table 4.20 Johnson family types and parameters for $(Y_1, Y_2, \ldots, Y_{10})$ in CLAY/10/7490 (Source: Table 5, Ching and Phoon 2014b).

Parameter	Family type	Johnson parameters			
		a_X	b_X	a_Y	b_Y
Y_1 [ln(LL)]	SU	1.636	−1.166	0.616	3.479
Y_2 [ln(PI)]	SU	1.433	−0.265	0.918	3.178
Y_3 (LI)	SU	1.434	−1.068	0.629	0.358
Y_4 [ln(σ_v'/P_a)]	SB	3.150	0.256	14.458	−7.010
Y_5 [ln(σ_p'/P_a)]	SB	4.600	21.548	576.785	−4.793
Y_6 [ln(s_u/σ_v')]	SU	2.039	−0.517	1.427	−1.461
Y_7 [ln(S_t)]	SU	2.393	−2.080	1.885	0.461
Y_8 (B_q)	SU	2.676	0.161	0.513	0.615
Y_9 (ln[$(q_t - \sigma_v)/\sigma_v'$])	SU	1.340	−0.572	0.659	1.476
Y_{10} (ln[$(q_t - u_2)/\sigma_v'$])	SU	2.134	−1.102	1.154	0.657

from 1 to 419 with an average 30 data points per study. The clay properties cover a wide range of OCR (but mostly 1–10), a wide range of S_t (insensitive to quick clays; sites with $S_t = 1 \sim$ tens or hundreds are fairly typical), and a wide range of PI (but mostly 8–100). There are a few data points for fissured clays as well as organic clays. Details are given in Ching and Phoon (2014a). Ten dimensionless clay parameters are compiled in this database: $Y_1 = \ln(LL)$, $Y_2 = \ln(PI)$, $Y_3 = LI$, $Y_4 = \ln(\sigma_v'/P_a)$, $Y_5 = \ln(\sigma_p'/P_a)$, $Y_6 = \ln(s_u/\sigma_v')$, $Y_7 = \ln(S_t)$, $Y_8 = B_q$, $Y_9 = \ln[(q_t - \sigma_v)/\sigma_v']$, and $Y_{10} = \ln[(q_t - u_2)/\sigma_v']$. For Y_6, the s_u values in the data points are all converted to the mobilized s_u values defined by Mesri and Huvaj (2007). The statistics are summarized in Table 4.19.

Due to its flexibility, the Johnson distribution is adopted in Ching and Phoon (2014b) to fit the univariate $(Y_1, Y_2, \ldots, Y_{10})$ data. The resulting Johnson parameters are summarized in Table 4.20. The fitted Johnson distributions are plotted in Figure 4.2 together with the empirical histograms constructed from Y data. For each bivariate data (Y_i, Y_j), Y_i and Y_j can be individually transformed to standard normal

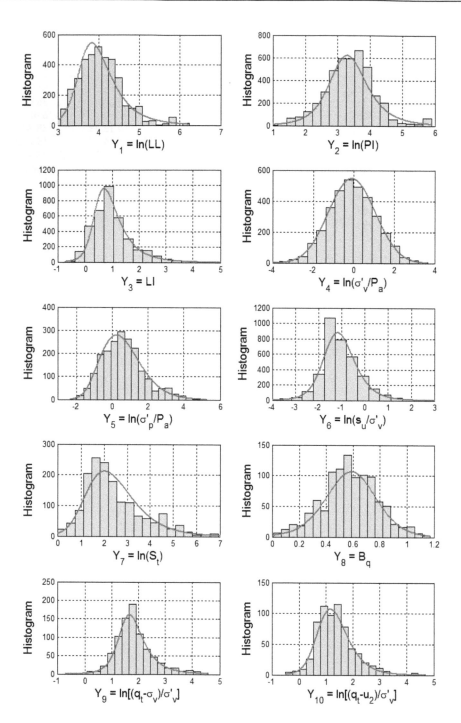

Figure 4.2 Empirical histograms constructed from Y data (solid curves are the fitted Johnson distributions) (Source: Figure 3, Ching and Phoon 2014b).

Table 4.21 Numbers of available (X_i, X_j) data pairs (Source: Table 4 in Ching and Phoon 2014b).

	Y_1	Y_2	Y_3	Y_4	Y_5	Y_6	Y_7	Y_8	Y_9	Y_{10}
Y_1 [ln(LL)]	3822	3822	3412	2084	1362	1835	1184	680	618	541
Y_2 [ln(PI)]		4265	3424	2169	1433	2173	1203	688	626	549
$Y_3 = $ LI		Index properties	3661	1999	1314	1709	1279	660	598	521
$Y_4 = \ln(\sigma_v'/P_a)$				3370	1944	2419	853	965	862	668
$Y_5 = \ln(\sigma_p'/P_a)$					2028	1423	554	780	691	543
$Y_6 = \ln(s_u/\sigma_v')$						3532	715	595	533	525
$Y_7 = \ln(S_t)$				Stresses and strengths			1589	240	230	190
$Y_8 = B_q$								1016	862	668
$Y_9 = \ln[(q_t - \sigma_v)/\sigma_v']$	Symmetric								862	590
$Y_{10} = \ln[(q_t - u_2)/\sigma_v']$								CPTU parameters		668

random variables (X_i, X_j) using the CDF transform given in Eq. 4.11 together with the distribution type/parameters in Table 4.20.

Table 4.21 shows the number of bivariate (X_i, X_j) data points. The diagonal entries in Table 4.21 are the numbers of data points with univariate information, which are identical to the numbers in the second column of Table 4.19. The off-diagonal numbers in Table 4.21 are the numbers of data points with bivariate information, i.e., two parameters are measured in close proximity. Data points with bivariate information are typically abundant: most (X_i, X_j) pairs are associated with more than 400 data points. The only exceptions are the (X_7, X_8), (X_7, X_9), and (X_7, X_{10}) pairs which are associated with about 200 data points (the entries in grey). It is apparently uncommon to measure S_t and CPTU data together in the literature. There are three diagonal boxes in Table 4.21, representing the parameters relating to (a) index properties, (b) stresses and strengths, and (c) CPTU parameters. It is clear that index properties have the most bivariate data points, while the CPTU parameters have the least in this CLAY/10/7490 database.

The correlation matrix C cannot be estimated using Eq. 4.6 because this equation requires genuine multivariate data. Instead, each correlation coefficient δ_{ij} can be estimated with an entry-by-entry bivariate manner based on a bivariate dataset (X_i, X_j) using Eq. 4.7 (same strategy as that adopted for CLAY/7/6310). The correlation matrix C constructed in the entry-by-entry bivariate manner is not guaranteed to be positive definite, but Ching and Phoon (2014b) proposed a bootstrap method to mitigate this issue. The resulting correlation matrix C for $(X_1, X_2, \ldots, X_{10})$ is shown in Table 4.22.

Simulation

One attractive feature of the multivariate normal distribution is that simulation is computationally simple. Let $\underline{U}$ be a 10×1 vector containing 10 independent samples of the standard normal random variable. Let L be the lower triangular Cholesky factor satisfying $C = L \times L^T$, in which C is the 10×10 correlation matrix given in Table 4.22. Samples of *correlated* standard normal random variables can be obtained easily using $\underline{X} = L \times \underline{U}$, in which $\underline{X}$ is the random *vector* $(X_1, X_2, \ldots, X_{10})$. Simulated samples of soil parameters $(Y_1, Y_2, \ldots, Y_{10})$ can be obtained using Eq. 4.13. Simulated samples

Table 4.22 Correlation matrix **C** for $(X_1, X_2, \ldots, X_{10})$ in CLAY/10/7490 (Source: Table 8, Ching and Phoon 2014b).

		X_1	X_2	X_3	X_4	X_5	X_6	X_7	X_8	X_9	X_{10}
	X_1 [for ln(LL)]	1.00	0.91	−0.25	−0.24	−0.30	0.10	−0.21	0.09	0.09	0.07
	X_2 [for ln(PI)]		1.00	−0.32	−0.21	−0.27	0.04	−0.25	0.11	0.00	−0.01
	X_3 (for LI)		Index properties	1.00	−0.49	−0.57	0.01	0.59	−0.05	0.06	−0.05
	X_4 [for ln(σ'_v/P_a)]				1.00	0.72	−0.50	0.00	0.20	−0.38	−0.32
C =	X_5 [for ln(σ'_p/P_a)]					1.00	0.01	0.06	−0.03	0.11	0.04
	X_6 [for ln(s_u/σ'_v)]						1.00	0.18	−0.24	0.73	0.63
	X_7 [for ln(S_t)]				Stresses & strengths			1.00	0.18	0.15	−0.08
	X_8 (for B_q)								1.00	−0.45	−0.63
	Y_9 (ln[($q_t − \sigma_v)/\sigma'_v$])	Symmetric								1.00	0.74
	Y_{10} (ln[($q_t − u_2)/\sigma'_v$])								CPTU		1.00
									parameters		

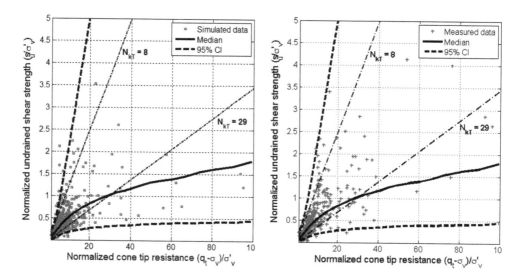

Figure 4.3 Comparison between (a) simulated and (b) measured data for the $s_u/\sigma'_v − (q_t − \sigma_v)/\sigma'_v$ correlation (Source: Figure 12, Ching and Phoon 2014b).

of other physical parameters, such as $\ln(\text{OCR})$, $\ln(s_u/\sigma'_p)$, and CPTU cone factor (N_{kT} and N_{kE}), can be derived from the samples of $(Y_1, Y_2, \ldots, Y_{10})$:

$$\ln(\text{OCR}) = \ln(\sigma'_p/P_a) − \ln(\sigma'_v/P_a) = Y_5 − Y_4$$

$$\ln(s_u/\sigma'_p) = \ln(s_u/\sigma'_v) + \ln(\sigma'_v/P_a) − \ln(\sigma'_p/P_a) = Y_6 + Y_4 − Y_5$$

$$\ln(N_{kT}) = \ln[(q_t − \sigma_v)/s_u] = \ln[(q_t − \sigma_v)/\sigma'_v] − \ln(s_u/\sigma'_v) = Y_9 − Y_6 \qquad (4.18)$$

$$\ln(N_{kE}) = \ln[(q_t − u_2)/s_u] = \ln[(q_t − u_2)/\sigma'_v] − \ln(s_u/\sigma'_v) = Y_{10} − Y_6$$

As an example, Figure 4.3 compares measured and simulated data for the correlation between s_u/σ'_v and $(q_t − \sigma_v)/\sigma'_v$. Most of the data points fall within the range

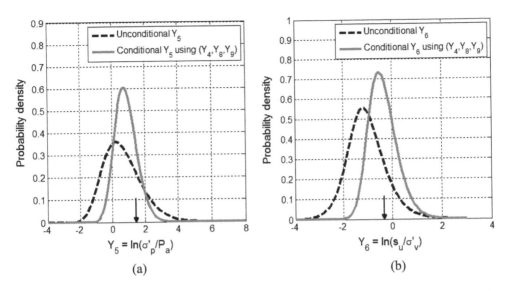

Figure 4.4 Conditional probability distributions based on $Y_4 = -0.577$, $Y_8 = 0.207$, and $Y_9 = 2.899$ for (a) normalized preconsolidation stress (Y_5) and (b) normalized undrained shear strength (Y_6) (arrow indicates actual measured value) (Source: Figure 15, Ching and Phoon 2014b).

$8 \leqslant N_{kT} \leqslant 29$ reported by Rad and Lunne (1988) for s_u obtained from triaxial compression strength $[N_{kT} = (q_t - \sigma_v)/s_u]$. The median and 95% confidence interval for the simulated data broadly agrees with the measured data in trend and scatter. More comparisons between the measured and simulated $(Y_1, Y_2, \ldots, Y_{10})$ data can be found in Ching and Phoon (2014b). The conclusions are similar: the simulated and measured data are consistent in trend and scatter.

Conditioning

Based on the multivariate probability distributions constructed in previous section, it is possible to update the marginal distribution of any one parameter or even the multivariate distribution of any group of parameters given information from other parameters covered by the multivariate probability distribution. This is called the Bayesian updating. Consider an example involving updating the normalized preconsolidation stress (Y_5) and the normalized undrained shear strength (Y_6) based on the normalized effective vertical stress (Y_4), the pore pressure ratio (Y_8) and the normalized cone tip resistance (Y_9). This is a realistic example as Y_8 and Y_9 are routinely measured simultaneously in piezocone soundings. For illustration, assume that the following piezocone information is available: $Y_4 = \ln(\sigma_v'/P_a) = \ln(0.562) = -0.577$, $Y_8 = B_q = 0.178$ and $Y_9 = \ln[(q_t - \sigma_v)/\sigma_v'] = \ln(17.78) = 2.878$. The unconditioned and conditioned distributions for Y_5 and Y_6 are shown in Figure 4.4 (details for obtaining the conditional distribution are described in Ching and Phoon 2014b). The arrows in the figure indicate the actual measured values of Y_5 and Y_6.

Table 4.23 Updated mean and COV of CIUC s_u for unstructured clays under various combinations of information (Based on Eqs. 22–28, Ching et al. 2010).

Information	Updated mean of CIUC s_u (in kPa)	Updated COV of s_u
OCR	$(\sigma'_v/P_a) \times OCR^{0.64} \times 43.474$	0.313
N_{60}	$N_{60}^{0.602} \times (\sigma'_v/P_a)^{0.243} \times 33.905$	0.282
$q_t - \sigma_v$	$[(q_t - \sigma_v)/P_a]^{0.976} \times 8.634$	0.346
OCR, N_{60}	$OCR^{0.373} \times N_{60}^{0.256} \times (\sigma'_v/P_a)^{0.685} \times 38.690$	0.182
OCR, $q_t - \sigma_v$	$OCR^{0.431} \times [(q_t - \sigma_v)/P_a]^{0.326} \times (\sigma'_v/P_a)^{0.674} \times 25.001$	0.196
$N_{60}, q_t - \sigma_v$	$N_{60}^{0.362} \times [(q_t - \sigma_v)/P_a]^{0.399} \times (\sigma'_v/P_a)^{0.146} \times 19.086$	0.218
OCR, $N_{60}, q_t - \sigma_v$	$OCR^{0.291} \times N_{60}^{0.200} \times [(q_t - \sigma_v)/P_a]^{0.220} \times (\sigma'_v/P_a)^{0.534} \times 27.322$	0.160

4.6 MULTIVARIATE NORMAL DISTRIBUTIONS CONSTRUCTED WITH *INCOMPLETE* BIVARIATE DATA

4.6.1 CLAY/4/BN

Ching et al. (2010) presented another clay database containing four soil parameters: $Y_1 = OCR$, $Y_2 = s_u$ from CIUC test, $Y_3 = q_t - \sigma_v$ (net cone resistance), and $Y_4 = N_{60}$ (SPT N corrected for energy efficiency). The range of OCR of this database is wider – from 1 to 50. However, only bivariate data on $(Y_1, Y_2) = (OCR, s_u)$, $(Y_3, Y_2) = (q_t - \sigma_v, s_u)$, and $(Y_4, Y_2) = (N_{60}, s_u)$ are available. Bivariate data on $(Y_1, Y_3) = (OCR, q_t - \sigma_v)$, $(Y_1, Y_4) = (OCR, N_{60})$, and $(Y_3, Y_4) = (q_t - \sigma_v, N_{60})$ are missing, i.e., the bivariate correlations $\{\delta_{ij}: i = 1, \ldots, n - 1, j = i + 1, \ldots, n\}$ are only partially known. Given that even complete bivariate information is not available, it is not possible to apply the aforementioned CDF transform approach directly. It is accurate to say that although it is common to measure more than two soil parameters in a site investigation, it is uncommon to establish correlations between *all possible pairs* of soil parameters. It is useful to point out that data limitation in geotechnical engineering is not confined to the quantity of data, but can extend to incompleteness in a deep way such as missing pairs of soil parameters. We have highlighted previously that even complete information on all pairs of soil parameters is incomplete in the multivariate sense.

To deal with this difficulty of incomplete bivariate correlations, Ching et al. (2010) constructed a multivariate normal distribution using a Bayesian network model which prescribed a dependency structure based on some postulated but reasonable conditional relationships between the soil parameters. Needless to say, this prescribed dependency structure has to be validated. They considered $Y_1 = OCR$ as a given number and the remaining soil parameters (Y_2, Y_3, Y_4) are lognormally distributed random variables. Hence, $\ln(Y_2) = \ln(s_u) = \lambda_2 + \xi_2 X_2$, $\ln(Y_3) = \ln(q_t - \sigma_v) = \lambda_3 + \xi_3 X_3$, and $\ln(Y_4) = \ln(N_{60}) = \lambda_4 + \xi_4 X_4$, in which X_i are standard normal random variables. The subsequent Bayesian analyses derived in Ching et al. (2010) lead to the transformation models in Table 4.23. It can be seen from the table that updated COV typically decreases with increasing amount of information. This again demonstrates numerically that uncertainty in a design soil parameter can be reduced by multivariate correlations between the design parameter and other available pieces of site investigation information.

Table 4.24 Correlation matrix **C** for (X_1, X_2, X_3, X_4) for the four selected parameters of unstructured clays (Source: Phoon et al. 2012, with permission from ASCE).

	X_1 (for OCR)	X_2 (for s_u)	X_3 (for $q_t - \sigma_v$)	X_4 (for N_{60})
X_1 (for OCR)	1.000	0.554	0.355	0.395
C = X_2 (for s_u)		1.000	0.642	0.714
X_3 (for $q_t - \sigma_v$)			1.000	0.458
X_4 (for N_{60})	Symmetric			1.000

Table 4.25 Updated mean and standard deviation of ϕ' for clean sands under various combinations of information (Based on Eqs. 11–17, Ching et al. 2012).

Information	Updated mean of ϕ' (in degrees)	Updated stdev (in degrees)
I_R	$3 \times I_R + \phi_{cv}$	1.960
$(N_1)_{60}$	$6.220 \times \ln[(N_1)_{60}] + 23.167$	3.086
q_{t1}	$7.819 \times \ln(q_{t1}) + 2.401$	3.919
$I_R, (N_1)_{60}$	$1.996 \times I_R + 0.665 \times \phi_{cv} + 2.081 \times \ln[(N_1)_{60}] + 7.751$	1.655
I_R, q_{t1}	$2.335 \times I_R + 0.778 \times \phi_{cv} + 1.735 \times \ln(q_{t1}) + 0.533$	1.753
$(N_1)_{60}, q_{t1}$	$3.840 \times \ln[(N_1)_{60}] + 2.993 \times \ln(q_{t1}) + 15.22$	2.423
$I_R, (N_1)_{60}, q_{t1}$	$1.814 \times I_R + 0.605 \times \phi_{cv} + 1.518 \times \ln[(N_1)_{60}] +$ $1.183 \times \ln(q_{t1}) + 6.015$	1.524

Based on the results of Ching et al. (2010), Phoon et al. (2012) further assumed OCR to be lognormal with a reasonable COV = 0.25, i.e., $\ln(Y_1) = \ln(OCR) = \lambda_1 + \xi_1 X_1$. Under this assumption, they showed that the underlying standard normal variables (X_1, X_2, X_3, X_4) have the correlation matrix shown in Table 4.24. The correlation matrix in Table 4.24 and the transformation models in Table 4.23 should be suitable for unstructured clays covering a fairly wide range of OCR.

4.6.2 SAND/4/BN

Ching et al. (2012) adopted the Bayes-net model proposed in Ching et al. (2010) for clean sands. The study was based on a database containing five selected parameters of normally consolidated clean sands: $Y_1 = \phi_{cv}$ (critical state friction angle), $Y_2 = I_R$ (dilatancy index, see Bolton 1986 and Ching et al. 2012), $Y_3 = \phi'$ (peak secant friction angle), $Y_4 = (q_t/P_a)/(\sigma_v'/P_a)^{0.5} = q_{t1}$ (corrected cone resistance), and $Y_5 = (N_1)_{60}$ (SPT N corrected for energy efficiency and overburden stress). The authors considered $Y_1 = \phi_{cv}$ and $Y_2 = I_R$ as given numbers and the remaining soil parameters (Y_3, Y_4, Y_5) are random variables: Y_3 is normal, while Y_4 and Y_5 are lognormal. Technically, this means the results presented in Ching et al. (2012) is *conditional* on ϕ_{cv} and I_R. Hence, $Y_3 = \phi' = \mu_3 + \sigma_3 X_3$, $\ln(Y_4) = \ln(q_{t1}) = \lambda_4 + \xi_4 X_4$, and $\ln(Y_5) = \ln[(N_1)_{60}] = \lambda_5 + \xi_5 X_5$, in which X_i are standard normal random variables. Bayesian analyses produced the transformation models in Table 4.25. It can be seen from the table that the updated standard deviation typically decreases with increasing

Table 4.26 Correlation matrix **C** for $(X_1, X_2, X_3, X_4, X_5)$ for the five selected parameters of clean sands.

	X_1 (for ϕ_{cv})	X_2 (for I_R)	X_3 (for ϕ')	X_4 (for q_{t1})	X_5 [for $(N_1)_{60}$]
X_1 (for ϕ_{cv})	1.000	0.000	0.642	0.491	0.536
X_2 (for I_R)		1.000	0.642	0.491	0.536
C = X_3 (for ϕ')			1.000	0.764	0.835
X_4 (for q_{t1})		Symmetric		1.000	0.638
X_5 [for $(N_1)_{60}$]					1.000

amount of information. This again illustrates that uncertainty in a design soil parameter can be reduced by multivariate correlations between the design parameter and other available pieces of site investigation information.

If we further assume ϕ_{cv} and I_R are normal with reasonable standard deviations of 3° and 1°, respectively, i.e., $Y_1 = \phi_{cv} = \mu_1 + 3X_1$ and $Y_2 = I_R = \mu_2 + X_2$, and also assume independence between ϕ_{cv} and I_R, it can be shown that the underlying standard normal variables $(X_1, X_2, X_3, X_4, X_5)$ has the correlation matrix shown in Table 4.26. The correlation matrix in Table 4.26 and the transformation models in Table 4.25 should be suitable for normally consolidated clean sands.

4.7 MULTIVARIATE DISTRIBUTIONS CONSTRUCTED WITH THE COPULA THEORY

As an alternative to the multivariate normal distribution, this section presents a copula-based approach for modelling the multivariate distribution of multiple soil parameters. The copula theory has been widely used for constructing bivariate distributions in the geotechnical literature (e.g., Li et al. 2012, 2013, 2015; Li and Tang 2014; Tang et al. 2013, 2015; Wu 2013; Zhang et al. 2014; Huffman and Stuedlein 2014). However, there are limited studies applying copulas to multivariate distributions with n dimensions (n > 2), because only the elliptical copulas have practical n-dimensional generalizations. This section demonstrates the use of two elliptical copulas (i.e., Gaussian and t copulas) to construct the multivariate distribution for the Clay/5/345 database (Table 4.1; also Section 4.4.1).

4.7.1 Copula theory

As mentioned in Section 4.3, the multivariate normal distribution involves converting non-normally distributed soil parameters $\underline{Y} = (Y_1, Y_2, \ldots, Y_n)'$ into standard normal random variables $\underline{X} = (X_1, X_2, \ldots, X_n)'$ using the CDF transform $X = \Phi^{-1}[F(Y)]$. An alternate approach is to convert $\underline{Y}$ into standard uniform random variables $\underline{U} = (U_1, U_2, \ldots, U_n)'$ by $U = F(Y)$. This implies that the multivariate CDF $F(y_1, y_2, \ldots, y_n)$ for $\underline{Y}$ can be represented by a CDF of coupled standard uniform variables $(u_1, u_2, \ldots, u_n)$, namely $C(u_1, u_2, \ldots, u_n)$. The function $C(u_1, u_2, \ldots, u_n)$ is called a copula function.

According to Sklar's theorem (e.g., Nelsen 2006), the following relationship between $F(y_1, y_2, \ldots, y_n)$ and $C(u_1, u_2, \ldots, u_n)$ holds:

$$F(y_1, y_2, \ldots, y_n) = C(u_1, u_2, \ldots, u_n) = C(F_1(y_1), F_2(y_2), \ldots, F_n(y_n)) \quad (4.19)$$

where $F_i(\cdot)$ is the marginal (univariate) CDF of Y_i. It can be seen from Eq. 4.19 that the multivariate CDF $F(y_1, y_2, \ldots, y_n)$ is converted into a copula function $C(u_1, u_2, \ldots, u_n)$ with $u_i = F_i(y_i)$. It is understood that the relationship $u_i = F_i(y_i)$ characterizes the univariate probability distribution for each individual soil parameter, while the copula function C describes the multivariate correlation among all standard uniform variables ($u_1, u_2, \ldots, u_n$). Therefore, the key tasks for the multivariate copula approach are (a) to determine each marginal (univariate) distribution $F_i(\cdot)$ and (b) to select an appropriate copula that provides reasonable fit to the correlation structure among ($U_1, U_2, \ldots, U_n$). Note that the above two tasks can be decoupled. The multivariate PDF of $\underline{Y} = (Y_1, Y_2, \ldots, Y_n)$ can be derived from the multivariate CDF by taking derivatives of Eq. 4.19 (e.g., McNeil et al. 2005):

$$f(y_1, y_2, \ldots, y_n) = \frac{\partial^n C(F_1(y_1), F_2(y_2), \ldots, F_n(y_n))}{\partial F_1(y_1) \ldots \partial F_n(y_n)} \prod_{i=1}^{n} \frac{\partial F_i(y_i)}{\partial y_i}$$

$$= c(F_1(y_1), F_2(y_2), \ldots, F_n(y_n)) \prod_{i=1}^{n} f_i(y_i)$$

$$= c(u_1, u_2, \ldots, u_n) \prod_{i=1}^{n} f_i(y_i) \quad (4.20)$$

where $c(u_1, u_2, \ldots, u_n) = \partial^n C(u_1, u_2, \ldots, u_n)/\partial u_1 \ldots \partial u_n$ is the copula density function; $f_i(\cdot)$ is the PDF of Y_i.

4.7.2 Elliptical copulas (Gaussian and t copulas)

The Gaussian copula and the t copula are derived from the multivariate normal distribution and the multivariate t distribution, respectively. Specifically, the multivariate Gaussian copula has the following copula function $C^{Ga}(u_1, u_2, \ldots, u_n; C)$ and copula density function $c^{Ga}(u_1, u_2, \ldots, u_n; C)$ (McNeil et al. 2005):

$$C^{Ga}(u_1, u_2, \ldots, u_n; C) = \Phi_n(\Phi^{-1}(u_1), \Phi^{-1}(u_2), \ldots, \Phi^{-1}(u_n); C) \quad (4.21)$$

$$c^{Ga}(u_1, u_2, \ldots, u_n; C) = |C|^{-1/2} \exp\left(-\frac{1}{2} \underline{X}'(C^{-1} - I)\underline{X}\right) \quad (4.22)$$

where C is the correlation matrix defined in Eq. 4.4; $\Phi_n(., \ldots, .; C)$ is the standardized multivariate normal distribution with correlation matrix C; $|C|$ is the determinant of C; I is the identity matrix; $\underline{X} = (\Phi^{-1}(u_1), \Phi^{-1}(u_2), \ldots, \Phi^{-1}(u_n))'$ is the vector of standard normal random variables. Substituting the Gaussian copula density function in Eq. 4.22 into Eq. 4.20 leads to the multivariate standard normal PDF in Eq. 4.4. Therefore, the multivariate normal distribution adopted in the previous sections is a combination of the Gaussian copula and standard normal univariate distributions.

The multivariate t copula has the following copula function $C^t(u_1, u_2, \ldots, u_n; \mathbf{C}, \nu)$ and copula density function $c^t(u_1, u_2, \ldots, u_n; \mathbf{C}, \nu)$ (McNeil et al. 2005):

$$C^t(u_1, u_2, \ldots, u_n; \mathbf{C}, \nu) = t_n(t_\nu^{-1}(u_1), t_\nu^{-1}(u_2), \ldots, t_\nu^{-1}(u_n); \mathbf{C}, \nu) \tag{4.23}$$

$$c^t(u_1, u_2, \ldots, u_n; \mathbf{C}, \nu) = |\mathbf{C}|^{-1/2} \frac{\Gamma\left(\frac{\nu+n}{2}\right)\left[\Gamma\left(\frac{\nu}{2}\right)\right]^{n-1}}{\left[\Gamma\left(\frac{\nu+1}{2}\right)\right]^n} \frac{\left[1 + \frac{1}{\nu}\underline{T}'\mathbf{C}^{-1}\underline{T}\right]^{-(\nu+n)/2}}{\prod\limits_{i=1}^{n}\left(1 + \frac{t_i^2}{\nu}\right)^{-(\nu+1)/2}} \tag{4.24}$$

where $t_n(., \ldots, .; \mathbf{C}, \nu)$ is the standard multivariate Student's t CDF with correlation matrix $\mathbf{C}$ and ν degrees of freedom; $t_\nu^{-1}(\cdot)$ is the inverse of univariate CDF of Student's t distribution with ν degrees of freedom; $\underline{T} = (t_\nu^{-1}(u_1), t_\nu^{-1}(u_2), \ldots, t_\nu^{-1}(u_n))'$ is a vector of Student's t random variables with ν degrees of freedom; $t_i = t_\nu^{-1}(u_i)$ is the i-th entry in $\underline{T}$; Γ is the Gamma function. The multivariate Gaussian copula is the limiting case of the multivariate t copula when ν approaches infinity. The additional parameter ν in the multivariate t copula characterizes the degree of non-normality in the multivariate soil data. The degree of non-normality increases with decreasing ν.

4.7.3 Kendall rank correlation

Correlation measures are essential to multivariate modelling. Section 4.2 introduced the commonly-used correlation measure, i.e., the product-moment (Pearson) correlation. Note that the Pearson correlation measures the degree of linear correlation only. When the transformation model in Eq. 4.1 is not linear (For example, $Y_1 = a + bY_2^3 + \varepsilon$), the Pearson correlation coefficient ρ_{12} between Y_1 and Y_2 will not be equal to ± 1 even if $s_\varepsilon = 0$. Furthermore, the Pearson correlation is invariant only under strictly monotonic linear transformations. Since the CDF transform $X = \Phi^{-1}[F(Y)]$ for most distributions is nonlinear, the Pearson correlation between Y_1 and Y_2 is not equal to that between X_1 and X_2. Therefore, when Y is transformed to X or X to Y, the Pearson correlation needs to be re-evaluated. To overcome the aforementioned limitation underlying the Pearson correlation, this section introduces the Kendall rank correlation that only depends on the ranks rather than the actual numerical values.

Unlike the Pearson correlation, the Kendall correlation measures the degree of "concordance" between Y_1 and Y_2. The concept of concordance is simple. Let (Y_1, Y_2) and (Y_1', Y_2') be a pair of data points. Then, (Y_1, Y_2) and (Y_1', Y_2') are concordant if $(Y_1 - Y_1')(Y_2 - Y_2') \geqslant 0$ and are discordant if $(Y_1 - Y_1')(Y_2 - Y_2') < 0$. Consider a bivariate dataset (Y_i, Y_j) of sample size N. There are $0.5 \times N \times (N-1)$ possible pairs of data, namely (Y_{im}, Y_{jm}) versus (Y_{in}, Y_{jn}) $(m < n)$. For each pair (Y_{im}, Y_{jm}) and (Y_{in}, Y_{jn}), the concordance is judged by the sign of $(Y_{im} - Y_{in})(Y_{jm} - Y_{jn})$. Then, the Kendall correlation coefficient (τ_{ij}) is defined as the portion of concordance minus the portion of discordance (e.g., Nelsen 2006):

$$\tau_{ij} = \frac{\sum\limits_{m<n} \text{sgn}[(y_{im} - y_{in})(y_{jm} - y_{jn})]}{0.5 \times N \times (N-1)} \tag{4.25}$$

where sgn($\cdot$) is calculated by

$$\text{sgn} = \begin{cases} 1 & (y_{im} - y_{in})(y_{jm} - y_{jn}) \geqslant 0 \ (\text{concordant}) \\ -1 & (y_{im} - y_{in})(y_{jm} - y_{jn}) < 0 \ (\text{discordant}) \end{cases} \quad m < n = 1, 2, \ldots, N \quad (4.26)$$

The numerator in Eq. 4.25 is the number of concordant pairs minus the number of discordant pairs, and the denominator is the total number of pairs. The Kendall rank correlation is invariant with respect to strictly monotonic transformations. This means that the Kendall correlation between Y_1 and Y_2 remains unchanged after being transformed into standard normal random variables X_1 and X_2. This advantage will simplify the calibration of the correlation matrix C in the Gaussian and t copulas, as demonstrated in the following section.

4.7.4 Estimating C using Pearson and Kendall correlations

This section first demonstrates the application of the Gaussian copula to the Clay/5/345 database. The application of the t copula will be demonstrated in a later section. As shown in Eqs. 4.21 and 4.22, the Gaussian copula is characterized by the correlation matrix C. It is the correlation matrix for the underlying standard normal variables $\underline{X} = (\Phi^{-1}(u_1), \ \Phi^{-1}(u_2), \ \ldots, \ \Phi^{-1}(u_n))'$. The (i, j) entry in the C matrix is denoted by δ_{ij}, which is the Pearson correlation between (X_i, X_j). It will be clear that δ_{ij} can be estimated using the Pearson or Kendall correlation between (Y_i, Y_j). The focus in this section is on the comparison of the Pearson and Kendall correlations as correlation measures for estimating δ_{ij}. The marginal (univariate) distributions for the five soil parameters in the Clay/5/345 database are summarized in Table 4.4. It can be seen that all the five soil parameters are roughly lognormally distributed. With the marginal distributions determined, the next step is to calibrate the correlation matrix C in the Gaussian copula. Note that δ_{ij} can be determined by matching either Pearson or Kendall correlation. In general, δ_{ij} in the matrix C and the Pearson correlation between Y_i and Y_j (denoted by ρ_{ij}) have the following relationship (Li and Tang 2014):

$$\rho_{ij} = \int_{-\infty}^{\infty} \int_{-\infty}^{\infty} \left(\frac{y_i - \mu_i}{\sigma_i} \right) \left(\frac{y_j - \mu_j}{\sigma_j} \right) \frac{f_i(y_i)f_j(y_j)}{\sqrt{1 - \delta_{ij}^2}} \exp\left\{ -\frac{\delta_{ij}^2 x_i^2 - 2\delta_{ij} x_i x_j + \delta_{ij}^2 x_j^2}{2(1 - \delta_{ij}^2)} \right\} dy_i dy_j$$

$$(4.27)$$

where μ_i and σ_i are the mean and standard deviation of Y_i; $x_i = \Phi^{-1}[F_i(y_i)]$ is standard normal. It is nontrivial to solve the above integral equation for δ_{ij}. However, because Y_i and Y_j are both lognormally distributed, Eq. 4.27 has the following explicit analytical form:

$$\delta_{ij} = \frac{\ln\left(1 + \rho_{ij} \times \sqrt{\exp(\xi_i^2) - 1} \times \sqrt{\exp(\xi_j^2) - 1}\right)}{\xi_i \times \xi_j}$$

$$(4.28)$$

where ξ_i (shown in the last column of Table 4.4) is the standard deviation of $\ln(Y_i)$. From here on, the method of estimating δ_{ij} described in Eqs. 4.27 & 4.28 is referred

Table 4.27 Pearson and Kendall correlations for $(Y_1, Y_2, Y_3, Y_4, Y_5)$ and the corresponding **C** matrix, together with the AIC and BIC scores for the Gaussian copula.

	Method P	Method K
Pearson correlation among Y (ρ_{ij})	$\begin{bmatrix} 1 & 0.053 & -0.500 & -0.060 & -0.208 \\ 0.053 & 1 & 0.169 & 0.844 & 0.567 \\ -0.500 & 0.169 & 1 & 0.303 & 0.417 \\ -0.060 & 0.844 & 0.303 & 1 & 0.725 \\ -0.208 & 0.567 & 0.417 & 0.725 & 1 \end{bmatrix}$	
Kendall correlation among Y (τ_{ij})	$\begin{bmatrix} 1 & -0.103 & -0.620 & -0.113 & -0.175 \\ -0.103 & 1 & 0.177 & 0.747 & 0.592 \\ -0.620 & 0.177 & 1 & 0.209 & 0.279 \\ -0.113 & 0.747 & 0.209 & 1 & 0.662 \\ -0.175 & 0.592 & 0.279 & 0.662 & 1 \end{bmatrix}$	
Correlation matrix **C** (δ_{ij}) in Gaussian copula	$\begin{bmatrix} 1 & 0.065 & -0.914 & -0.077 & -0.261 \\ 0.065 & 1 & 0.249 & 0.881 & 0.635 \\ -0.914 & 0.249 & 1 & 0.414 & 0.533 \\ -0.077 & 0.881 & 0.414 & 1 & 0.780 \\ -0.261 & 0.635 & 0.533 & 0.780 & 1 \end{bmatrix}$	$\begin{bmatrix} 1 & -0.161 & -0.827 & -0.176 & -0.272 \\ -0.161 & 1 & 0.274 & 0.922 & 0.802 \\ -0.827 & 0.274 & 1 & 0.323 & 0.424 \\ -0.176 & 0.922 & 0.323 & 1 & 0.863 \\ -0.272 & 0.802 & 0.424 & 0.863 & 1 \end{bmatrix}$
AIC score	83.4	-1462.4
BIC score	121.8	-1424.0

to as Method P. Note that Eq. 4.28 is applicable only when Y's are all lognormally distributed. If they are not lognormal, Eq. 4.27 should be used in Method P to solve for δ_{ij}.

There is another method of estimating δ_{ij} through the Kendall correlation τ_{ij} (Li and Tang 2014):

$$\delta_{ij} = \sin\left(\frac{\pi \times \tau_{ij}}{2}\right) \tag{4.29}$$

This method of estimating δ_{ij} described in Eq. 4.29 is referred to as Method K. Note that Method K is applicable regardless of the distribution types of Y.

Table 4.27 shows the Pearson correlations among $(Y_1, Y_2, Y_3, Y_4, Y_5)$ (ρ_{ij}), where ρ_{ij} is estimated by Eq. 4.7 (X is replaced by Y), and Kendall correlations among $(Y_1, Y_2, Y_3, Y_4, Y_5)$ (τ_{ij}), where τ_{ij} is estimated by Eq. 4.25. Given ρ_{ij} and τ_{ij}, δ_{ij} can be determined using either Method P or Method K. The results are shown in Table 4.27. Note that the δ_{ij} estimated by Method P and Method K are quite different. To quantify the goodness of fits to the Clay/5/345 database for the two calibrated Gaussian copulas, both the Akaike Information Criterion (AIC) and Bayesian Information Criterion (BIC) are adopted (Li and Tang 2014). The AIC and BIC scores for the two Gaussian copulas are summarized in Table 4.27. It is clear that the Gaussian copula using Method K (matching Kendall correlation) provides better fit to the multivariate data in the sense that AIC and BIC scores are significantly smaller (smaller is better).

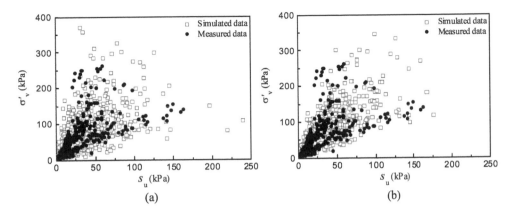

Figure 4.5 Comparison between simulated and measured data for the $s_u - \sigma'_v$ correlation produced by (a) Method P and (b) Method K.

In addition to AIC and BIC, the fitness of the calibrated Gaussian copulas can be verified by comparing their simulated samples with the measured data. Taking the correlation between s_u and σ'_v as an example, Figure 4.5 shows the measured data and the simulated data from Method P and Method K. It can be seen that both methods can generate samples that broadly agree with the measured data in trend and scatter. However, the samples from Method P exhibit larger scattering about the main body of the measured data. On the other hand, the samples from Method K provide better fit to the main body of the measured data. The reason is that the δ_{ij} estimated from Method P is 0.635, which is smaller than the δ_{ij} estimated from Method K ($\delta_{ij} = 0.802$). Therefore, Method P may underestimate the correlation between s_u and σ'_v in multivariate modelling.

The reason why Method K outperforms Method P can be explained as follows: Method P is applicable only when (Y_1, Y_2, Y_3, Y_4, Y_5) are lognormally distributed. In the case that (Y_1, Y_2, Y_3, Y_4, Y_5) are only *roughly* lognormal, the errors for the distribution fitting can affect the estimated δ_{ij}. On the other hand, Method K is generally applicable regardless of the distribution types for (Y_1, Y_2, Y_3, Y_4, Y_5). Even if there are errors for distribution fitting, these errors do not affect the estimated δ_{ij}. Therefore, Method K is superior to Method P for calibrating δ_{ij} in the Gaussian copula, owing to both its simplicity and robustness.

It is also noteworthy that both Method P and Method K adopted in this section match the correlations for the physical parameters (Y_1, Y_2, Y_3, Y_4, Y_5). This is the common method for estimating the C matrix in the Gaussian copula. The reason is that only the correlation matrix for $\underline{Y}$ instead of the $\underline{Y}$ data is typically available in the geotechnical practice. If the $\underline{Y}$ data is available, the data for $\underline{Y}$ can be firstly converted to $\underline{X}$ and then estimate C using the $\underline{X}$ data. This method was adopted in Sections 4.4 and 4.5 and will be denoted by Method XP below, because it computes the Pearson correlation for $\underline{X}$.

Among Methods P, K, and XP, it is possible that Method K is the most robust method of estimating the C matrix, because the accuracy in the estimated Kendall rank correlation is independent of the CDF transform from $\underline{Y}$ to $\underline{X}$. The only assumption for Method K is that the underlying $\underline{X}$ is multivariate normal, which is the assumption for Eq. 4.29. Method XP requires the same assumption. However, Method XP estimates the Pearson correlation for $\underline{X}$. The estimated Pearson correlation will be inaccurate if the converted $\underline{X}$ exhibits nonlinear correlations. Therefore, it is possible that Method XP is less robust than Method K. For the Clay/5/345 dataset, $\underline{X}$ exhibits fairly linear correlations. As a result, the $\mathbf{C}$ matrix estimated by Method XP (Table 4.5) is similar to the one by Method K (Table 4.27). It is possible that Method P is the least robust, because not only Method P requires the same assumption (underlying $\underline{X}$ is multivariate normal) but Eq. 4.28 also requires the estimation for ξ_i and ξ_j. If ξ_i and ξ_j are estimated inaccurately, Method P may perform poorly. Both Methods K and XP are not afflicted by this issue. The robustness of these methods under the multivariate normal framework has been recently systematically studied by Ching et al. (2016). To our knowledge, the robustness of these methods has not been studied systematically using simulated multivariate non-normal data. This is a fruitful topic for future research.

4.7.5 Comparison between the Gaussian and t copulas

In the previous section, the Gaussian copula is adopted. However, the Gaussian copula may not be the best choice for the Clay/5/345 database. Thus, this section further compares the suitability of the t copula against the Gaussian copula.

The calibration method adopted here is different from that adopted in the previous section. In the previous section, δ_{ij} is estimated by either Method P or Method K. These methods are "methods of moments". However, the degrees of freedom ν in the t copula cannot be calibrated using these methods. In this section, the Maximum Likelihood Estimation (MLE) (McNeil et al. 2005) is adopted to determine δ_{ij} and ν in the t copula. For consistency, the MLE is also adopted to determine δ_{ij} in the Gaussian copula.

The resulting MLE's for δ_{ij} and ν are shown in Table 4.28. For comparison, the AIC and BIC scores for the calibrated Gaussian and t copulas are also given in the table. First of all, the correlation matrix $\mathbf{C}$ in the Gaussian copula based on MLE should be compared with that based on Method K (Table 4.27). They are fairly consistent, but the AIC and BIC scores based on MLE are slightly smaller (MLE fits slightly better). Note that MLE is considered as a robust method for parameter estimation and that Method

Table 4.28 Calibrated correlation matrices, degrees of freedom, AIC and BIC scores for the Gaussian and t copulas using MLE.

	Gaussian copula					t copula				
Correlation matrix **C** (δ_{ij}) in Gaussian or t copula	$\begin{bmatrix} 1 & -0.115 & -0.818 & -0.161 & -0.277 \\ -0.115 & 1 & 0.260 & 0.910 & 0.769 \\ -0.818 & 0.260 & 1 & 0.332 & 0.425 \\ -0.161 & 0.910 & 0.332 & 1 & 0.838 \\ -0.277 & 0.769 & 0.425 & 0.838 & 1 \end{bmatrix}$					$\begin{bmatrix} 1 & -0.112 & -0.832 & -0.159 & -0.293 \\ -0.112 & 1 & 0.250 & 0.913 & 0.763 \\ -0.832 & 0.250 & 1 & 0.323 & 0.429 \\ -0.159 & 0.913 & 0.323 & 1 & 0.833 \\ -0.293 & 0.763 & 0.429 & 0.833 & 1 \end{bmatrix}$				
Degrees of freedom ν	∞					15				
AIC score	−1474.3					**−1489.9**				
BIC score	−1435.9					**−1447.7**				

K performs similarly with MLE. This indicates that Method K should be considered as a robust method of estimating δ_{ij} in the Gaussian copula as well.

Recall that the degrees of freedom ν in the t copula characterize the degree of non-normality. With MLE for $\nu = 15$ (relatively high), it is expected that the degree of non-normality in the Clay/5/345 database is insignificant. This explains why the calibrated correlation matrices C for the Gaussian and t copulas (Table 4.28) are quite similar. The AIC and BIC scores also give the same conclusion: the scores for the Gaussian and t copulas are not very different (the t copula performs slightly better). Therefore, the Gaussian copula can be considered as an appropriate copula for the Clay/5/345 database.

Recall that the Gaussian copula is uniquely characterized by the correlation matrix C, and C can be determined in both the full multivariate manner and the entry-by-entry bivariate manner. Therefore, the Gaussian copula is applicable to both genuine multivariate data and incomplete multivariate data with multiple sets of bivariate data. It has been noted that the correlation matrix produced by multiple sets of bivariate data may not be positive definite, but a bootstrap method proposed by Ching and Phoon (2014b) can be applied to mitigate this issue. On the other hand, the t copula is governed by both the correlation matrix C and the degrees of freedom ν. Since ν cannot be calibrated from the correlations for Y, the t copula is only applicable to genuine multivariate data with the aid of MLE.

4.8 CONCLUSIONS

It is possible to construct the multivariate probability distributions of soil parameters based on the databases in the literature. Several such multivariate databases and distributions are demonstrated in this chapter. The main challenge lies in the fact that genuine multivariate soil data are rarely available in the literature. Nonetheless, with the techniques discussed in this chapter, it is still possible to construct the multivariate distribution based on bivariate correlation data in the literature. Moreover, it is possible to go beyond the multivariate normal distribution with the aid of the copula theory. The copula theory decomposes a multivariate distribution into a copula function and multiple marginal (univariate) distributions. As a result, the modelling of a multivariate distribution is simplified into the selection of an appropriate copula and marginal distributions.

With the multivariate distribution of soil parameters, it is possible to update the probability distributions for the target parameters (e.g., s_u) based on the information of multiple soil parameters (e.g., Atterberg's limits and CPTU parameters) through Bayesian analysis. Such Bayesian analysis can be done for both the aforementioned multivariate normal distribution framework and the copula framework. By incorporating multiple soil information, the COVs in the target parameters can be further reduced. Chapter 6 will discuss simplified RBD methods that are sufficiently responsive to a wide range of COVs. With such simplified RBD methods, the multiple soil information can be converted into more economical design outcomes (e.g., a shorter pile). By doing this, the linkage between site investigation efforts and geotechnical RBD outcomes can be constructed in an explicit and defensible way. The construction of this linkage is a unique topic in geotechnical engineering.

REFERENCES

Bjerrum, L. (1954) Geotechnical properties of Norwegian marine clays. *Geotechnique*, 4 (2), 49–69.

Bjerrum, L. & Simons, N.E. (1960) Comparison of shear strength characteristics of normally consolidated clays. In: *Proc. of Research Conference on Shear Strength of Cohesive Soils.* Boulder, ASCE. pp. 711–726.

Bolton, M.D. (1986) The strength and dilatancy of sands. *Geotechnique*, 36 (1), 65–78.

Chen, B.S.Y. & Mayne, P.W. (1996) Statistical relationships between piezocone measurements and stress history of clays. *Canadian Geotechnical Journal*, 33 (3), 488–498.

Chen, J.R. (2004) *Axial Behavior of Drilled Shafts in Gravelly Soils.* PhD Dissertation. Ithaca, NY, Cornell University.

Ching, J. & Phoon, K.K. (2012a) Modeling parameters of structured clays as a multivariate normal distribution. *Canadian Geotechnical Journal*, 49 (5), 522–545.

Ching, J. & Phoon, K.K. (2012b) Corrigendum: Modeling parameters of structured clays as a multivariate normal distribution. *Canadian Geotechnical Journal*, 49 (12), 1447–1450.

Ching, J. & Phoon, K.K. (2012c) Establishment of generic transformations for geotechnical design parameters. *Structural Safety*, 35, 52–62.

Ching, J. & Phoon, K.K. (2013) Multivariate distribution for undrained shear strengths under various test procedures. *Canadian Geotechnical Journal*, 50 (9), 907–923.

Ching, J. & Phoon, K.K. (2014a) Transformations and correlations among some clay parameters – The global database. *Canadian Geotechnical Journal*, 51 (6), 663–685.

Ching, J. & Phoon, K.K. (2014b) Correlations among some clay parameters – The multivariate distribution. *Canadian Geotechnical Journal*, 51 (6), 686–704.

Ching, J. & Phoon, K.K. (2015a) Constructing multivariate distributions for soil parameters. Chapter 1. In: Phoon, K.K. & Ching J. (eds.) *Risk and Reliability in Geotechnical Engineering.* London, Taylor & Francis.

Ching, J. & Phoon, K.K. (2015b) Reducing the transformation uncertainty for the mobilized undrained shear strength of clays. *ASCE Journal of Geotechnical and Geoenvironmental Engineering*, 141 (2), 04014103.

Ching, J., Phoon, K.K. & Chen, Y.C. (2010) Reducing shear strength uncertainties in clays by multivariate correlations. *Canadian Geotechnical Journal*, 47 (1), 16–33.

Ching, J., Chen, J.R., Yeh, J.Y. & Phoon, K.K. (2012) Updating uncertainties in friction angles of clean sands. *ASCE Journal of Geotechnical and Geoenvironmental Engineering*, 138 (2), 217–229.

Ching, J., Phoon, K.K. & Lee, W.T. (2013) Second-moment characterization of undrained shear strengths from different test modes. In: *Foundation Engineering in the Face of Uncertainty, Geotechnical Special Publication Honoring Professor F. H. Kulhawy.* ASCE. pp. 308–320.

Ching, J., Phoon, K.K. & Yu, J.W. (2014a) Linking site investigation efforts to final design savings with simplified reliability-based design methods. *ASCE Journal of Geotechnical and Geoenvironmental Engineering*, 140 (3), 04013032.

Ching, J., Phoon, K.K. & Chen, C.H. (2014b) Modeling CPTU parameters of clays as a multivariate normal distribution. *Canadian Geotechnical Journal*, 51 (1), 77–91.

Ching, J., Li, D.Q. & Phoon, K.K. (2016) Robust estimation of correlation coefficients among soil parameters under the multivariate normal framework, conditionally accepted by Structural Safety.

D'Ignazio, M., Phoon, K.K., Tan, S.A. & Länsivaara, T.T. (2016) Correlations for undrained shear strength of Finnish soft clays, *Canadian Geotechnical Journal*, in press.

Evans, J.D. (1996) *Straightforward Statistics for the Behavioral Sciences.* Brooks/Cole Publishing, Pacific Grove, CA.

Hatanaka, M. & Uchida, A. (1996) Empirical correlation between penetration resistance and internal friction angle of sandy soils. *Soils and Foundations,* 36 (4), 1–9.

Huffman, J.C. & Stuedlein, A.W. (2014) Reliability-based serviceability limit state design of spread footings on aggregate pier reinforced clay. *ASCE Journal of Geotechnical and Geoenvironmental Engineering,* 140 (10), 04014055.

Jamiolkowski, M., Ladd, C.C., Germain, J.T. & Lancellotta, R. (1985) New developments in field and laboratory testing of soils. In: *Proceeding of the 11th International Conference on Soil Mechanics and Foundation Engineering, San Francisco.* Vol. 1. pp. 57–153.

Jensen, F.V. (1996) An Introduction to Bayesian Networks. New York, Springer.

Kulhawy, F.H. & Mayne, P.W. (1990) *Manual on Estimating Soil Properties for Foundation Design.* Report EL-6800. Palo Alto, Electric Power Research Institute. Available online at EPRI.COM.

Ladd, C.C. & Foott, R. (1974) New design procedure for stability in soft clays. *ASCE Journal of the Geotechnical Engineering Division,* 100 (7), 763–786.

Li, D.Q. & Tang, X.S. (2014) Modeling and simulation of bivariate distribution of shear strength parameters using copulas. Chapter 2. In: *Risk and Reliability in Geotechnical Engineering.* Boca Raton, CRC Press. pp. 77–128.

Li, D.Q., Tang, X.S., Zhou, C.B. & Phoon, K.K. (2012) Uncertainty analysis of correlated non-normal geotechnical parameters using Gaussian copula. *Science China Technological Sciences,* 55 (11), 3081–3089.

Li, D.Q., Tang, X.S., Phoon, K.K., Chen, Y.F. & Zhou, C.B. (2013) Bivariate simulation using copula and its application to probabilistic pile settlement analysis. *International Journal for Numerical and Analytical Methods in Geomechanics,* 37 (6), 597–617.

Li, D.Q., Zhang, L., Tang, X.S., Zhou, W., Li, J.H., Zhou, C.B. & Phoon, K.K. (2015) Bivariate distribution of shear strength parameters using copulas and its impact on geotechnical system reliability. *Computers and Geotechnics,* 68, 184–195.

Liu, S., Zou, H., Cai, G., Bheemasetti, B.V., Puppala, A.J. & Lin, J. (2016) Multivariate correlation among resilient modulus and cone penetration test parameters of cohesive subgrade soils, *Engineering Geology,* 209, 128–142.

Locat, J. & Demers, D. (1988) Viscosity, yield stress, remoulded strength, and liquidity index relationships for sensitive clays. *Canadian Geotechnical Journal,* 25, 799–806.

Marcuson III, W.F. & Bieganousky, W.A. (1977) SPT and relative density in course sands. *ASCE Journal of the Geotechnical Engineering Division,* 103 (11), 1295–1309.

Mayne, P.W., Christopher, B.R. & DeJong, J. (2001) *Manual on Subsurface Investigations.* National Highway Institute Publication No. FHWA NHI-01-031. Washington, DC, Federal Highway Administration.

McNeil, A.J., Frey, R. & Embrechts, P. (2005) *Quantitative Risk Management: Concepts, Techniques and Tools.* Princeton, Princeton University Press.

Mesri, G. (1975) Discussion on "New design procedure for stability of soft clays". *ASCE Journal of the Geotechnical Engineering Division,* 101 (4), 409–412.

Mesri, G. (1989) A re-evaluation of $s_u(mob) = 0.22\sigma_p$ using laboratory shear tests. *Canadian Geotechnical Journal,* 26 (1), 162–164.

Mesri, G. & Huvaj, N. (2007) *Shear Strength Mobilized in Undrained Failure of Soft Clay and Silt Deposits.* ASCE Geotechnical Special Publication 173, Geo-Denver.

Mitchell, J.K. (1993) *Fundamentals of Soil Behaviour.* 2nd edition. New York, John Wiley and Sons.

Müller, R., Larsson, S. & Spross, J. (2014) Extended Multivariate Approach for Uncertainty Reduction in the Assessment of Undrained Shear Strength in Clays, *Canadian Geotechnical Journal,* 51 (3), 231–245.

NAVFAC (1982) *Soil Mechanics DM7.1*. Naval Facilities Engineering Command, Alexandria.

Nelsen, R.B. (2006) *An Introduction to Copulas*. 2nd edition. New York, Springer.

Ng, I.T., Yuen, K.V. & Dong, L. (2016) Nonparametric estimation of undrained shear strength for normally consolidated clays. *Marine Georesources and Geotechnology*, 34 (2), 127–137.

Phoon, K.K. & Ching, J. (2013) Multivariate model for soil parameters based on Johnson distributions. In: Withiam, J.L., Phoon, K.K. & Hussein, M.H. (eds.) *Foundation Engineering in the Face of Uncertainty: Honoring Fred H. Kulhawy (GSP 229)*. Reston, ASCE. pp. 337–353.

Phoon, K.K., Ching, J. & Huang, H.W. (2012) Examination of multivariate dependency structure in soil parameters. In: *GeoCongress 2012 – State of the Art and Practice in Geotechnical Engineering (GSP 225)*. Reston, ASCE. pp. 2952–2960.

Rad, N.S. & Lunne, T. (1988) Direct correlations between piezocone test results and undrained shear strength of clay. In: *Proc. International Symposium on Penetration Testing, ISOPT-1, Orlando*. Vol. 2. Rotterdam, Balkema. pp. 911–917.

Robertson, P.K. & Campanella, R.G. (1983) Interpretation of cone penetration tests: Part I – Sands. *Canadian Geotechnical Journal*, 20 (4), 718–733.

Salgado, R., Bandini, P. & Karim, A. (2000) Shear strength and stiffness of silty sand. *ASCE Journal of Geotechnical and Geoenvironmental Engineering*, 126 (5), 251–462.

Slifker, J.F. & Shapiro, S.S. (1980) The Johnson system: Selection and parameter estimation. *Technometrics*, 22 (2), 239–246.

Stas, C.V. & Kulhawy, F.H. (1984) Critical evaluation of design methods for foundations under axial uplift and compressive loading. Report EL-3771. Palo Alto, Electric Power Research Institute.

Tang, X.S., Li, D.Q., Rong, G., Phoon, K.K. & Zhou, C.B. (2013) Impact of copula selection on geotechnical reliability under incomplete probability information. *Computers and Geotechnics*, 49, 264–278.

Tang, X.S., Li, D.Q., Zhou, C.B. & Phoon, K.K. (2015) Copula-based approaches for evaluating slope reliability under incomplete probability information. *Structural Safety*, 52, 90–99.

Wroth, C.P. & Wood, D.M. (1978) The correlation of index properties with some basic engineering properties of soils. *Canadian Geotechnical Journal*, 15 (2), 137–145.

Wu, X.Z. (2013) Probabilistic slope stability analysis by a copula-based sampling method. *Computational Geosciences*, 17 (5), 739–755.

Zhang, J., Huang, H.W., Juang, C.H. & Su, W.W. (2014) Geotechnical reliability analysis with limited data: Consideration of model selection uncertainty. *Engineering Geology*, 181, 27–37.

Chapter 5

Statistical characterization of model uncertainty

*Mahongo Dithinde, Kok-Kwang Phoon, Jianye Ching,
Limin Zhang, and Johan V. Retief*

ABSTRACT

One of the key elements for practical implementation of RBD in geotechnical engineering is the characterization of calculation model uncertainty. The calculation models used in the analysis and design of geotechnical structures are in general incomplete and inexact as a result of lack of knowledge or simplification for mathematical convenience. The objective of this chapter is to present available methodologies for characterizing calculation model uncertainty for geotechnical structures. Although the examples given are limited, the methodologies presented are applicable to a wide range of geotechnical structures. First, the general statistical characterization entailing (a) exploratory data analysis, (b) outliers detection and correction of anomalous values, (c) using the corrected data to compute the sample moments (mean, standard deviation, skewness, and kurtosis), (d) verification of the randomness of the model factor, (e) determining the appropriate probability distribution for M and (f) removal of statistical dependencies is presented. This is followed by a presentation of published model factor statistics for geotechnical structures at ultimate and serviceability limit states. It is concluded that significant data on model uncertainty statistics is now available as prior information to enable reliability calibration of partial factors required for semi-probabilistic design approach.

5.1 INTRODUCTION

Calculation model uncertainty is one of the important sources of geotechnical design uncertainties with a significant influence on reliability analysis and code calibration of partial factors for semi-probabilistic design. It arises from imperfections of analytical models for representing geological conditions and predicting engineering behaviour. Generally, the mathematical modelling of physical processes entails making unrealistic assumptions and simplifications physically and geometrically just to create a useable and oftentimes an analytically tractable model. For example, Boussinesq's solution for stresses beneath a point load assumes that the soil is semi-infinite, homogeneous, isotropic and elastic. Obviously, stresses obtained by equations based on these assumptions may differ from stresses obtained in real soil by a significant margin. Even the derivation of the classical Terghazi's bearing capacity equation makes the assumption that the soil beneath the foundation is a homogeneous semi-infinite mass. But as demonstrated in Chapter 3, soils exhibit inherent spatial variability, making the assumption of homogeneity unrealistic. Therefore inevitably, calculation models for resistance and load effects are an over simplifications of complex real world phenomena. Consequently, there is uncertainty in the calculation model predictions even if the model inputs are known with certainty. It is for this reason that the Probabilistic Model Code of the Joint Committee on Structural Safety (JCSS 2001) considers model uncertainty to be a random variable accounting for the effects neglected in the models and simplifications in the mathematical relations.

In the August 2016 revised version of JCSS (2001), the following "indicative" model statistics are recommended for the following geotechnical structures in Table 3.7.5.1:

1. Embankment slope stability based on failure arc analysis (e.g., Bishop, Spencer, etc.) or 2-D FEM
    –   Homogeneous soils (mean = 1.1, standard deviation = 0.05)
    –   Non homogeneous soils (mean = 1.1, standard deviation = 0.10)
2. Stability of retaining (sheet piled) walls based on Brinch Hansen, or Blum, Elastic/ plastic spring supported beam model (mean = 1.0, standard deviation = 0.10)
3. Shallow foundations stability based on Brinch Hansen
    –   Homogeneous soil profile (mean = 1.0, standard deviation = 0.15)
    –   Non homogeneous soil profile (mean = 1.0, standard deviation = 0.20)
    –   Settlement: prediction (mean = 1.0, standard deviation = 0.20–0.30)
4. Foundation piles (driven) based on CPT based empirical design rules
    –   Point bearing capacity (mean = 1.0, standard deviation = 0.25)
    –   Shaft resistance (mean = 1.0, standard deviation = 0.15)
5. Embankment settlement prediction (mean = 1.0, standard deviation = 0.20)

However, it is not explained if these indicative statistics were estimated from load test databases or were gathered primarily from engineering experience.

The current practice in quantifying model uncertainty in various fields including geotechnical engineering, involves comparing results produced by the theoretical/ computational model with physical test results. Hence, model uncertainty is generally represented in terms of the ratio of the measured to predicted values. In this Chapter, this ratio is referred to as the model factor. Mathematically the model factor (M) is expressed as:

$$M = \frac{R_m}{R_c} \qquad (5.1)$$

where R_m = measured or real response estimated from test results and R_c = calculated response based on the theoretical/computational model (called "calculation model" from hereon). Eq. (5.1) is identical to Eq. 3.9.3, Section 3.9 of JCSS (2001).

In Eq. (5.1), it should be noted that measured response can also be affected by testing uncertainties. However, well conducted tests should produce fairly reliable results. Ideally, robust model uncertainty statistics can only be evaluated using: (1) realistically large-scale prototype tests, (2) a sufficiently large and representative database, and (3) reasonably high quality testing where extraneous uncertainties are well-controlled.

With the possible exception of foundations, insufficient test data are available to perform robust characterization of model uncertainties in many geotechnical calculation models. Furthermore, it is important to note that the model factor applies to a specific set of conditions (e.g., failure mode, calculation model, local conditions and experience base, etc.). Therefore, a proliferation of model factors can be expected.

It should be noted that a set of observations of the model factor will take on a range of values representing a sample from the population of interest. Such a raw dataset does not convey much information and therefore needs to be reduced to manageable forms to facilitate its interpretation. It is therefore natural to consider M as a random variable following some probability distribution function. To estimate the statistical properties of this random variable, the following steps are taken: (a) exploratory data analysis, (b) outlier detection and correction of anomalous values, (c) using the corrected data to compute the sample moments (mean, standard deviation, skewness, and kurtosis), (d) verification of the randomness of M, and (e) determining the appropriate probability distribution for M. If M is not random, additional steps are needed to remove its dependency on some underlying parameters.

This Chapter first presents the general methodologies for deriving model uncertainty statistics. Published model factor statistics for geotechnical structures, mostly foundations, are next presented. These include (1) laterally loaded rigid bored piles (ULS), (2) axially loaded piles (ULS), (3) shallow foundations (ULS), (4) axially loaded piles (SLS), (4) limiting tolerable displacement (SLS); (5) factor of safety of a slope calculated by limit equilibrium method and (6) base heave for excavation in clays.

5.2   EXPLORATORY DATA ANALYSIS

The first stage in any data analysis is to explore the data collected in order to reveal general patterns/features of the dataset. Accordingly, the compiled database of model factor observations needs to be subjected to exploratory data analysis, which generally involves examination of graphical outputs (e.g., histograms and normality tests) and descriptive statistics. Figure 5.1 presents an illustration of an output of exploratory data analysis for model factors of pile capacities. The histograms of two datasets consisting of N observations are compared to the probability density functions for a lognormal distribution based on the sample mean and standard deviation (S.D.) in this example. The Anderson-Darling test probability p_{AD} for goodness of fit to normality ($\ln(M)$ is normal if M is lognormal) is presented as well. A probability less than 0.05 means the hypothesis of normality can be rejected.

It is apparent from Figure 5.1 that the graphical display uncovers the following hidden or at least not readily noticed features in the dataset:

- Immediate impression of the range of the data, its most frequently occurring values, and the degree to which it is scattered about the mean,
- Outlying observations which somehow do not fit the overall pattern of the data,
- The exhibition of two or more peaks which may imply an inhomogeneous mixture of data from different samples,
- Whether the data is symmetric or asymmetric,
- Indication of the underlying theoretical distribution for the data.

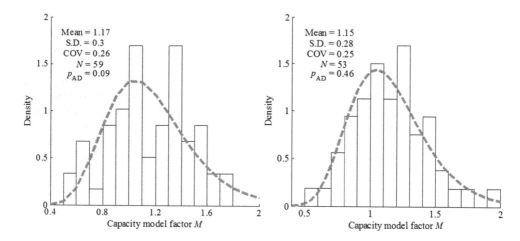

Figure 5.1 Exploratory data analysis of model factors of pile capacities (After Dithinde et al. (2011) with permission from ASCE).

For reliability analysis and design, the key statistics are the mean and the standard deviation, among others. In addition to the measure of centrality and dispersion, the sample mean (m_M) and sample standard deviation (s_M) of the model factor are considered as indicators of the accuracy and precision of the calculation method. An accurate and precise method gives $m_M = 1$ and $s_M = 0$ respectively, which means that for each case, the calculated capacity equals the measured capacity (an ideal case). However, due to presence of uncertainties, the ideal case cannot be attained in practice. Therefore in reality, a calculation method is considered better when m_M is close to 1 and s_M is close to 0. In general, when $m_M > 1$, the calculated capacity is less than the actual capacity, which is conservative and safe; whereas when $m_M < 1$, the calculated capacity is greater than the actual capacity, which is unconservative and unsafe.

Ideally, a calculation model should capture the key features of the physical system, and the remaining difference between the model and reality should be random in nature because it is caused by numerous minor factors that were left out of the model. The statistics of the model factor should capture these random differences resulting from model idealisations. In practice, the ratio between the measured result and the calculated result may not be random in the sense that it is systematically affected by input parameters such as the problem geometry. It is incorrect to model M as a random variable in this situation. The simplest approach to remove dependency is by linear regression of $\ln(R_m)$ on $\ln(R_c)$. If Eq. (5.1) is applicable, the gradient of the line should be close to 1. The statistical treatment of model factor data is closely related to the field of design assisted by testing for which guidance is provided by ISO 2394:2015 Annex C and EN 1990:2002 Annex D. In fact, Eq. (5.1) is identical to Eq. (C.14) in ISO 2394:2015 Annex C. A general approach towards the classification and statistical treatment of model uncertainty is presented by Holický et al. (2015) and implementation is demonstrated by Dithinde et al. (2011).

5.3   DETECTION OF DATA OUTLIERS

Data outliers are extreme values (high or low) that appear to deviate markedly from the main body of a data set. In general, outliers in data may be attributed to human error, instrument error and/or natural deviations in populations. The presence of outliers may greatly influence any calculated statistics leading to biased results. For instance, they may increase the variability of a sample and decrease the sensitivity of subsequent statistical tests (McBean and Rovers, 1998). Therefore, prior to further numerical treatment of samples and application of statistical techniques for assessing the parameters of the population, it is important to identify extreme values and correct erroneous ones. However, it is important to note that there is also a possibility that what appears to be an outlier is a correct observation representing the true state of nature. Therefore, the data point suspected to be an outlier must be carefully scrutinised for errors to justify its exclusion from subsequent analysis.

A number of procedures have been developed to detect outliers. The procedures can be divided into univariate and bivariate approaches. In a univariate approach, screening data for outliers is carried out on each variable while in the bivariate approach, variables are considered simultaneously. In this regard the sample z-score and box plots constitute a basic univariate approach while scatter plots of predicted versus actual performance constitute a common bivariate approach. Since there may be some correlation between the variables, the bivariate approach is considered to be statistically superior as it considers more information (Robinson et al. 2005).

5.3.1   Sample z-score method

The z score is a measure of the number of standard deviations that an observation is above or below the mean. A positive z-score indicates that the observation is above the mean while a negative z-score denotes that the observation is below the mean. The z-score of an observation in a given data set is given by the expression:

$$z = \frac{x - \bar{x}}{s} \tag{5.2}$$

where: x = original data value; $\bar{x}$ = the sample mean; s = the sample standard deviation; z = the z-score corresponding to x.

According to Chebychev's rule, in any distribution, the proportion of scores between the mean and k standard deviation contains at least $1 - 1/k^2$ scores. This rule implies at least 75% of the scores lie between the mean plus/minus two standard deviations ($\pm 2s$), and 89% of the scores would lie between the mean plus/minus three standard deviations ($\pm 3s$). Another commonly adopted rule based on frequency expectations produced by the normal distribution is the following: approximately 68% of the z-scores reside between mean and $\pm 1s$, approximately 95% of the scores resides between mean and $\pm 2s$, and approximately 99% of the scores reside between mean and $\pm 3s$. Both rules have led to the general expectation that almost all the observations in a data set will have z-score less than 3 in absolute value. This implies that all the observation will fall within the interval ($\bar{x} - 3s$ to $\bar{x} + 3s$). Therefore the observation with z-score greater than ± 3 is considered an outlier.

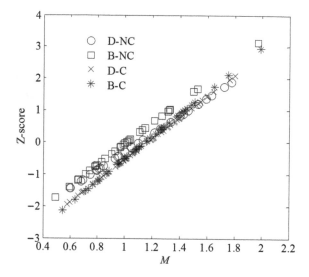

Figure 5.2 Sample z-scores for model factors from 4 load test databases: D-NC (driven piles in non-cohesive soils), B-NC (bored piles in non-cohesive soils), D-C (driven piles in cohesive soils) and B-C (bored piles in cohesive soils) (After Dithinde and Retief 2013).

To apply the above principle to the model factor dataset, first the z-score for each data point is determined and the z-scores are then plotted against the model factors (original data values). As an illustration, plots of z-score versus the model factors of pile capacities are presented in Figure 5.2 for four load test databases. Two observations located at $z = 3$ can be identified as potential outliers.

5.3.2 Box plot method

The box plot method is a more formalised statistical procedure for detecting outliers in a data set. A box plot displays a 5-number summary in a graphical form. The 5-number summary consists of; the most extreme values in the data set (the maximum and minimum values), the lower and upper quartiles, and the median. These values are presented together and ordered from lowest to highest: minimum value, lower quartile, median value, upper quartile, and largest value. Each of these values describe a specific part of a data set: the median identifies the centre of a data set; the upper and lower quartiles span the middle half of a data set; and the highest and lowest observations provide additional information about the actual dispersion of the data.

In using the box plot to identify outliers in the data set, the inter-quartile range (IQR) is required. The inter-quartile range is the difference between the upper quartile and the lower quartile. Any observation that is more than 1.5 IQR beyond the upper and lower quartiles is regarded as an outlier. Typical examples of box plots are presented in Figure 5.3, indicating observations 53 and 156 to be classified as outliers from load test databases B-NC and B-C, respectively.

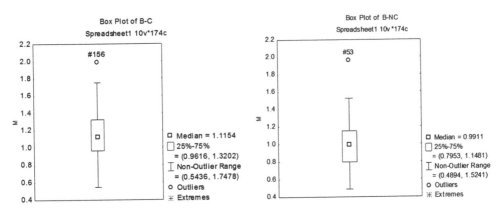

Figure 5.3   Box Plot methods for model factors from load test databases: B-C (bored piles in cohesive soils) and B-NC (bored piles in non-cohesive soils) (Dithinde and Retief 2013).

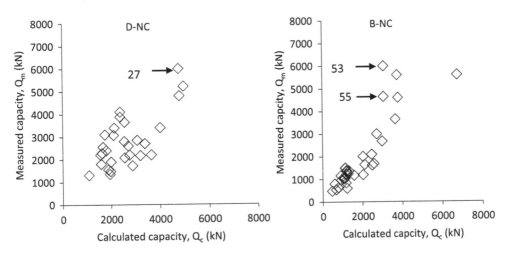

Figure 5.4   Scatter plots of measured capacity versus calculated capacity for driven in non-cohesive soils (D-NC) and bored piles in non-cohesive soils (B-NC) (After Dithinde et al. 2011) with permission from ASCE.

5.3.3   Scatter plot method

The main variables in the computation of model factor realisations are the calculated and measured results. It is reasonable to expect the calculated and measured results to be positively correlated, i.e. the calculated result should be large when the measured result is large and vice-versa. Situations that do not follow this expectation would appear as a point lying a significant distance from the general trend of the data in the scatter plots of measured capacity (Q_m) vs calculated capacity (Q_c). Typical examples for pile foundation are presented in Figure 5.4. It can be seen that two data points in the bored piles in non-cohesive soils dataset (case no. 53 and 55) are potential outliers.

5.4 PROBABILISTIC MODEL FOR M

The theory of reliability is based on a general principle that the basic variables (actions, material properties and geometric data) can be modeled as random variables having appropriate types of probability distribution. Accordingly, one of the key objectives of statistical characterization of the model factor is to determine its distribution function. This is customarily interpreted as the "actual" probability distribution of the random variable under consideration and therefore extends beyond the available sample (i.e. the distribution of the entire population). Once the probability distribution function is known, inferences based on the statistical properties of the distribution can be made. Needless to say, in the presence of a finite sample size, the "actual" probability distribution cannot be identified with certainty. To be more specific, every sample quantile (the collection of all sample quantiles forms the empirical cumulative distribution function) is subject to statistical uncertainty. It is well-known that the statistical uncertainty increases as the sample quantile value decreases. The sample mean and standard deviation are subject to statistical uncertainty as well. This statistical uncertainty aspect is not covered in this Chapter.

For reliability calibration and related studies, the most commonly applied distributions to describe actions, materials properties and geometric data are the normal and lognormal distributions (Holický, 2009; Allen et al. 2005). Therefore, it is reasonable to test goodness-of-fit to the normal and lognormal distribution before considering more complicated distributions. The goodness-of-fit can be examined through (i) a cumulative distribution function (CDF) plotted using a standard normal variate, z as the vertical axis, and (ii) direct distribution fitting to the data.

The cumulative distribution function is a common tool for statistical characterization of random variables used in reliability calibration (e.g., Allen et al. 2005). In the context of the model factor analysis, the CDF is a function that represents the probability that a value of M less than or equal to a specified value. The CDF should be familiar to geotechnical engineers, because it is identical to the grain size gradation curve. This probability can be transformed to the standard normal variable (or variate), z, and plotted against M values (on x-axis) for each data point. This approach is equivalent to plotting the model factor values and their associated probability values on a normal probability paper. An important property of a CDF plotted in this manner is that normally distributed data plot as a straight line while lognormally distributed data on the other hand will plot as a curve. Examples of CDF plots are presented in Figure 5.5. A further characterization entails fitting predicted normal and lognormal distributions to the CDF of the data sets. These theoretical distributions are also shown in Figure 5.5. Both distributions seem to fit the data reasonably well, although the lognormal distribution seems to provide a slightly better fit at the tails.

In the direct distribution fitting method, normal and lognormal probability density functions based on the sample moment parameters are fitted to the histogram of M as illustrated in Figure 5.6. The graphical comparison indicates the degree to which the alternative distributions provide a smoothed representation of the M data. The uneven nature of the histogram is expected for a finite sample size. A large sample size produces less unevenness. It is not possible to judge if the degree of unevenness is explainable by statistical uncertainties associated with a finite sample size by inspection.

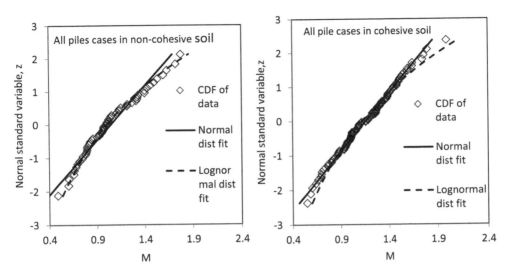

Figure 5.5 CDF plots with normal and lognormal fit (After Dithinde and Retief 2013).

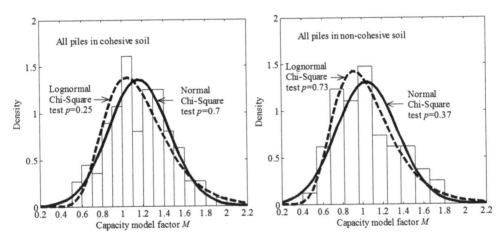

Figure 5.6 Normal and lognormal distribution fit to the data (After Dithinde and Retief 2013).

The quantitative assessment of the difference between the empirical data frequencies and the assumed distributions is achieved through a goodness-of-fit test. In Figure 5.6, the Chi-Square goodness-of-fit test was used. In these tests, the p-value is a measure of the goodness of fit, with larger values indicating a better fit. Just like the Anderson-Darling goodness-of-fit test was used in Figure 5.1, it is customary to conclude that there is no evidence to reject a hypothesized distribution if the p-value is larger than 0.05. However, based on past studies and practical considerations, a lognormal distribution is found to be a more suitable probability distribution for the model factor.

5.5 VERIFICATION OF RANDOMNESS OF THE MODEL FACTOR

Reliability analysis is based on the assumption of randomness of the basic variables including the model factor. Generally, the randomness of the model factor is verified by investigating the presence or absence of correlation with related input parameters (e.g., material properties and geometric data) in the database. The presence of correlation between M and deterministic variations in the input parameters would indicate that:

- The calculation model does not fully take the effects of the input parameters into account.
- The assumption that M is a random variable is not valid.

The measure of the degree of association between variables is the correlation coefficient. The basic and most widely used type of correlation coefficient is Pearson r, also known as the linear or product-moment correlation. The correlation can be negative or positive. When it is positive, the dependent variable tends to increase as the independent variable increases; when it is negative, the dependent variable tends to decrease as the independent variable increases. The numerical value of r lies between the limits -1 and $+1$. A high absolute value of r indicates a high degree of association whereas a small absolute value indicates a small degree of association. When the absolute value is 1, the relationship is said to be perfect and when it is zero, the variables are independent. These observations strictly apply to normal random variables. They are approximately true for distributions close to normal. For strongly non-normal distributions, the Spearman rank correlation coefficient is a more robust measure. For values lying between these limits, a critical question is "when is the numerical value of the correlation coefficient considered significant"? Several authors in various fields have suggested guidelines for the interpretation of the correlation coefficient. The following interpretation suggested by Franzblau (1958) seems to be popular:

- Range of r: 0 to ± 0.2 – indicate no or negligible correlation.
- Range of r: ± 0.2 to ± 0.4 – indicate a low degree of correlation.
- Range of r: ± 0.4 to ± 0.6 – indicate a moderate degree of correlation.
- Range of r: ± 0.6 to ± 0.8 – indicate a marked degree of correlation.
- Range of r: ± 0.8 to ± 1 – indicate a high correlation.

The statistical significance of the correlation is determined through hypothesis testing and presented in terms of the usual p-value. In this test, the null hypothesis is that there is no correlation between M and the given input parameter (indicative of statistical independence). A small p-value ($p < 0.05$) indicates that the null hypothesis is not valid and should be rejected. The correlations between the model factor of a pile capacity and inputs parameters (shaft length and diameter) are presented in Figure 5.7.

5.5.1 Removal of statistical dependencies

In order to correct for the calculation model uncertainty, the common practice is to apply a model factor as an independent random variable on the calculated capacity. This is only valid if the model factor does not vary systematically with some underlying

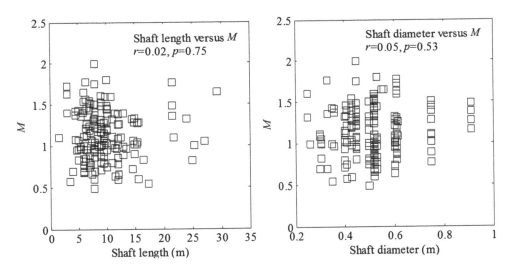

Figure 5.7 Correlation between the model factor of a pile capacity and input parameters (shaft length and diameter) (Adapted from Dithinde 2007).

factors. However, if there is some degree of correlation, then the calculated reliability index is affected unless the correlation is explicitly accounted for. Even though correlation can be incorporated into the reliability analysis, it complicates the calculations as it involves transforming the original variables to a set of uncorrelated variables. Therefore, to apply the model factor as an independent random variable in reliability analysis, the statistical dependencies need to be removed. The two approaches to the treatment of correlation are further discussed in the subsequent subsections.

5.5.1.1 *Generalised model factor approach*

The generalised model factor approach entails performing regression using the calculated values as the predictor variable. It was alluded to in section 5.4 that in general the appropriate probabilistic model for the model factor is taken to be the lognormal probability distribution. Taking pile foundations as an example, the generalised model factor is derived from the regression of $\ln(Q_m)$ on $\ln(Q_c)$ where Q_m is the measured capacity and Q_c is the calculated capacity. The resulting functional relationship between $\ln(Q_m)$ and $\ln(Q_c)$ is given by a general regression model of the form:

$$\ln(Q_m) = a + b\ln(Q_c) + \varepsilon \tag{5.3}$$

in which a and b are regression constants and ε is a normal random variable with zero mean and non-zero variance.

Taking antilog on both sides of equation 5.3 yields;

$$Q_m = \exp(a) \cdot \exp(\varepsilon) \cdot Q_c^b \tag{5.4}$$

The regression model in the form of Eq. 5.3 or 5.4 removes systematic effects and the remaining component tends to appear random (Phoon and Kulhawy 2005).

Eq. (5.4) can be re-written as:

$$Q_m = \exp(a + \varepsilon)Q_c^b \qquad (5.5)$$

Let

$$\exp(a + \varepsilon) = M \qquad (5.6)$$

Then;

$$Q_m = MQ_c^b \qquad (5.7)$$

Eq. (5.7) is the generalised representation of the model factor M. This equation is immediately recognised as being of the same form as that for the conventional model factor (i.e. $Q_m = MQ_c$). In fact, the conventional model factor is a special case of the generalised model factor with $b = 1$.

In equation (5.6), ε is a random variable and therefore M will likewise be random. Assuming M is lognormally distributed, its mean and variance are as follows:

$$\mu_M = \exp(a + 0.5\xi^2) \qquad (5.8)$$
$$\sigma_M^2 = \mu_M^2[\exp(\xi^2) - 1] \qquad (5.9)$$

in which ξ is the standard deviation of ε. The generalised model factor as presented in Eq. (5.7) is not dimensionless in contrast to the conventional model factor equation. The force unit (i.e. kN) adopted for the measured and predicted capacity is applicable to Eq. (5.7). To make the generalised model factor dimensionless, both the measured and calculated capacity need to be normalised. Dithinde (2007) investigated the following normalization schemes for the generalised model factor for piles:

- Scheme 1: dividing $\ln(Q_m)$ and $\ln(Q_c)$ by area of pile base $\times$ atmospheric pressure $(A_b P_a)$
- Scheme 2: dividing $\ln(Q_m)$ and $\ln(Q_c)$ by volume of water displaced by the pile (i.e. volume of piles $\times$ unit weight of water (V_w))
- Scheme 3: dividing $\ln(Q_m)$ and $\ln(Q_c)$ by weight of pile shaft (W_s)

The dimensionless generalised model factor is then obtained from the regression analyses of normalised $\ln(Q_m)$ on normalised $\ln(Q_c)$. Figure 5.8 presents two typical regression results. The regression results presented in Figure 5.8 are then used in conjunction with Eq. 5.8 and Eq. 5.9 to compute the required generalised model factor statistics. The regression parameters and the ensuing generalised model factor statistics for pile foundations are summarised in Table 5.1. It is evident from Table 5.1 that for a given pile class, the model factor statistics corresponding to the three normalisation schemes are comparable. Note that the statistics and the regression parameter "b" are used as inputs into the performance function for computation of the reliability index.

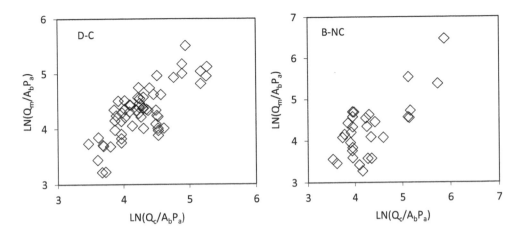

Figure 5.8 Regression of normalized ln(Q_m) and ln(Q_c) (Adapted from Dithinde, 2007).

Table 5.1 Generalised model factor statistics for various normalisation schemes (Adapted from Dithinde 2007).

Case	Normalisation Scheme	Regression parameters				Generalised M statistics		
		R^2	a	b	ξ	μ	σ	COV
D-NC	A_bP_a	0.88	−0.116	1.03	0.293	0.93	0.28	0.30
	V_w	0.78	0.103	0.985	0.294	1.16	0.35	0.30
	W_s	0.55	0.527	0.887	0.291	1.77	0.53	0.30
B-NC	A_bP_a	0.76	−0.183	1.03	0.264	0.86	0.23	0.27
	V_w	0.77	−0.019	0.993	0.264	1.02	0.27	0.27
	W_s	0.74	−0.425	1.094	0.262	0.68	0.18	0.27
D-C	A_bP_a	0.68	0.223	0.977	0.273	1.30	0.36	0.28
	V_w	0.74	0.670	0.89	0.267	2.02	0.55	0.27
	W_s	0.74	0.571	0.89	0.267	1.83	0.50	0.27
B-C	A_bP_a	0.79	0.298	0.961	0.244	1.39	0.34	0.25
	V_w	0.80	0.277	0.967	0.244	1.36	0.34	0.25
	W_s	0.80	0.248	0.967	0.244	1.32	0.33	0.25

5.5.1.2 Verification of removal of systematic dependency

The correlation of practical significance is that between the model factor and the calculated performance. To verify the removal of such correlation, the regression error (ε) which represents the model factor is plotted against the normalised ln(Q_c). For a given pile class, the model error for a given data point is determined from Eq. (5.3). As an example the model error for normalisation scheme 1 is given by:

$$\varepsilon = \ln(Q_m/A_bp) - a - b\ln(Q_c/A_b) \tag{5.10}$$

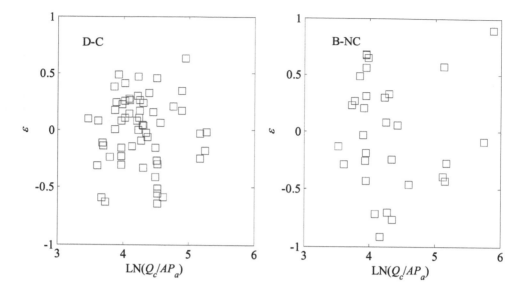

Figure 5.9   Scatter plots of ε versus ln(Q_c/AP) (Adapted from Dithinde 2007).

Illustrative scatter plots of the model error (ε) versus the calculated capacity are presented in Figure 5.9. Visual inspection of the scatter plots shows that there is no correlation between M (or ε) and $\ln(Q_c/A_b)$.

5.5.2   Model factor as a function of input parameters

The generalised model factor approach is purely empirical. Statistical dependencies are removed in a gross manner by regressing the measured capacity against the calculated capacity. There is no physical insight on the sources of these statistical dependencies. The practical limitation that comes immediately to mind is how general are the generalised model factors and their statistics? A related limitation is that the scope of applicability (for example, the range of pile lengths or soil strengths) is not explicit, although one could argue that if the database covers most practical scenarios, the generalized model factors characterized from this database would be useful even in the absence of an explicit scope of applicability.

Over-simplification is deliberately adopted at times to reduce the solution to a simple analytical form. Although one may suspect various input parameters to be the explanatory variables behind statistical dependencies between the model factor and the calculated capacity, it is not easy to remove these dependencies in a more physical way by regressing the model factor against each input parameter (in contrast to regressing the model factor against the gross predicted capacity described in section 5.5.1.1) because the values of these input parameters cannot be varied systematically in a load test database for regression analysis.

Zhang et al. (2015) studied the calculation of cantilever retaining wall deflections in undrained clay using the mobilized strength design (MSD) method proposed by

Osman and Bolton (2004). The displacement model factor M is defined as:

$$\delta_m = M \times \delta_c \tag{5.11}$$

where δ_m = measured wall top displacement either from case histories in the field or from model tests in the laboratory and δ_c = calculated displacement. The authors found that this standard definition cannot be applied directly in the case of the MSD calculation method, because M is a function of six input parameters: (1) excavation width, (2) excavation depth, (3) wall thickness, (4) at-rest lateral earth pressure coefficient, (5) undrained shear strength ratio at mid-depth, and (6) ratio between undrained Young's modulus and undrained shear strength at mid-depth.

It is not possible to remove these dependencies from M using field data, because the values of these input parameters cannot be varied systematically for regression analysis. The authors proposed a rather novel approach consisting of: (1) removing these dependencies using the finite element method (FEM) where the input parameters can be freely varied and (2) characterizing the displacement model factor for the finite element method which is unlikely to suffer from the same dependency problem, given that it is mechanically more consistent. Step (1) is carried out by defining the ratio between the FEM wall top displacement δ_{c_FEM} and the corresponding MSD calculated displacement δ_{c_MSD}:

$$\delta_{c\_FEM} = \eta \times \delta_{c\_MSD} \tag{5.12}$$

The correction factor η in Eq. (5.12) can be decomposed into a systematic part (f) that is determined using multivariate regression and a residual random factor η^* (regression error) as follows:

$$\eta = f \times \eta^* \tag{5.13}$$

This regression can be carried out, because a large number of design scenarios described by different combinations of the six input parameters can be analyzed using FEM and MSD. No field data is involved in Eq. (5.12). Field data is involved in Step (2) where the model factor for FEM (M_{FEM}) is characterized in the usual way:

$$\delta_m = M_{FEM} \times \delta_{c\_FEM} \tag{5.14}$$

Zhang et al. (2015) showed that M_{FEM} is indeed not plagued by the dependency problem. Combining Steps (1) and (2), it is quite clear that the model factor for MSD (M) is:

$$M = M_{FEM} \times \eta^* \times f \tag{5.15}$$

The critical observation here is that M is not a random variable because of the deterministic function f, although it follows the standard model factor definition. More specifically, M is the product of a random variable ($M_{FEM} \times \eta^*$) and a deterministic function f. One can also take the alternate view that the calculated displacement from MSD should be modified by f. By rearranging Eq. (5.15), it is easy to see that the model factor for this modified MSD method ($\delta_{c\_MSD} \times f$) will be a random variable $M^* = (M_{FEM} \times \eta^*)$.

Other geotechnical problems have been studied using the same framework:

- Bearing capacity of strip footings under combined positive loading (Phoon & Tang 2015a), combined negative loading and combined general loading (including positive and negative loading) (Phoon & Tang 2015b).
- Bearing capacity of circular footings on dense sand (Tang & Phoon 2016a).
- Uplift capacity of helical anchors in clay (Tang & Phoon 2016b).

The results are summarized in Table 5.2. The main differences in these studies from Zhang et al. (2015) are that: (a) FEM is replaced by finite element limit analysis (FELA) and (b) ultimate limit state is considered rather than serviceability limit state. It should be pointed out that the size of the load test database adopted by Phoon and Tang (2015a, 2015b) was larger than what is commonly available in practice: 120 load tests for combined positive loading and 72 load tests for combined negative loading. Because the size of a load test database is usually much smaller, Phoon and Tang (2015c) examined the effect of the load test database on the model statistics by randomly drawing smaller databases from the original large parent databases.

5.6 AVAILABLE MODEL FACTOR STATISTICS

Available model factor statistics are primarily restricted to simple calculation methods for foundations. This section presents model statistics for a variety of foundations at the ultimate and serviceability limit states (ULS and SLS): Section 5.6.1 – laterally loaded rigid bored piles (ULS); Section 5.6.2 – axially loaded piles (ULS); Section 5.6.3 – shallow foundations (ULS); Section 5.6.4 – axially loaded foundations (SLS); Section 5.6.5 – limiting tolerable displacement (SLS). Section 5.6.6 presents model statistics for the factor of safety of a slope calculated by limit equilibrium method while Section 5.6.7 presents model statistics for the basal heave factor of safety in excavation in clays.

5.6.1 Laterally loaded rigid bored piles (ultimate limit state)

The capacity of a laterally loaded pile is typically predicted using conventional ultimate lateral soil stress models. The statistics of the model factor for different ultimate lateral soil stress models under undrained and drained loading modes are presented in Table 5.3.

5.6.2 Axially loaded piles (ultimate limit state)

Tables 5.4 and 5.5 present the model factor statistics for the capacity of axially loaded piles for various calculation methods, soil conditions, and failure interpretation methods. See Paikowsky et al. (2004) for NCHRP Report 507. The model factors are defined following Eq. (5.1).

5.6.3 Shallow foundations (ultimate limit state)

Model factor statistics for shallow foundations are scarce compared to pile foundations. Model statistics for different loading modes (vertical eccentric loading, inclined

Table 5.2 Summary of model statistics with response modified by f.

Problem	Variables		$\ln f = b_0 + \Sigma b_i x_i$		M^* Mean	COV	Notation
Strip footings on sand under positive combined loading (Phoon and Tang 2015a)	x_1	$\gamma D/p_a$	b_0	0.28	1.04	0.1	D foundation width
	x_2	ξ	b_1	-5.05			γ unit weight of sand
	x_3	$\tan\phi_a$	b_2	11.4			p_a atmospheric pressure
	x_4	d/B	b_3	-0.26			d embedment depth
	x_5	α/ϕ_a	b_4	-0.09			ϕ_a repose angle of sand
	x_6	e/B	b_5	0.21			e load eccentricity
	x_7	$(e/B)(\alpha/\phi_a)$	b_6	-1.12			α load inclination
			b_7	-0.98			
Strip footings on sand under negative combined loading (Phoon and Tang 2015b)	x_1	$\gamma D/p_a$	b_0	0.1	1.07	0.1	ξ empirical parameter
	x_2	ξ	b_1	-4.5			$\xi = 0.02\sim0.12$
	x_3	$\tan\phi_a$	b_2	10.4			
	x_4	d/B	b_3	-0.25			
	x_5	α/ϕ_a	b_4	-0.12			
	x_6	e/B	b_5	-1.03			
	x_7	$(e/B)(\alpha/\phi_a)$	b_6	-0.45			
			b_7	-1.81			
Strip footings on sand under general combined loading (Phoon and Tang 2015b)	x_1	$\gamma D/p_a$	b_0	0.1	1.06	0.13	
	x_2	ξ	b_1	-4.5			
	x_3	$\tan\phi_a$	b_2	10.25			
	x_4	d/B	b_3	-0.15			
	x_5	α/ϕ_a	b_4	0.05			
	x_6	e/B	b_5	-0.93			
	x_7	$(e/B)(\alpha/\phi_a)$	b_6	-0.05			
			b_7	-2.53			
Circular footings on dense sand (Tang and Phoon 2016a)	x_1	$\tan\phi_{cv}$	b_0	1.97	1.02	0.15	D foundation diameter
	x_2	D_R	b_1	-3.12			D_R relative density of sand
	x_3	$\gamma D/p_a$	b_2	2.23			ϕ_{cv} critical state friction angle
			b_3	-0.68			
Helical anchors in clay under tension loading (Tang and Phoon 2016b)	x_1	n	b_0	0.75	0.95	0.16	n number of helix plates
	x_2	S/D	b_1	-0.05			S plate spacing
	x_3	H/D	b_2	-0.11			D diameter of helix plate
	x_4	$\gamma H/s_u$	b_3	-0.03			H depth of top helix
			b_4	-0.11			s_u undrained shear strength
Cantilever retaining wall deflections in undrained clay (Zhang et al. 2015)	x_1	$2D/B$	b_0	0.89	1.02	0.26	EI wall stiffness
	x_2	$\ln H_c/B$	b_1	-0.13			B wall width
	x_3	$\ln \gamma D^4/EI$	b_2	0.43			E_{ur} soil stiffness
	x_4	$1/K_0$	b_3	0.12			σ'_v effective vertical stress
	x_5	s_u/σ'_v	b_4	0.69			K at-rest lateral earth coefficient
	x_6	E_{ur}/s_u	b_5	-0.74			D wall depth
			b_6	-7×10^{-4}			H_c excavation depth

Table 5.3 Model factors for rigid bored piles based on the hyperbolic capacity (Source: Phoon and Kulhawy 2005).

Calculation Model[a]	Statistics of Model Factor	
Undrained:	*Number of load tests = 74*	
Reese (1958)	Range	0.75–2.72
	Mean	1.42
	COV	0.29
Hansen (1961)	Range	0.86–3.61
	Mean	1.92
	COV	0.29
Broms (1964a)	Range	1.08–4.49
	Mean	2.28
	COV	0.37
Stevens and Audibert (1979)	Range	0.55–2.13
	Mean	1.11
	COV	0.29
Randolph and Houlsby (1984)	Range	0.67–2.52
	Mean	1.32
	COV	0.29
Drained:	*Number of load tests = 77*	
Reese et al. (1974)	Range	0.40–3.35
	Mean	1.19
	COV	0.43
Hansen (1961)	Range	0.55–2.33
	Mean	0.98
	COV	0.33
Broms (1964b)	Range	0.85–3.40
	Mean	1.80
	COV	0.38
Simplified Broms (1964b)	Range	0.59–2.62
	Mean	1.30
	COV	0.38

[a] Model and reference details given in Phoon & Kulhawy (2005).

eccentric loading) are reported in NCHRP 651 (Paikowsky et al. 2010). Tables 5.6 through 5.10 present the summary model factor statistics for the different loading conditions.

5.6.4 Axially loaded pile foundations (serviceability limit state)

Limit state design requires that the occurrence of both ultimate and serviceability limit states are sufficiently improbable. For consistency, it is imperative that serviceability limit state verification be based on reliability principles. Tables 5.11 and 5.12 present values of the model factor for driven steel H-piles and bored piles, respectively. The model factor here refers to the ratio of the measured pile settlement at the working

Table 5.4 Model factors for driven piles (Source: NCHRP Report 507, Dithinde et al. 2011 with permission from ASCE).

Calculation method	No. of cases	Pile type	Soil type	Mean	COV	Source
β-method	4	H-pile	Clay	0.61	0.61	NCHRP Report 507
λ-method	16			0.74	0.39	
α-Tomlinson	17			0.82	0.40	
α-API	16			0.90	0.41	
SPT-97 mob	8			1.04	0.39	
λ-method	18	Concrete piles	Clay	0.76	0.39	
α-API	17			0.81	0.36	
β-method	8			0.81	0.31	
α-Tomlinson	18			0.87	0.48	
α-Tomlinson	18	Pipe piles	Clay	0.64	0.50	
α-API	19			0.79	0.54	
β-method	12			0.45	0.60	
λ-method	19			0.67	0.55	
SPT-97 mob	12			0.39	0.62	
Nordlund	19	H-pile	Sand	0.94	0.4	
Meyerhof	18			0.81	0.38	
β-method	19			0.78	0.51	
SPT-97 mob	18			1.35	0.43	
Nordlund	36	Concrete piles	Sand	1.02	0.48	
β-method	35			1.1	0.44	
Meyerhof	36			0.61	0.61	
SPT-97 mob	36			1.21	0.47	
Nordlund	19	Pipe piles	Sand	1.48	0.52	
β-method	20			1.18	0.62	
Meyerhof	20			0.94	0.59	
SPT-97 mob	19			1.58	0.52	
α-Tomlinson/Nordlund/Thurman	20	H-pile	Mixed soils	0.59	0.39	
α-API/Nordlund/Thurman	34			0.79	0.44	
β-method/Thurman	32			0.48	0.48	
SPT-97 mob	40			1.23	0.45	
α-Tomlinson/Nordlund/Thurman	33	Concrete piles	Mixed soils	0.96	0.49	
α-API/Nordlund/Thurman	80			0.87	0.48	
β-method/Thurman	80			0.81	0.38	
SPT-97 mob	71			1.81	0.50	
FHWA CPT	30			0.84	0.31	
α-Tomlinson/Nordlund/Thurman	13	Pipe piles	Mixed soils	0.74	0.59	
α-API/Nordlund/Thurman	32			0.8	0.45	
β-method/Thurman	29			0.54	0.48	
SPT-97 mob	33			0.76	0.38	
Static formula	28	Concrete piles	Sand	1.11	0.33	Dithinde et al. 2011
Static formula	59		Clay	1.17	0.26	
Meyerhof	24		Sand	1.22	0.54	*FHWA-HI-98-032*

Table 5.5 Model factors for bored piles (Source: Dithinde et al. 2011 with permission from ASCE, Zhang & Chu 2009a, NCHRP Report 507).

Calculation method[1]	Constr. Method[2]	No. of cases	Soil type	Mean	COV	Source
Static formula	Mixed	30	Sand	0.98	0.24	Dithinde et al. 2011
Static formula		53	Clay	1.15	0.25	
FHWA (1999)	Casing	11	Sand/silt	0.6	0.58	Zhang and Chu 2009a
FHWA (Hong Kong data)	Casing	17	Sand/silt	1.06	0.28	
FHWA (1999)	RCD	15	Rocks	0.48	0.52	
COP (BD 2004)	RCD	15	Rocks	2.57	0.31	
FHWA (1999)	Mixed	32	Sand	1.71	0.60	NCHRP Report 507
	Casing	12		2.27	0.46	
	Slurry	9		1.62	0.74	
R&W	Mixed	32		1.22	0.67	
	Casing	12	Sand	1.45	0.5	
	Slurry	9		1.32	0.62	
FHWA (1999)	Mixed	53	Clay	0.9	0.47	
	Casing	14		0.84	0.50	
	Dry	30		0.88	0.48	
FHWA (1999)	Mixed	44	Clay + Sand	1.19	0.30	
	Casing	21		1.04	0.29	
	Dry	12		1.32	0.28	
	Slurry	10		1.29	0.27	
R&W	Mixed	44		1.09	0.35	
	Casing	21	Clay + Sand	1.01	0.42	
	Slurry	12		1.2	0.32	
	Slurry	10		1.16	0.25	
C&K	Mixed	46	Rock	1.23	0.40	
	Dry	29		1.29	0.34	
IGM	Mixed	46	Rock	1.3	0.34	
	Dry	29		1.35	0.31	

[1] Model and reference details given by source references.
[2] Casing = pile bore excavation assisted by steel casing; RCD = reverse circulation drilling in rocks; slurry = excavation assisted by mineral slurry; dry = excavation above groundwater.

Table 5.6 Vertical-eccentric loading using the effective foundation width B' (Source: NCHRP 651).

Tests[1]	No. cases	Minimum slope criterion			Two-slope criterion		
		Mean	Std	COV	Mean	Std	COV
DEGEBO – radial load path	17 (15)[2]	2.22	0.754	0.340	2.04	0.668	0.328
Montrasio(1994)/Gottardi (1992) – radial load path	14	1.71	0.399	0.234	1.52	0.478	0.313
Perau (1995) – radial load path	12	1.43	0.337	0.263	1.19	0.470	0.396
All cases	34 (41)[2]	1.83	0.644	0.351	1.61	0.645	0.400

[1] Model and reference details given by NHCRP 651.
[2] Number of cases for two-slope criterion.

Table 5.7 Vertical-eccentric loading using the full foundation width B (Source: NCHRP 651).

Tests[1]	No. cases	Minimum slope criterion			Two-slope criterion		
		Mean	Std	COV	Mean	Std	COV
DEGEBO – radial load path	17 (15)[2]	1.30	0.464	0.358	1.20	0.425	0.355
Montrasio (1994)/Gottardi (1992) – radial load path	14	0.97	0.369	0.38	0.86	0.339	0.396
Perau (1995) – radial load path	12	0.79	0.302	0.383	0.64	0.296	0.464
All cases	34 (41)[2]	1.05	0.441	0.420	0.92	0.423	0.461

[1] Model and reference details given by NHCRP 651.
[2] Number of cases for two-slope criterion.

Table 5.8 Inclined-eccentric loading when using the effective foundation width B′ (Source: NCHRP 651).

Tests[1]	No. cases	Minimum slope criterion			Two-slope criterion		
		Mean	Std	COV	Mean	Std	COV
DEGEBO/Gottardi (1992) – radial load path	8	2.06	0.813	0.394	1.78	0.552	0.310
Montrasio (1994)/ Gottardi (1992)	6	2.13	0.496	0.234	2.12	0.495	0.233
Perau (1995) – Positive eccentricity	8	2.16	1.092	0.506	2.15	1.073	0.500
Step-like load path Perau (1995) – Negative eccentricity	7	3.43	1.792	0.523	2.29	1.739	0.713
All step-like load cases	21	2.57	1.352	0.526	2.56	1.319	0.516
All cases	29	2.43	1.234	0.508	2.34	1.201	0.513

[1] Model and reference details given by NHCRP 651.

Table 5.9 Inclined-eccentric loading when using the full foundation width B (Source: NCHRP 651).

Tests[1]	No. cases	Minimum slope criterion			Two-slope criterion		
		Mean	Std	COV	Mean	Std	COV
DEGEBO/Gottardi (1992) – radial load path	8	1.07	0.448	0.417	0.94	0.365	0.387
Montrasio (1994)/ Gottardi (1992)	6	1.18	0.126	0.106	1.18	0.125	0.106
Perau (1995) – Positive eccentricity	8	0.70	0.136	0.194	0.70	0.135	0.194
Step-like load path Perau (1995) – Negative eccentricity	7	1.09	0.208	0.191	1.08	0.208	0.193
All step-like load cases	21	0.97	0.267	0.276	0.96	0.267	0.277
All cases	29	1.00	0.322	0.323	0.96	0.290	0.303

[1] Model and reference details given by NHCRP 651.

Table 5.10 Statistics for the ratio of measured (q_{L2}) to calculated bearing capacity (q_{ult}) for all foundations on rock using the Carter and Kulhawy (1988) method (Source: NCHRP 651).

Cases	n	No. of sites	m_λ	σ_λ	COV
All (measured q_u)	119	78	8.00	9.92	1.240
Measured discontinuity spacing (s')	83	48	8.03	10.27	1.279
Fractured with measured discontinuity spacing (s')	20	9	4.05	2.42	0.596
All non-fractured	99	60	8.80	10.66	1.211
Non-fractured with measured discontinuity spacing (s')	63	39	9.29	11.44	1.232
Non-fractured with s' based on AASHTO (2007)	36	21	7.94	9.22	1.161

n = number of case histories, m_λ = mean of biases, σ_λ = standard deviation, COV = coefficient of variation, q_u = uniaxial compressive strength of intact rock, q_{L2} = measured capacity interpreted using the L2 method.

Table 5.11 Model factors for driven piles at the working load level defined as one half of Davisson's capacity (Source: Zhang et al. 2008).

Calculation method	No. of cases	Soil type	Mean	COV
Vesic (1977)	34	Sand/silt	1.02	0.23
Fleming et al. (1992)	34	Sand/silt	0.66	0.22
Load transfer method	34	Sand/silt	1.34	0.22
Vesic (1977)	30	Rocks	0.96	0.27
Fleming et al. (1992)	30	Rocks	0.81	0.28
Load transfer method	30	Rocks	1.16	0.24

Table 5.12 Model factors for large-diameter bored piles at the working load level defined as one half of Davisson's capacity (Adapted from Zhang and Chu 2009b).

Calculation method	Construction method	No. of cases	Soil type	Mean	COV
Vesic (1977)	Casing	20	Sand/silt	0.24	0.38
Mayne and Harris (1993)	Casing	12	Sand/silt	0.64	0.22
Reese and O'Neill (1989)	Casing	19	Sand/silt	1.80	0.31
Vesic (1977)	RCD	14	Rocks	0.87	0.30
Kulhawy and Carter (1992)	RCD	14	Rocks	1.01	0.24
Load transfer method using correlation with RQD	RCD	14	Rocks	1.21	0.30

load level and the calculated settlement at the same load level. The working load level is defined as one half of Davisson's capacity.

Model factor statistics for an allowable settlement = 25 mm from another SLS study are presented in Table 5.13. In this study, the SLS model statistics were derived by fitting measured load-settlement data to a hyperbolic equation (Eq. 5.16). At the ultimate limit state, a consistent load test interpretation procedure should be used to produce a single "measured capacity" from each measured load-displacement curve. The ratio of the measured capacity to the calculated capacity is called a model factor as defined in Eq. (5.1). The same approach applies to the serviceability limit state

Table 5.13 SLS model factor statistics for allowable settlement = 25 mm (Source: Dithinde et al. 2011, with permission from ASCE).

| | | Actual statistics | | | | Statistics from first-order second-moment approximations | | | |
| | | M_s | | M_sM | | M_s | | M_sM | |
Case	N	μ	COV	μ	COV	μ	COV	μ	COV
D-NC	28	1.084	0.077	1.202	0.351	1.076	0.082	1.195	0.340
B-NC	30	1.079	0.083	1.057	0.238	1.094	0.114	1.072	0.266
D-C	59	1.082	0.047	1.259	0.260	1.083	0.049	1.267	0.265
B-C	53	1.077	0.063	1.236	0.250	1.073	0.059	1.234	0.257

D-NC = driven piles in non-cohesive soils, B-NC = bored piles in non-cohesive soils, D-C = driven piles in cohesive soils, B-C = bored piles in cohesive soils, M_s = model factor for SLS, M_sM = combined statistics.

(SLS). The capacity is replaced by an allowable capacity that depends on the allowable displacement. The distribution of the SLS model factor is established from a load test database in the same way. Notice that the SLS model factor has to be re-evaluated when a different allowable displacement is prescribed. Hence, it is important to note that Tables 5.11 to 5.13 only apply to an allowable settlement = 25 mm. If the allowable settlement is treated as a random variable in the serviceability limit state, a more general approach involving fitting measured load-displacement data to a normalized hyperbolic curve is recommended as detailed below:

$$\frac{Q}{Q_m} = \frac{y}{a + by} \tag{5.16}$$

in which Q = applied load, Q_m = failure load or capacity interpreted from a measured load-displacement curve, "*a*" and "*b*" = curve-fitting parameters, and y = pile butt displacement. Note that the curve-fitting parameters are physically meaningful, with the reciprocals of "*a*" and "*b*" equal to the initial slope and asymptotic value of the hyperbolic curve, respectively. The curve-fitting equation is empirical and other functional forms can be considered (Phoon and Kulhawy 2008). However, the important criterion is to apply a curve-fitting equation that produces the least scatter in the measured normalized load-displacement curves. Each measured load-displacement curve is thus reduced to two curve-fitting parameters. Based on "*a*" and "*b*" statistics estimated from the load test database (Table 5.14), one can construct an appropriate bivariate probability distribution for (*a, b*) that can reproduce the scatter in the normalized load over the full range of displacements. Details are given in Phoon and Kulhawy (2008). Table 5.11 is in fact produced using this general curve-fitting approach. It is evident that this approach can be used in conjunction with a random allowable settlement. This approach has been applied to various foundation types (Phoon et al. 2006; Phoon et al. 2007; Akbas and Kulhawy 2009a; Dithinde et al. 2011; Stuedlein and Reddy 2013; Huffman and Stuedlein 2014; Huffman et al. 2015).

Table 5.14 Statistics for hyperbolic parameters [Source: ACIP under axial compression (Phoon et al. 2006) with permission from ASCE; Spread foundation, drilled shaft, pressure-injected footing under axial uplift (Phoon et al. 2007) with permission from ASCE; Driven piles in non-cohesive soil, Bored piles in non-cohesive soils, Driven piles in cohesive soils, and Bored piles in cohesive soil (Dithinde et al. 2011) with permission from ASCE; Augered Cast-In-Place Piles in Granular Soils (Stuedlein and Reddy, 2013); Spread footing on clay (Huffman et al. 2015)].

Augered cast-in-place pile (compression)	Spread footing (uplift)
No. tests = 40	No. tests = 85
a: Mean = 5.15 mm, SD = 3.07 mm, COV = 0.60	a: Mean = 7.13 mm, SD = 4.66 mm, COV = 0.65
b: Mean = 0.62, SD = 0.16, COV = 0.26	b: Mean = 0.75, SD = 0.14, COV = 0.18
Correlation = −0.67	Correlation = −0.24
Augered cast-in-place pile in granular soils	Spread footing on clay
No. tests = 87	No. tests = 30
a (k2): Mean = 3.40 mm, COV = 0.49	a(k2): Mean = 0.70 mm, COV = 0.16
b (k1): Mean = 0.16 mm, COV = 0.23	b(k1): Mean = 0.013, COV = 0.53
Bored pile (uplift)	Pressure injected footing (uplift)
No. tests = 48	No. tests = 25
a: Mean = 1.34 mm, SD = 0.73 mm, COV = 0.54	a: Mean = 1.38 mm, SD = 0.95 mm, COV = 0.68
b: Mean = 0.89, SD = 0.063, COV = 0.07	b: Mean = 0.77, SD = 0.21, COV = 0.27
Correlation = −0.59	Correlation = −0.73
Driven piles in non-cohesive soils (compression)	Bored piles in non-cohesive soils (compression)
No. tests = 28	No. tests = 30
a: Mean = 5.55 mm, SD = 3.00 mm, COV = 0.54	a: Mean = 4.10 mm, SD = 3.20 mm, COV = 0.78
b: Mean = 0.71, SD = 0.10, COV = 0.14	b: Mean = 0.77, SD = 0.16, COV = 0.21
Correlation = −0.778	Correlation = −0.876
Driven piles in cohesive soils (compression)	Bored piles in non-cohesive soils (compression)
No. tests = 59	No. tests = 53
a: Mean = 3.58 mm, SD = 2.04 mm, COV = 0.57	a: Mean = 2.79 mm, SD = 2.04 mm, COV = 0.57
b: Mean = 0.78, SD = 0.09, COV = 0.11	b: Mean = 0.82, SD = 0.09, COV = 0.11
Correlation = −0.886	Correlation = −0.801

SD = standard deviation, COV = coefficient of variation.

For SLS the relationship of interest is that between allowable load (Q_a) and the resulting permissible settlement (y_a) given by:

$$Q_a = \frac{y_a}{a + by_a} Q_m \tag{5.17}$$

Let

$$\frac{y_a}{a + by_a} = M_s \tag{5.18}$$

Then,

$$Q_a = M_s Q_m \tag{5.19}$$

where M_s is the SLS model factor and the other symbols are as defined previously. The actual statistics for M_s are shown in columns 3 and 4 of Table 5.13. On the basis of first-order second moment analysis, the mean (μ_{M_s}) and COV (COV_{M_s}) of M_s can be estimated as follows (Phoon and Kulhway 2008):

$$\mu_{MS} = \frac{y_a}{\mu_a + \mu_b y_a} \tag{5.20}$$

$$COV_{MS} = \frac{\sqrt{\sigma_a^2 + y_a^2 \sigma_b^2 + 2 y_a \rho_{a,b} \sigma_a \sigma_b}}{\mu_a + \mu_b y_a} \tag{5.21}$$

where μ_a and μ_b = mean of a and b respectively, and σ_a and σ_b = standard deviation of a and b, respectively. Using Eq. 5.20 and 5.21 in conjunction with the hyperbolic parameter statistics (e.g., Table 5.14) as well as their correlations, M_s statistics can be computed for a given allowable settlement. For routine building structures with an allowable settlement of 25 mm, the estimated SLS model uncertainty statistics are presented in columns 7 and 8 of Table 5.13. The results show that the actual and estimated statistics are close, implying that Eqs. (5.20) and (5.21) are reasonable approximations. It is important to distinguish between M_s and M. The statistics for the former are meant for SLS and they are functions of the permissible settlement (y_a), while the statistics for the latter are meant for ULS.

Since Q_m is generally unavailable at the design stage then Eq. (5.19) needs to be modified as follows for reliability calibration:

$$Q_a = M_s(MQ_c) \tag{5.22}$$

The important point here is that uncertainties in ULS (manifested in M) must be included if Q_a is calculated from Q_c. Assuming that M_s and M are uncorrelated, then the combined statistics ($M_s M$) can be estimated using first-order second moment analysis as follows:

$$\mu_{M_s M} = \mu_s \mu_M \tag{5.23}$$

$$COV_{M_s M} = \sqrt{COV_{M_s}^2 + COV_M^2} \tag{5.24}$$

The ensuing combined statistics are presented Table 5.13 as follows: actual statistics in columns 5 & 6 and estimated statistics in columns 9 & 10. Even for the combined statistics, the estimated and actual values are quite close.

This approach is practical and grounded realistically on the load test database with minimal assumptions. The mean values in Tables 5.11 to 5.13 are different, because the calculation methods are different. This is to be expected. It is more interesting to observe that the COVs are comparable, despite the diverse variety of calculation methods. It is worth mentioning that Akbas and Kulhawy (2009b) have suggested a probabilistic approach to address *differential* settlement of footings on cohesionless soils.

5.6.5 Limiting tolerable displacement (serviceability limit state)

Another important serviceability consideration is limiting tolerable displacements of structures. The limiting tolerable displacements of a structure are affected by many

Table 5.15 Statistics of intolerable settlement and limiting tolerable settlement of buildings (Source: Table 3, Zhang & Ng 2007, with permission from ASCE).

	No. of cases	Observed intolerable settlement (mm)		Limiting tolerable settlement (mm)	
		Mean	Standard deviation	Mean	Standard deviation
Foundation type					
All	221	328	265	156	118
Shallow foundations	165	321	280	218	185
Deep foundations	52	349	218	106	55
Structural type					
All	185	296	220	134	109
Frame structures	115	278	236	148	126
With load-bearing wall	52	303	257	112	48
Soil type					
All	182	311	270	165	159
Clay	126	357	290	169	131
Sand and fill	56	207	151	86	56
Usage of building					
All	164	269	247	150	144
Mill structure	29	308	193	183	156
Office structure	135	255	265	121	64

factors, including the type and size of the structure, the intended usage of the structure, substructure-superstructure interactions, the properties of the structural materials and the subsurface soils, and the rate and uniformity of settlement (Zhang & Ng 2005, 2007). For a full reliability-based design for serviceability limit states, it is preferable to obtain the probability distributions of limiting tolerable displacement. Tables 5.15 and 5.16 provide statistics of intolerable settlement and angular distortion, and limiting tolerable settlement and angular distortion of buildings based on records of the displacements of 380 buildings. The intolerable or limiting tolerable displacements are shown to follow the lognormal distribution.

5.6.6 Factor of safety of a slope calculated by limit equilibrium method

Factor of safety (FS) is commonly used to quantify the safety level of a slope. The most popular method of determining the FS of a slope is the limit equilibrium method (LEM). Due to the uncertainty and variability involved in ground conditions and analytical methods, the calculated FS of a slope is not exact. In general, FS calculated by LEM depends on the way that the input soil strength is determined (e.g., unconfined compression test or vane shear test) and the calculation method (e.g., the Bishop simplified or Spencer method). The model factor for a FS is defined as the actual FS divided by the calculated FS.

Wu (2009) investigated a collection of undrained slope case histories analyzed by LEM with circular slip surface assumption (e.g., simplified Bishop). He demonstrated that if unconfined compression or vane shear test is the way of determining the input

Table 5.16 Statistics of intolerable and limiting tolerable angular distortion (Source: Table 4, Zhang & Ng 2007, with permission from ASCE).

Statistics	No. of cases	Observed intolerable angular distortion (radian)		Limiting tolerable angular distortion (radian)	
		Mean	Standard deviation	Mean	Standard deviation
Foundation type					
All	120	0.012	0.012	0.003	0.003
Shallow foundations	63	0.013	0.011	0.006	0.006
Deep foundations	57	0.008	0.011	0.002	0.002
Structural type					
All	191	0.012	0.014	0.004	0.005
Frame structures	152	0.011	0.015	0.005	0.004
With load-bearing wall	39	0.015	0.011	0.004	0.002
Soil type					
All	126	0.011	0.013	0.006	0.014
Clay	103	0.011	0.011	0.005	0.014
Sand and fill	23	–	–	–	–
Usage of building					
All	83	0.015	0.013	0.005	0.003
Mill structure	17	0.032	0.026	0.006	0.003
Office structure	66	0.013	0.013	0.003	0.002

undrained shear strength (s_u), the model factor for the resulting FS has a mean value roughly equal to 1.0 and COV ranging from 0.13 to 0.24.

Travis et al. (2011a) collected 301 FSs calculated by two-dimensional (2D) LEM for 157 failed slopes, and Travis et al. (2011b) further conducted statistical analysis for the database. They showed that the FSs for the failed slopes have sample mean shown in Table 5.17. The sample standard deviation for the ln(FS) is also shown in the table. They found that the type of LEM has the main effect on the mean and standard deviation. Therefore, Table 5.17 presents the FS statistics for four types of LEM: (a) direct method, including infinite slope, ordinary method of slice, Swedish circle, etc.; (b) Bishop method, the simplified Bishop method; (c) force method, including the Janbu and Lowe-Karafiath methods; (d) complete method, including the Spence, Morgenstern-Price, and Chen-Morgenstern methods. If one accepts the view that the actual FS for a failed slope is 1, the model factor is then simply $M = $ (actual FS)/(calculated FS) $= 1/FS$. Let us denote the mean of FS by μ and its COV by δ. If FS is lognormal, let us further denote the mean value of ln(FS) by λ and its variance by ξ^2:

$$\lambda = \ln(\mu) - 0.5 \times \xi^2 \qquad \xi^2 = \ln(1 + \delta^2) \tag{5.25}$$

The inverse relation is

$$\mu = \exp(\lambda + 0.5 \times \xi^2) \qquad \delta = [\exp(\xi^2) - 1]^{0.5} \tag{5.26}$$

Table 5.17 Statistics of the FS calculated by four types of LEM for failed slopes (revised from Travis et al. 2011b).

	Direct	Bishop	Force	Complete
Number of cases (n)	83	134	43	41
Mean of FS (μ)	0.98	1.04	1.10	1.05
Standard deviation of ln(FS) (ξ)	0.21	0.20	0.20	0.15
Mean of the model factor M [$=\exp(\xi^2)/\mu$]	1.07	1.00	0.95	0.97
COV of the model factor M [$=[\exp(\xi^2) - 1]^{0.5}$]	0.21	0.20	0.20	0.15

Table 5.18 Statistics of the FS calculated by two types of LEM for failed slopes (revised from Bahsan et al. 2014).

LEM method	Man-made slopes (n = 34)				Natural slopes (n = 9)	
	Fill slopes (n = 27)		Cut slopes (n = 7)			
	λ (mean of M)	ξ (COV of M)	λ (mean of M)	ξ (COV of M)	λ (mean of M)	ξ (COV of M)
Simplified Bishop	−0.068 (1.11)	0.28 (0.28)	0.158 (0.89)	0.28 (0.28)	0.001 (1.41)	0.83 (1.00)
Spencer	−0.137 (1.19)	0.27 (0.27)	0.140 (0.90)	0.26 (0.26)	−0.124 (1.57)	0.81 (0.96)

It is clear that the mean value of ln(1/FS) has mean $= -\lambda$ and variance $= \xi^2$. It follows from Eq. (5.26) that $M = 1/FS$ has mean value $= \exp(-\lambda + 0.5 \times \xi^2) = \exp(\xi^2)/\mu$ and COV $= [\exp(\xi^2) - 1]^{0.5}$. Table 5.17 shows that the mean and COV of M for various LEM methods. In general, the mean for the model factor M ranges from 0.95 to 1.07 and COV ranges from 0.15 to 0.21. This is consistent to the range summarized by Wu (2009).

Bahsan et al. (2014) collected 43 case histories of failed undrained slopes. They re-analyzed all cases using the simplified Bishop and Spencer methods. In the analysis, they transformed the input s_u value to the mobilized s_u defined by Mesri and Huvaj (2007). They modelled the vertical spatial variability for s_u by adopting thin horizontal clay layers in LEM and they also modelled the tension crack in LEM. Table 5.18 presents the statistics of ln(FS) for the simplified Bishop and Spence methods. The mean of ln(FS) is denoted by λ, and standard deviation is denoted by ξ. They found that the statistics for man-made slopes (fills and cuts) are quite different from those for natural slopes: the variability of FS calculated from failed natural slopes is very high (very large ξ). Again, if one accepts the view that the actual FS for a failed slope is 1, the model factor M is simply 1/FS. According to Eq. (5.26), the mean value of M is equal to $\exp(\lambda + 0.5 \times \xi^2)$ and the COV of M is equal to $[\exp(\xi^2) - 1]^{0.5}$. Table 5.18 shows the mean and COV for the model factor in the parenthesis. For man-made slopes, the mean value of M ranges from 0.89 to 1.19 and its COV ranges from 0.26 to 0.28. For natural slopes, the mean value ranges from 1.41 to 1.57 and its COV ranges from 0.96 to 1.00.

Table 5.19 Statistics of the model factor for base heave FS (revised from Wu et al. 2014).

	Modified Terzaghi	Bjerrum-Eide	Slip circle
Mean of M	1.02	1.09	1.27
COV of M	0.157	0.147	0.221

5.6.7 Base heave for excavation in clays

Wu et al. (2014) collected 24 case histories for excavation in clays. Among the 24 cases, 8 cases totally failed by base heave, 7 cases nearly failed, and 9 cases did not fail. Based on this database, they estimated the mean and COV of the model factors (M) for three well known methods for calculating base-heave FS for excavation in clays: (modified) Terzaghi method, Bjerrum-Eide method, and slip circle method. Here, the model factor (M) for FS is defined as the actual FS divided by the calculated FS. Table 5.19 summarizes the estimated mean and COV of the model factor.

5.7 CONCLUSIONS

The acceptance of reliability analysis and design in geotechnical practice calls for a concerted effort on characterization of calculation model uncertainty. Model uncertainty is generally represented in terms of the ratio of the measured to calculated values termed model factor (M), considered as a random variable following some probability distribution function. The model factor applies to a specific set of conditions (e.g., failure mode, calculation model, local conditions and experience base, etc.). Therefore, a proliferation of model factors can be expected.

With the current state of knowledge, the statistics of M are derived following well established statistical data analysis procedure comprising of (a) exploratory data analysis, (b) outlier detection and correction of anomalous values, (c) using the corrected data to compute the sample moments (mean, standard deviation, skewness, and kurtosis), (d) verification of the randomness of M, and (e) determining the appropriate probability distribution for M.

Reliability based design is based on the notion of randomness of the basic variables including the model factor. Therefore if M depicts some statistical dependency with deterministic variations in the database, such statistical dependencies need to be removed. Accordingly two approaches namely the "generalised model factor" and "model factor as a function of input parameters" have been presented in this Chapter.

Available model factor statistics are primarily restricted to simple calculation methods for foundations (both ultimate and serviceability limit state). A comprehensive survey of model statistics for foundations is conducted in this Chapter. Some model statistics on the factor of safety for slope stability and basal heave are also available. More research is needed to characterize the model uncertainties in other common geotechnical systems (such as retaining walls and ground improvement methods).

ACKNOWLEDGMENTS

The authors are grateful to the assistance of Dr. Chong Tang in preparing some of the figures.

REFERENCES

Allen, T.M., Nowak, A.S. & Bathurst, R.J. (2005) *Calibration to Determine Load and Resistance Factors for Geotechnical and Structural Design.* Washington, DC, Transport and Research Board.

Akbas, S.O. & Kulhawy, F.H. (2009a) Axial compression of footings in cohesionless soil. I: load-settlement behavior. *Journal of Geotechnical & Geoenvironmental Engineering*, ASCE, 135 (11), 1562–1574.

Akbas, S.O. & Kulhawy, F.H. (2009b) Reliability-based design approach for differential settlement of footings on cohesionless soils. *Journal of Geotechnical & Geoenvironmental Engineering*, ASCE, 135 (12), 1779–1788.

AASHTO (2007) *AASHTO LRFD Bridge Design Specifications*, 4th ed., AASHTO, Washington, DC.

Bahsan, E., Liao, H.J. & Ching, J. (2014) Statistics for the calculated safety factors of undrained failure slopes. *Engineering Geology*, 172, 85–94.

Carter, J.P. & Kulhawy, F.H. (1988) *Analysis and Design of Foundations Socketed into Rock.* Report No. EL-5918. New York, Empire State Electric Engineering Research Corporation and Electric Power Research Institute. p. 158.

Dithinde, M. (2007) *Characterisation of Model Uncertainty for Reliability Based Design of Pile Foundations.* PhD Dissertation submitted to the University of Stellenbosch. Available from: http://www.hdl.handle.net/10019.1/12612.

Dithinde, M. & Retief, J.V. (2013) Pile design practice in southern Africa I: Resistance statistics. *Journal of the South African Institution of Civil Engineering*, 55 (1), 60–71.

Dithinde, M., Phoon, K.K., De Wet, M. & Retief, J.V. (2011) Characterization of model uncertainty in the static pile design formula. *ASCE Journal of Geotechnical and Geoenvironmental Engineering*, 137 (1), 70–85.

EN 1990 (2002) *Eurocode: Basis of Structural Design.* Brussels, Committee for Standardization (CEN).

FHWA (2001) *Load and Resistance Factor Design (LRFD) for Highway Bridge Substructures.* Publication No. FHWA- HI-98-032.

Fleming, W.G.K., Weltman, A.J., Randolph, M.F. & Elson, W.K. (1992) *Pile Engineering.* New York, John Wiley & Sons.

Franzblau, A. (1958) *A Primer for Statistics for Non-Statistician.* New York, NY, Harcourt Brace & World.

Holický, M. (2009) *Reliability Analysis for Structural Design.* Stellenbosch, SUNMeDIA Press. ISBN: 978-1-920338-11-4.

Holický, M., Retief, J.V. & Sykora, M. (2015) Assessment of model uncertainty for structural resistance. *Probabilistic Engineering Mechanics Journal.* Available from: http://dx.doi.org/10.1016/j.probengmech.2015.09.008.

Huffman, J.C. & Stuedlein, A.W. (2014) Reliability-based serviceability limit state design of spread footings on aggregate pier reinforced clay, *Journal of Geotechnical and Geoenvironmental Engineering*, ASCE, 140 (10), 04014055

Huffman, J.C., Strahler, A.W. & Stuedlein, A.W. (2015) Reliability-based serviceability limit state design for immediate settlement of spread footings on clay. *Soils and Foundations* 55 (4), 798–812.

ISO 2394:2015. *General Principles of Reliability for Structures*. Geneva, International Organisation for Standardisation.

JCSS (2001) *Probabilistic Model Code*. The Joint Committee on Structural Safety. ISBN: 978-3-909386-79-6.

Kulhawy, F.H. & Carter, J.P. (1992) Socketed foundation in rock masses. In: Bell, F.H. (ed.) *Engineering in Rock Masses*. Oxford, Butterworth-Heinemann. pp. 509–529.

Mayne, P.W. & Harris, D.E. (1993) *Axial Load-Displacement Behavior of Drill Shaft Foundations in Piedmont Residuum*. FHWA Publication No. 41-30-3175. Atlanta, Georgia Institute of Technology.

McBean, E.A. & Rovers, F.A. (1998) *Statistical Procedures for Analysis of Environmental Monitory Data and Risk Assessment*. New Jersey, Prentice-Hall.

Mesri, G. & Huvaj, N. (2007) Shear strength mobilized in undrained failure of soft clay and silt deposits. In: DeGroot, D.J., Vipulanandan, C., Yamamuro, J.A., Kaliakin, V.N., Lade, P.V., Zeghal, M., El Shamy, U., Lu, N., Song, C.R. (eds.) *Proceedings of Advances in Measurement and Modeling of Soil Behavior (GSP 173)*. Denver, CO, ASCE. p. 1.

Osman, A.S. & Bolton, M.D. (2004) A new design method for retaining walls in clay. *Canadian Geotechnical Journal*, 41 (3), 451–466.

Paikowsky, S.G., Birgisson, B., McVay, M., Nguyen, T., Kuo, C., Baecher, G.B., Ayyub, B., Stenersen, K., O'Malley, K., Chernauskas, L. & O'Neill, M. (2004) *Load and Resistance Factors Design for Deep Foundations*. NCHRP Report 507. Washington, DC, Transportation Research Board of the National Academies.

Paikowsky, S.G., Canniff, M.C., Lesny, K., Kisse, A., Amatya, S. & Muganga, R. (2010) *LRFD Design and Construction of Shallow Foundations for Highway Bridge Structures*. NCHRP Report 651. Washington, DC, Transportation Research Board of the National Academies.

Phoon, K.K. & Kulhawy, F.H. (2005) Characterization of model uncertainties for laterally loaded rigid drilled shafts. *Geotechnique*, 55 (1), 45–54.

Phoon, K.K. & Kulhawy, F.H. (2008) Serviceability limit state reliability-based design. In: Phoon, K.K. (ed.) *Reliability-Based Design in Geotechnical Engineering: Computations and Applications*. London, Taylor & Francis. pp. 344–383.

Phoon, K.K. & Tang, C. (2015a) *Model Uncertainty for the Capacity of Strip Footings Under Positive Combined Loading*. Geotechnical Special Publication in honour of Wilson. H. Tang, ASCE, in press.

Phoon, K.K. & Tang, C. (2015b) Model uncertainty for the capacity of strip footings under negative and general combined loading. In: *12th International Conference on Applications of Statistics and Probability in Civil Engineering, ICASP 12, Vancouver, Canada, July 12–15*.

Phoon, K.K. & Tang, C. (2015c) Effect of load test database size on the characterization of model uncertainty. In: *Symposium on Reliability of Engineering Systems, Taipei, Taiwan*.

Phoon, K.K., Chen, J.-R. & Kulhawy, F.H. (2006) Characterization of model uncertainties for augered cast-in-place (ACIP) piles under axial compression. In: *Foundation Analysis & Design: Innovative Methods (GSP 153)*. Reston, ASCE. pp. 82–89.

Phoon, K.K., Chen, J.-R. & Kulhawy, F.H. (2007) Probabilistic hyperbolic models for foundation uplift movements. In: *Probabilistic Applications in Geotechnical Engineering (GSP 170)*. Reston, ASCE. CDROM.

Reese, L.C. & O'Neill, M.W. (1989) New design method for drilled shaft from common soil and rock tests. In: Kulhawy, F.H. (ed.) *Foundation Engineering: Current Principles and Practices*. Vol. 2. New York, ASCE. pp. 1026–1039.

Robinson, R.B., Cox, C.D. & Odom, K. (2005) Identifying outliers in correlated water quality data. *Journal of Environmental Engineering*, 131 (4), 651–657.

Studlein, A.W. & Reddy, S.C. (2013) Factors affecting the reliability of augered cast-in-place piles in granular soils at the serviceability limit state. *The Journal of the Deep Foundations Institute*, 7 (2), 46–57

Tang, C. & Phoon, K.K. (2016a) Model uncertainty of Eurocode 7 approach for the bearing capacity of circular footings on dense sand, *International Journal of Geomechannics*, ASCE, in press.

Tang, C. & Phoon, K.K. (2016b) Model uncertainty of cylindrical shear method for calculating the uplift capacity of helical anchors in clay, *Engineering Geology*, 207, 14–23.

Travis, Q.B., Schmeeckle, M.W. & Sebert, D.M. (2011a) Meta-analysis of 301 slope failure calculations. II: Database analysis. *ASCE Journal of Geotechnical and Geoenvironmental Engineering*, 137 (5), 471–482.

Travis, Q.B., Schmeeckle, M.W. & Sebert, D.M. (2011b) Meta-analysis of 301 slope failure calculations. I: Database description. *ASCE Journal of Geotechnical and Geoenvironmental Engineering*, 137 (5), 453–470.

Vesic, A.S. (1977) *Design of Pile Foundations*. National Cooperative Highway Research Program Synthesis of Practice No. 42. Washington, DC, Transportation Research Board.

Wu, T.H. (2009) Reliability of geotechnical predictions. In: *Geotechnical Risk and Safety, Proceedings of the 2nd International Symposium on Geotechnical Safety and Risk*. Gifu, Japan, CRC Press, Taylor & Francis Group. pp. 3–10.

Wu, S.H., Ou, C.Y. & Ching, J. (2014) Calibration of model uncertainties for basal heave stability of wide excavations in clay. *Soils and Foundations*, 54, 1159–1174.

Zhang, L.M. & Ng, A.M.Y. (2005) Probabilistic limiting tolerable displacements for serviceability limit state design of foundations. *Geotechnique*, 55 (2), 151–161.

Zhang, L.M. & Ng, A.M.Y. (2007) Limiting tolerable settlement and angular distortion for building foundations. Geotechnical Special Publication No. 170. In: Phoon, K.K., Fenton, G.A., Glynn, E.F., Juang, C.H., Griffiths, D.V., Wolff, T.F. & Zhang, L.M. (eds.) *Probabilistic Applications in Geotechnical Engineering*. Reston, ASCE. Available in CD ROM.

Zhang, L.M. & Chu, L.F. (2009a) Calibration of methods for designing large-diameter bored piles: Ultimate limit state. *Soils and Foundations*, 49 (6), 883–896.

Zhang, L.M. & Chu, L.F. (2009b) Calibration of methods for designing large-diameter bored piles: Serviceability limit state. *Soils and Foundations*, 49 (6), 897–908.

Zhang, L.M., Xu, Y. & Tang, W.H. (2008) Calibration of models for pile settlement analysis using 64 field load tests. *Canadian Geotechnical Journal*, 45 (1), 59–73.

Zhang, D.M., Phoon, K.K., Huang, H.W. & Hu, Q.F. (2015) Characterization of model uncertainty for cantilever deflections in undrained clay. *ASCE Journal of Geotechnical and Geoenvironmental Engineering*, 141 (1), 04014088.

Chapter 6

Semi-probabilistic reliability-based design

Kok-Kwang Phoon and Jianye Ching

ABSTRACT

Geotechnical design codes, be it reliability-based or otherwise, must cater to diverse local site conditions and diverse local practices that grew and adapted over the years to suit these conditions. One obvious example is that the COVs of geotechnical parameters can vary over a wide range, because diverse property evaluation methodologies exist to cater to these diverse practice and site conditions. Another example is that deep foundations are typically installed in layered soil profiles that vary from site to site. These diverse design settings do not surface in structural engineering. If the performance of geotechnical RBD were to be measured by its ability to achieve a more uniform level of reliability than that implied in existing allowable stress design over these diverse settings (which is recommended in Section D.5, ISO2394:2015), then LRFD and comparable simplified RBD formats widely used in structural design codes are not adequate. While it is understandable for geotechnical RBD to adopt structural LRFD concepts at its initial stage of development over the past decades, it is timely for the geotechnical design code community to look into how we can improve our state of practice in simplified geotechnical RBD. This chapter demonstrates that improved formats such as the Quantile Value Method coupled with effective random dimension (ERD-QVM) exist that can cater to a more realistic and diverse range of design scenarios. Specifically, ERD-QVM can maintain an acceptably uniform level of reliability over a wide range of COVs of geotechnical parameters and a wide range of layered soil profiles. It can achieve this while retaining the simplicity of an algebraic design check similar to the traditional factor of safety format and LRFD. ERD-QVM is a step in the right direction to develop geotechnical RBD for geotechnical engineers. More research is urgently needed for geotechnical RBD to gain wider acceptance among practitioners.

6.1 INTRODUCTION

ISO2394:2015 contains a new informative Annex D on "Reliability of Geotechnical Structures". The need to achieve consistency between geotechnical and structural reliability-based design is explicitly recognized for the first time in ISO2394:2015 with the inclusion of Annex D. The emphasis in Annex D is to identify and characterize critical elements of the geotechnical reliability-based design (RBD) process, while respecting the diversity of geotechnical engineering practice. These elements are applicable to any implementations of RBD, be it in a simplified format such as the Partial Factor Approach (PFA), the Load and Resistance Factor Design (LRFD), the Multiple Resistance and Load Factor Design (MRFD) (Phoon et al. 2003a), the Robust LRFD (R-LRFD) (Gong et al. 2016), and the Quantile Value Method (QVM) (Ching and Phoon 2011), or in a full probabilistic form such as the expanded RBD approach (Wang et al. 2011).

Clause 4.4.1 of ISO2394:2015 states that RBD can be applied in place of full risk assessments "when the consequences of failure and damage are well understood and within normal ranges". The objective of RBD is to adjust a set of design parameters such that a prescribed target probability of failure is achieved or at least not exceeded. For example, the depth of a bored pile is a practical design parameter that can be adjusted readily. In principle, it is possible to adjust the shaft diameter but it is less practical to constantly change the diameter of a rotary auger within a single site. This constructability consideration applies to the current allowable stress design (ASD) method. The trial-and-error adjustment of a design parameter such as the depth of a bored pile is common to RBD and ASD. The only difference is the design objective. The former considers a design to be satisfactory if a target probability of failure, say one in a thousand, is achieved. The latter considers a design to be satisfactory if a target global factor of safety, say three, is achieved. The advantages of using the probability of failure (or the reliability index) in place of the global factor of safety have been discussed elsewhere (Phoon et al. 2003b, 2003c).

Clause 4.4.1 of ISO2394:2015 also states that RBD can be further simplified "when in addition to the consequences also the failure modes and the uncertainty representation can be categorized and standardized". This simplified RBD approach is referred to as a semi-probabilistic approach. Simplified RBD formats in the form of PFA, LRFD, and MRFD are popular because practitioners can produce designs complying with the target probability of failure (or target reliability index), albeit approximately, while retaining the simplicity of performing one *algebraic check* per trial design. No tedious Monte Carlo simulations or more sophisticated probabilistic analyses are needed. From the perspective of a practitioner, there is no difference between applying a simplified RBD format, say LRFD, and the prevailing factor of safety format, other than multiplying a set of resistance and load factors to the corresponding resistance and load components (nominal or characteristic values) mandated in such codes. The key difference is that the numerical values of these resistance and load factors are not determined purely on experience or precedents, but calibrated by the code developer using reliability analysis to achieve a desired target reliability index. Once these resistance and load factors are made available in a design code, the practitioner can use them for design without having to perform reliability analysis. To the authors' knowledge, this simplified RBD approach is adopted in all geotechnical RBD codes to date. The obvious limitation associated with replacing reliability analysis with an algebraic design check is that the target reliability index cannot be achieved exactly. We note in passing that it is possible to achieve a desired factor of safety exactly for any design scenario, but the factor of safety concept is known to be inconsistent in many ways.

It is possible to achieve the target reliability index exactly under any design scenario if the full probabilistic approach is adopted. Hence, it is easy to achieve a completely uniform level of reliability under this approach. This is discussed in Chapter 7.

However, the prevalence of simplified RBD formats in all existing design codes implies that practitioners are not prepared to accept full probabilistic analyses at this point in time. For simplified RBD formats that are easy to use, Section D.5 "Implementation issues in geotechnical RBD" clarifies that the "key goal in geotechnical RBD is to achieve a more uniform level of reliability than that implied in existing allowable stress design". Section D.5 further highlights that reliability calibration of these formats are challenging in geotechnical engineering. There are many reasons why simplified geotechnical RBD formats are harder to calibrate than those in structural engineering. One reason is that these formats must cover a wide range of coefficients of variation (COVs) resulting from different soil property evaluation methodologies (Phoon 2015). Another source of challenge is that a simplified geotechnical RBD should be flexible enough to cover a range of soil profiles encountered in locales within the ambit of the design code.

The performance of a simplified RBD format should be measured by its ability to produce designs achieving a desired target reliability index within an acceptable error margin. When a simplified RBD format is first introduced into a design code, it should preferably produce designs comparable to those produced by the factor of safety method for continuity with past practice and experience. In fact, the target reliability index is commonly prescribed to comply with this judicious continuity principle. However, the primary goal must be to maintain a uniform level of reliability – this is the key basis for switching to RBD in the first place. For a simplified RBD format, the ability to maintain a uniform level of reliability is primarily related to the range of design scenarios covered by the code and the number of available factors that can be "tuned" during the reliability calibration process.

The authors recommend that a simplified RBD format should reveal the maximum deviation from the target reliability index among the range of design scenarios appearing in the calibration domain. In principle, application of a simplified RBD format to a design scenario lying outside the calibration domain can produce a reliability index far from the target value. Hence, it is important to state the salient features of the underlying calibration domain (e.g., range of pile diameters, pile lengths, statistics of geotechnical parameters, etc.) explicitly in association with any simplified RBD format to avoid conveying the impression that it can be applied to any design scenario, which is unlikely to be true. As highlighted above, one noteworthy feature of this calibration domain that is distinctive to geotechnical engineering is that COVs of geotechnical parameters can vary over a wide range, because of diverse property evaluation methodologies to cater for diverse practice and site conditions (refer to Section D.1, ISO2394:2015). It is easy to envisage that a single resistance or partial factor is unable to achieve a uniform reliability index if the range of COVs is sufficiently large, say between 10% and 70% for undrained shear strength estimated using different methods as shown in Table 3.7. Phoon and Ching (2013) demonstrated that a uniform reliability index is even more difficult to achieve in the presence of layered profiles.

LRFD was originally proposed for structural steel design (Ravindra and Galambos 1978). Geotechnical structures, particularly piles, have been mostly treated as a component comparable to columns in structural RBD codes. It is inappropriate to treat a pile in the same way as a structural column in the context of LRFD. The resistance provided by a column depends on the quality of concrete and steel. These structural materials are manufactured and their properties are assured by quality control. The ultimate strength of cast in-situ concrete may be impacted by the curing temperature, which in turn is affected by the ambient temperature. However, this site-specific issue can be mitigated by a variety of measures and it is significantly less influential than say a pile being installed in a stratified ground that changes from location to location. The variability of the ground also changes from location to location. The method of estimating design soil properties and other aspects of local practice may also change from location to location. In contrast, the testing methods for structural materials are highly standardized. In spite of these evident differences, geotechnical RBD continues to adopt the structural LRFD framework that does not explicitly consider site-specific effects in the evaluation of the factored resistance. To the authors' knowledge, geotechnical RBD was studied as a topic separate from structural RBD only in the nineties or thereabouts (e.g., Barker et al. 1991; Phoon et al. 1995). In the same vein, geotechnical aspects are explicitly considered only in ISO2394:2015, although the first edition of ISO2394 was published in 1973. It is understandable for geotechnical RBD to adopt structural LRFD concepts at its initial stage of development over the past decades. However, sufficient studies have been carried out to demonstrate that the traditional LRFD and PFA do not meet the needs of geotechnical engineering practice. The authors believe that it is timely for the geotechnical design code community to look into how we can improve our state of practice in simplified geotechnical RBD.

The key challenge in geotechnical RBD can be stated as follows. It is to calibrate a set of resistance factors, soil partial factors, or other factors such as quantiles that would produce designs satisfying the target reliability index approximately over a realistic range of design scenarios within the ambit of the design code, which must include diverse local site conditions and diverse local practices that grew and adapted over the years to suit these conditions. The factor of safety can be viewed as the reciprocal of a resistance factor. For example, a resistance factor $= 0.5$ is equivalent to a factor of safety $= 2$ if load factors are equal to 1. Based on this observation, it is clear that applying a resistance factor $= 0.5$ will produce a range of reliability indices as wide as applying a factor of safety $= 2$! Can we do better than this? Table B2 in EN1990:2002 recommends the minimum reliability index for three reliability classes (RCs) for the ultimate limit state. For RC1, RC2, and RC3, the minimum reliability indices for a 50 year reference period are 3.3, 3.8, and 4.3, respectively. In this case, one would require the simplified RBD format to maintain reliability indices of the designs to within a band of ± 0.5 or smaller. It may be argued that a single resistance factor value of say 0.5 is still applicable if one allows the engineer to adjust the nominal/characteristic resistance judiciously to accommodate site-specific conditions. This approach is similar to the existing allowable stress design approach where a relatively constant factor of safety is applied to a nominal resistance that should be suitably chosen by the engineer. However, from a RBD perspective, it is more difficult to calibrate a nominal resistance than a dimensionless resistance factor to maintain a relatively uniform level of reliability, unless the nominal resistance is well defined in the probabilistic sense. The QVM presented in Section 6.5 essentially exploits this possibility. We note in passing that geotechnical RBD also needs to address the limited availability of data (soil data, load test data, field monitoring data, etc.) in geotechnical practice. This challenge is discussed in Chapter 3. To the authors' knowledge, statistical uncertainty arising from small sample size has not been systematically incorporated in simplified RBD, say by applying a reduction factor to the resistance factor in LRFD or to the quantile in QVM.

This chapter reviews some common reliability-based calibration methods and their limitations. Two important limitations are illustrated numerically using a friction pile installed in a single layer soil profile and a two-layer soil profile. It is shown that the QVM (Ching and Phoon 2011) can maintain a relatively uniform level of reliability over a wide range of COVs for the unit side resistance in a single layer soil profile. However, QVM cannot maintain a uniform of reliability if it has to cater to sites containing different number of soil layers. This limitation is important for a design code covering deep foundations that are installed in layered soil profiles. The number of layers typically is not a constant given the geographical coverage of a design code and varying foundation lengths to carry different loads. This is an example of a site-specific condition commonly encountered in geotechnical practice. Ching et al. (2015) introduced the concept of an "effective random dimension" (ERD) into QVM to extend its reach to layered soil profiles (ERD-QVM). An additional gravity retaining wall is presented to show that the concept of ERD is related to the issue of redundancy and hence simplified RBD formats that do not cover layered soil profiles must still contend with variable ERD or variable redundancy during code calibration.

6.2 SURVEY OF CALIBRATION METHODS

6.2.1 Basic Load Resistance Factor Design (LRFD)

In geotechnical engineering, the most popular simplified RBD format in North America is the Load and Resistance Factor Design (LRFD) format (e.g., Paikowsky, 2004; Paikowsky et al. 2010). In its simplest form, the LRFD equation is:

$$\eta F_n \leq \psi Q_n \tag{6.1}$$

where η is the load factor (≥ 1) and ψ is the resistance factor (≤ 1), and therefore the name "load-and-resistance-factor-design" (LRFD). The parameters, F_n and Q_n, are the nominal load and nominal capacity (or resistance), respectively. This LRFD format is typically calibrated by assuming that the actual capacity (Q) can be modeled as the product of a bias factor (b_Q) and a nominal capacity typically obtained from a calculation procedure (Q_n):

$$Q = b_Q Q_n \tag{6.2}$$

The bias factor is similar to the model factor discussed in Chapter 5. The difference is that the bias factor is based on a nominal capacity that can be a conservative estimate mandated by a design code while the model factor is preferably based on a calculated capacity that is a best estimate. The bias factor is considered as a lognormal random variable in the reliability calibration process. Hence, it follows from Eq. (6.2) that Q is a lognormal random variable. The random load F can be related to the nominal load

(F_n) in the same way. The reason for doing this is to produce an analytical expression for the resistance factor:

$$\psi = \frac{\eta\,(F_n/\mu_F)\,\sqrt{\left(1 + V_F^2\right)\big/\left(1 + V_Q^2\right)}}{(Q_n/\mu_Q)\exp\left\{\beta_T\sqrt{\ln\left[\left(1 + V_F^2\right)\left(1 + V_Q^2\right)\right]}\right\}} \tag{6.3}$$

where μ_Q and μ_F are the mean value of Q and F, respectively; V_Q and V_F are the coefficient of variation (COV) of Q and F, respectively; and β_T = target reliability index.

The statistics of the bias factor is estimated from a load test database. It is evident that the bias factor is essentially a "lumped" factor capturing both systematic bias arising from the calculation model and random effects arising from parametric/model uncertainties. By virtue of its lumped nature, the statistics of the bias factor are theoretically a function of the design parameters (e.g., geometrical and soil parameters). In the most ideal case, the statistics of the bias factor are completely insensitive to the design parameters, i.e. they can be applied to all possible problem geometries, geologic formations, and soil properties. In this ideal case, statistics estimated from a load test database are robust and can be applied quite confidently to the full range of design scenarios encountered in practice. In the worst case, the statistics are very sensitive to one or more design parameters. For example, statistics for short piles may not be the same as statistics for long piles. This can arise because of physical reasons (e.g., side resistance dominates total resistance in long piles) or statistical reasons (e.g., spatial averaging of soil strength is more significant in long piles). In this case, it is debatable if the statistics derived from a load test database can be applied confidently to problems not covered by the database. Kulhawy and Phoon (2002) have highlighted this potential problem. Paikowsky (2002) has provided actual statistics to demonstrate that the statistics of bias factors are generally dependent on some design parameters. The statistics of the model factor can be afflicted by the same dependency issue and care must be taken to ensure that bias or model factor statistics are sufficiently general over the ambit of the design code. Details are given in Section 5.5.1.3.

If a particular design parameter, say length to diameter ratio of a pile (L/B), is influential, it is important to divide the range of the parameter into two or more segments and estimate different statistics over different segments. For example, it is possible to divide the range of L/B into two segments, say less than 10 for short piles and greater than 10 for long piles. The number of segments is clearly dependent on the sensitivity of the statistics to that particular design parameter, which is problem dependent. This segmentation procedure is a reasonable and practical solution to the above dependency problem. However, there is a more subtle but rarely appreciated problem in estimating sensitive statistics from a load test database. The problem is that if a statistics is sensitive, it would be important to ensure that the calibration examples are fairly uniformly distributed over any one segment of the parameter range. This is very difficult to do in a load test database as examples are usually collected from the literature rather than a single comprehensive research program. For example, for L/B > 10, it could be that L/B is predominantly between 30 and 50 in a particular database.

6.2.2 Extended LRFD and Multiple Resistance and Load Factor Design (MRFD)

Because of the various limitations in the basic LRFD approach, Phoon et al. (1995) proposed alternatives to this methodology that focused on realistic treatment of basic geotechnical issues than conformance with established LRFD procedures that evolved from structural engineering practice. The proposed approach simply allowed the "best" geotechnical calculation models to be used directly in the reliability calibration process, rather than artificially simplifying the capacity into a single lognormal random variable to fit the requirement of a closed-form reliability formula. The available calculation models were examined and compared with available load test data from the field and laboratory, as well as numerical simulations. These analyses led to a "best" calculation model (most accurate, least variability, essentially no bias) that was used in the reliability calibrations.

As part of this process, the key design parameters, such as the effective stress friction angle, the undrained shear strength, and the coefficient of horizontal soil stress, are modeled directly as random variables. A major advantage of this approach is that the range of an influential design parameter, and its variability, can be segmented and calibration points within each segment or domain can be selected to ensure uniform coverage of the variables during the calibration process (Fig. 6.1). The disadvantages are: (1) the closed-form reliability formula for lognormal random variables cannot be applied, and a more involved reliability calculation method such as the First-Order Reliability Method (FORM) is needed, and (2) it is necessary to adjust the resistance factor over each segment by an optimization procedure to minimize the deviation from the target reliability at each calibration point. However, once the calibration process

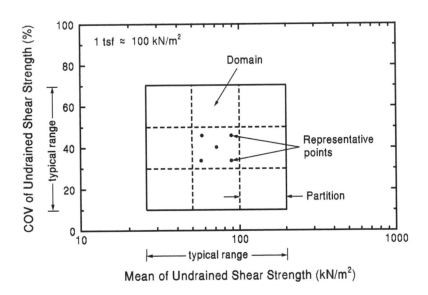

Figure 6.1 Partitioning of parameter space for calibration of resistance factors (Source: Figure D.3, ISO2394:2015. Reproduced with permission from the International Organization for Standardization (ISO). All rights reserved by ISO).

Table 6.1 Undrained ultimate uplift resistance factors for drilled shafts designed by $F_{50} = \psi_u Q_{un}$ or $F_{50} = \psi_{su} Q_{sun} + \psi_{tu} Q_{tun} + \psi_w W$ (Phoon et al. 1995).

Clay	COV of s_u (%)	ψ_u	ψ_{su}	ψ_{tu}	ψ_w
Medium	10–30	0.44	0.44	0.28	0.50
(mean $s_u = 25$–$50\,kN/m^2$)	30–50	0.43	0.41	0.31	0.52
	50–70	0.42	0.38	0.33	0.53
Stiff	10–30	0.43	0.40	0.35	0.56
(mean $s_u = 50$–$100\,kN/m^2$)	30–50	0.41	0.36	0.37	0.59
	50–70	0.39	0.32	0.40	0.62
Very Stiff	10–30	0.40	0.35	0.42	0.66
(mean $s_u = 100$–$200\,kN/m^2$)	30–50	0.37	0.31	0.48	0.68
	50–70	0.34	0.26	0.51	0.72

Note: Target reliability index $= 3.2$; $F_{50} = 50$-year return period load; $Q_{un} =$ nominal uplift capacity; $Q_{sun} =$ nominal uplift side resistance; $Q_{tun} =$ nominal uplift tip resistance; $W =$ weight of foundation.

is complete, the user never has to perform any reliability calculations or do any factor optimization, as described below.

Based on extensive studies of the optimization process, and detailed evaluation of typical ranges in the key design parameters and the COV ranges that should be encountered for them, it was found that a 3×3 segmenting was sufficient for practical calibration (Phoon et al. 1995). The typical ranges of COV to be expected are shown in Table 3.7. As can be seen, low variability corresponds to good quality direct lab or field measurements, medium is typical of most indirect correlations, while high represents strictly empirical correlations.

Once the calibration is complete, Eq. (6.1) is used directly with resistance factors such as those values shown under ψ_u in Table 6.1. Use of this table is straightforward with an appropriate site investigation. First, the mean value of the key design parameter is determined. For the undrained uplift example shown in Table 6.1, the key design parameter is the undrained shear strength (s_u). It is sufficient to determine the clay type (medium, stiff, or very stiff clay) based on the mean value of s_u. Second, the COV is determined either by direct measurements or from experience using Table 3.7. Note that it is not necessary to pin-point the exact COV value. It is sufficient to determine which variability tier (low, medium, high) or COV range is appropriate for the information at hand. Finally, the resistance factor corresponding to the clay type and variability tier is selected from Table 6.1. For example, for stiff clay and high variability, $\psi_u = 0.39$. If variability is reduced to "low" as a result of conducting more tests, the engineer can adopt a larger $\psi_u = 0.43$ in design. Table 6.1 represents the first attempt to link site investigation efforts to design explicitly. Although better methods have been developed in recent years to quantify the value of site investigation for design, it is important to view Table 6.1 as the minimum standard in geotechnical reliability-based design. The basic LRFD approach that prescribes one resistance factor value without reference to site investigation efforts is not appropriate for geotechnical engineering. The desirability of adopting LRFD purely for the purpose of harmonizing geotechnical and structural design at the design *format* level is debatable. Geotechnical and structural design should be harmonized by adopted a common basis in the form of reliability principles.

As noted previously, in LRFD (basic or extended), the ψ value is applied to the total geotechnical capacity, which is composed of distinctly different components. For example, the uplift capacity of a drilled shaft during undrained loading is composed of the side resistance, tip suction, and self-weight. These components, in turn, generally are nonlinear functions of more fundamental design parameters, such as the foundation depth, diameter, and weight and the soil undrained shear strength. The relative contribution of each component to the overall capacity is not constant, and the degrees of uncertainty associated with each component are different. For example, the shaft weight is almost deterministic in comparison to the undrained side resistance, because the COV of the unit weight of concrete is significantly smaller than the COV of the undrained shear strength.

This issue has been examined in detail (e.g., Phoon et al. 1995, 2003a), and it was found that a more consistent result in achieving a constant target reliability index was obtained using the following equation:

$$\eta F_n \le \psi_{su} Q_{sun} + \psi_{tu} Q_{tun} + \psi_w W \tag{6.4}$$

in which the ψ values are calibrated for each distinct term in the geotechnical capacity equation, as given in columns 4 through 6 of Table 6.1 for illustration, Q_{sun} is the nominal uplift side resistance, Q_{tun} is nominal uplift tip resistance, and W is the weight of the foundation. Eq. (6.4) is defined as the "multiple-resistance-and-load-factor-design" (MRFD) for a foundation in undrained uplift.

The LRFD and MRFD formats were compared by evaluating how closely the actual achieved values of β were to the target value β_T. For the extended LRFD format, $(\beta - \beta_T)/\beta_T$ was about plus or minus 5 to 10%. However, for the MRFD format, the $(\beta - \beta_T)/\beta_T$ values were only about 1/2 to 2/3 those from the extended LRFD format, indicating significant improvement over the LRFD format when each distinct term is assigned a resistance factor. For the basic LRFD, only one value is prescribed for the resistance factor. It is not specified or known whether the calibration was done for low, medium, or high variability. Assuming that calibration was done for a medium variability scenario, β would be greater than β_T for a low variability scenario and would be less than β_T for a high variability scenario, by amounts exceeding 10% at the limits.

From these results, it should be clear that the MRFD format is preferred over LRFD. Not only is the β_T more closely achieved, it is being done using proper geotechnical design equations where the relative weighting of each term is being addressed explicitly. And there is direct recognition of data quality in assessing the property variability. With these tools, an experienced engineer should have no trouble in selecting the appropriate ψ values for an improved design that is related to site investigation efforts.

6.2.3 Robust LRFD (R-LRFD)

A fundamental challenge in RBD is to address uncertainty rationally in analysis and design in the presence of limited data. A well-known problem is that the probability models constructed to handle uncertainty are themselves uncertain because data are insufficient to characterize the statistics and distributions exactly. Bayesian methods have been applied to address this statistical uncertainty (see Chapter 3), but Juang et al. (2013) took a different track to address this challenge by proposing a new design philosophy called robust geotechnical design (RGD). The essence of robust design is to derive a design that accounts for the effect of the variation in "noise factors" while simultaneously considers the safety and cost efficiency. The term noise factor refers to input parameters that are hard-to-control (i.e., cannot be easily adjusted by the designer) and hard-to-characterize (i.e., the uncertainty is recognized but difficult to quantify due to insufficient data). The original RGD approach retains the reliability index as a measure of safety (Juang et al. 2013; Juang and Wang 2013), and the robustness of a design is measured in terms of the variation of the failure probability (which is caused by the uncertainty in the probability distributions of noise factors). For practical applications and being consistent with the design code of LRFD, the reliability-based RGD has since been simplified to a format similar to LRFD called the Robust LRFD (R-LRFD) (Gong et al. 2016). The theoretical basis of R-LRFD is not probabilistic (because it assumes there are insufficient data for characterization of probability distributions) and safety is not defined by the reliability index.

It is noted that the traditional robustness measures that are based on probabilistic analyses, such as the variation of failure probability (Juang et al. 2013), signal-to-noise-ratio (SNR) (Taguchi 1986; Phadke 1989; Park et al. 2006; Gong et al. 2014), and feasibility robustness (Juang et al. 2013; Juang and Wang 2013), are suitable for a design with quantified uncertain parameters. They are, however, not applicable to the design with unquantified uncertain parameters. Thus, the robustness of a design (d) against the variation of noise factors (θ), in the context of R-LRFD, is measured with the concept of gradient.

The plots in Fig. 6.2(a) and 6.2(c) depict the gradient-based robustness concept: two designs (i.e., d_1 and d_2) with the same noise factors (θ_1) exhibit different patterns of system response; one (i.e., d_1 in Fig. 6.2a) yields high variation in the system response and the other (i.e., d_2 in Fig. 6.2c) yields low variation in the system response.

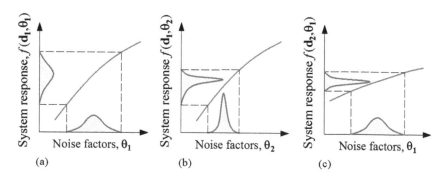

Figure 6.2 Conceptual illustration of the robustness of a design: (a) System response of a non-robust design d_1 with noise factors as its input; (b) Reducing the variation in the system response by reducing the variation in noise factors; (c) Reducing the variation in the system response by adopting a robust design d_2 without reducing the variation in noise factors (Gong et al. 2016).

The design that yields lower variation in the system response, which is by definition a more robust design, is shown to have a lower gradient. Thus, a robust design can be obtained by lowering the gradient of the system response to the noise factors without quantifying the uncertainties of noise factors. This characteristic of the gradient-based robustness measure is a perfect match with LRFD, in which the uncertainties in noise factors and the solution model are recognized but unquantified.

Within the framework of R-LFRD, the robustness of the resulting design (derived with fixed LRFD partial factors) against the unquantified uncertainty (due to diverse local site conditions and local practice) can be secured through a simplified multi-objective optimization scheme. In this simplified optimization scheme, safety that is evaluated with the traditional LRFD criterion is a compulsory constraint, while robustness and cost efficiency are the objectives to be optimized. Details of the R-LRFD methodology and application examples could be found in Gong et al. (2016).

6.2.4 LRFD for total settlement

The serviceability limit state (SLS) is the second limit state that is evaluated in foundation design. It often is the governing design criterion, particularly for large-diameter shafts and shallow foundations. Unfortunately, foundation movements are difficult to predict accurately, so reliability-based assessments of the SLS are not common. Ideally, the ultimate limit state (ULS) and the SLS should be checked using the same reliability-based design principle. However, the magnitude of uncertainties and the target reliability level for SLS are different from those of ULS, but these differences can be assessed consistently using reliability-calibrated deformation factors (analog of resistance factors).

Phoon et al. (1995) first examined this issue by employing large databases of foundation load-displacement data that could be normalized and evaluated. It was found that most databases could be best characterized by a two-parameter hyperbolic model, as illustrated in Fig. 6.3 for drilled shafts in uplift loading and as given below:

$$F/Q_u = y/(a + by) \qquad (6.5)$$

in which $F = $ load, $Q_u = $ uplift capacity, $y = $ displacement, and a and b are the curve-fitting parameters.

Recently, Phoon and Kulhawy (2008) summarized developments in SLS and noted that this model was most appropriate for the following foundation types: spread foundations in uplift (drained and undrained), drilled shafts in uplift and lateral-moment (drained and undrained), drilled shafts in compression (undrained), augered cast-in-place (ACIP) piles in compression (drained), and pressure-injected footings in uplift (drained). Drilled shafts in compression (drained) were fitted best by an exponential model. More recently, Akbas and Kulhawy (2009a) also showed that the hyperbolic model was appropriate for spread foundations in compression (drained).

The reliability of a foundation at the ULS is given by the probability of the capacity being less than the applied load. It is logical to follow the same approach for the SLS, where the capacity is replaced by an allowable capacity that depends on the allowable displacement (Phoon et al. 1995; Phoon and Kulhawy 2008). The nonlinearity of the load-displacement curve is captured by the two-parameter hyperbolic curve-fitting

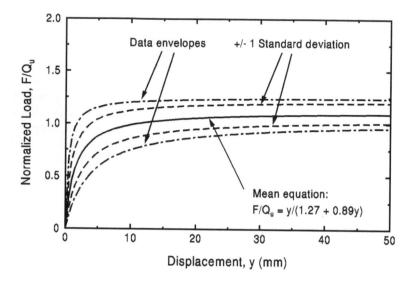

Figure 6.3 Load-displacement curves for drilled shafts in uplift (Phoon et al. 1995).

Table 6.2 Undrained uplift deformation factors for drilled shafts designed using $F_{50} = \psi_u Q_{uan}$ (Phoon et al. 1995).

Clay	COV of s_u (%)	ψ_u
Medium	10–30	0.65
(mean $s_u = 25$–$50\,kN/m^2$)	30–50	0.63
	50–70	0.62
Stiff	10–30	0.64
(mean $s_u = 50$–$100\,kN/m^2$)	30–50	0.61
	50–70	0.58
Very Stiff	10–30	0.61
(mean $s_u = 100$–$200\,kN/m^2$)	30–50	0.57
	50–70	0.52

Note: Target reliability index = 2.6.

equation. The uncertainty in the entire load-displacement curve is represented by a relatively simple bivariate random vector containing the hyperbolic parameters as its components, and the allowable displacement is introduced as a random variable for reliability analysis. The resulting LRFD equation is given below:

$$F_n = \psi_u Q_{uan} = \psi_u[Q_{un}y_a/(\mu_a + \mu_b y_a)] \tag{6.6}$$

where ψ_u is the uplift deformation factor given in Table 6.2, Q_{uan} is the nominal allowable uplift capacity, Q_{un} is the nominal uplift capacity, y_a is the allowable displacement, and μ_a and μ_b are the mean values of a and b, respectively. Note that the deformation factors are calibrated for a smaller β_T (or larger probability of failure) than for the ULS. Note that in reliability parlance, "failure" refers to any unsatisfactory performance such as foundation displacement exceeding an allowable value.

6.2.5 LRFD for differential settlement

In foundation design, the SLS for individual foundations is important, and it can be addressed as above. As long as the ground conditions are reasonably consistent, the differential settlements are likely to be minimal. However, for certain types of soil-foundation systems, such as spread footings on granular soils, the question of differential settlement can be very important. Conventional practices are empirical and commonly assume that the differential settlement is just some fixed percentage of the total computed settlement, typically ranging from 50 to 100%.

Akbas and Kulhawy (2009b) suggested a probabilistic approach to this problem to provide a more rational method of assessment. For illustration, they used the Burland and Burbridge (1985) settlement estimation method in the following form for estimating differential settlements:

$$\rho_{m1} - \rho_{m2} = (1/M)\zeta_s\zeta_i qB^{0.7}(I_{c1} - I_{c2}) \tag{6.7}$$

where ρ_{m1} and ρ_{m2} is the measured settlements for neighboring footings 1 and 2, M is the model factor (ratio of calculated-to-measured settlement), ζ_s is the shape factor, ζ_i is the depth of influence correction factor, q is the net increase in effective stress at foundation level, B is the footing width, and I_{c1} and I_{c2} are the compressibility index values for neighboring footings 1 and 2 ($=1.71/N_{60}^{1.4}$), with $N_{60} =$ standard penetration test N value corrected to an average energy ratio of 60%. The stress, model uncertainty, and geotechnical parameters were treated as random variables, including the I_c values that are correlated as a function of distance between the footings. Note that the definition of the model factor (M) in this study is the reciprocal of the more conventional definition presented in Chapter 5. The results of the study are presented in the following form:

$$q_d = \psi_D^{SLS} q_n = \psi_D^{SLS}[\rho_a/(\zeta_s\zeta_i B^{0.7}I_{cn})] \tag{6.8}$$

where ψ_D^{SLS} is the deformation factor for differential settlement, q_n is the nominal value of foundation applied stress, ρ_a is the allowable settlement limit, q_d is the revised design value of q_n, and I_{cn} is the nominal I_c calculated using mean N_{60}. The ψ_D^{SLS} values are given in a lengthy table and are a function of the allowable angular distortion (1/150, 1/300, 1/500), the COV of N (25 to 55%), and the center-to-center footing distance (3 to 9 m). For most parametric combinations, the deformation factors were less than 1.0, which is in contrast to some current practices for SLS. These practices may be unconservative.

6.2.6 First-order Reliability Method (FORM)

Annex C of EN 1990:2002 "Eurocode: Basis of Structural Design" discussed the application of a FORM design point method for calibration of partial factors. In essence, the performance function evaluated at the design point (or most probable failure point) is the simplified RBD equation. It is immediately clear that the design equation and the performance function are identical. The performance function is the best available physical model for estimation of the probability of failure. It can be known only implicitly in the form of a sophisticated numerical code. The design equation is typically a simple closed form equation provided in a code of practice for design purposes. The MRFD calibration approach does not require the performance function and the design equation to be linked. Another significant practical limitation of the FORM design point method is that only one design scenario rather than a range of representative design scenarios (see the dots in Fig. 6.1) can be selected for calibration.

6.2.7 Baseline technique

Ching and Phoon (2011) proposed a quantile-based reliability calibration approach that does not require the capacity to be lumped as a single lognormal random variable in LRFD nor does it require tedious segment by segment optimization of the resistance factors in MRFD. This Quantile Value Method (QVM) is discussed in Section 6.5. The approach shares some conceptual similarities with the above baseline technique (Task Committee on Structural Loadings of ASCE 1991). This technique essentially involves the matching of a suitably chosen nominal load and capacity to achieve a consistent level of reliability as shown below (Criswell and Vanderbilt 1987):

$$Q_\varepsilon = F_{50} \qquad\qquad (6.9)$$

where Q_ε is the ε exclusion limit of capacity (Q) and F_{50} is the 50-year return period load (F). The baseline calibration procedure involves adjusting the exclusion limit (ε) in Eq. (6.9) until a target annual probability of failure of 1% is achieved. From Fig. 6.4, it is clear that an exclusion limit of 5 to 10% should be used in Eq. (6.9) if the

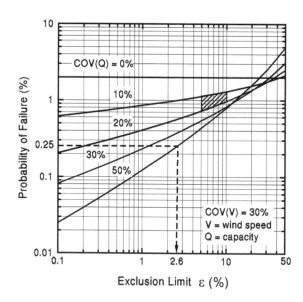

Figure 6.4 Reliability of structural components designed using $Q_\varepsilon = F_{50}$ (Source: Fig. 2, Phoon et al. 2003a, with permission from ASCE).

target probability of failure is 1% and the capacity COVs lie between 10 and 20%. The strength of the baseline RBD approach lies in its simplicity. First, a reasonably consistent target probability of failure of 1% can be achieved by simply matching a suitably chosen nominal load (F_{50}) and capacity (Q_5). No load or resistance factor is needed in Eq. (6.9). Second, the method makes use of familiar concepts, such as exclusion limits and return periods, which are already used widely in structural design. In addition, Fig. 6.4 shows that the probability of failure is fairly insensitive to changes in the capacity COV from 10 to 20% and exclusion limit from 5 to 10% (shaded area in Fig. 6.4).

The key differences between the Quantile Value Method (QVM) and the baseline method are: (1) a single quantile (synonymous with exclusion limit) is proposed for both capacity and load variables, (2) basic uncertain design parameters in the capacity model can be modeled as random variables with different probability distributions, rather than having to model the entire capacity as a single random variable (which is restrictive), and (3) a normalization scheme is applied to minimize the effects of the design parameters on the probabilistic characteristics of the safety ratio. The normalization scheme is the most critical component of QVM. It allows a *single* quantile to achieve consistent reliability, even over a wide range of COVs in the soil and load parameters. This is theoretically not possible for the FORM method. It is quite evident from Fig. 6.4 that the baseline method cannot maintain a single quantile for a COV of the capacity exceeding 20%. The mathematical details of the normalization scheme are furnished in Ching and Phoon (2011).

It is also worth emphasizing here that QVM bears no theoretical resemblance to the application of quantile in the definition of a characteristic value in the Eurocodes. The latter quantile is prescribed by design codes without reference to the target probability of failure. For example, a quantile between 5% and 10% is typically prescribed for the concrete compressive strength, f_{cu}, in structural design codes. The main purpose of this definition is to produce a suitably conservative compressive strength that varies consistently with the coefficient of variation of f_{cu}. The same quantile is applied to different performance functions, for example moment/shear capacity of a beam or compression capacity of a column. The quantile (ε) in QVM is fundamentally different. It is calibrated rather than prescribed to achieve a specific target probability of failure. It decreases in a relatively unique way with the target probability of failure for a given performance function as shown in Fig. 6.5. It is intrinsically related to the performance function. Hence, the quantile for a given soil property, say undrained shear strength, will vary when the property is applied within the context of different performance functions, even for the same target probability of failure.

6.2.8 Degree of understanding

The 2014 Canadian Highway Bridge Design Code (CAN/CSAS614:2014) calibrates ULS and SLS resistance factors based on the following three levels of understanding:

1. High understanding: extensive project-specific investigation procedures and/or knowledge are combined with prediction models of demonstrated quality to achieve a high level of confidence with performance predictions,

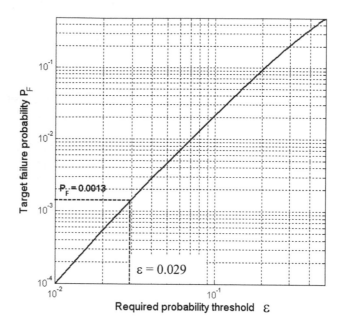

Figure 6.5 An example of ε vs. P_F relation for QVM (Ching and Phoon 2011, Figure 3).

2. Typical understanding: typical project-specific investigation procedures and/or knowledge are combined with conventional prediction models to achieve a typical level of confidence with performance predictions,
3. Low understanding: limited representative information (e.g., previous experience, extrapolation from nearby and/or similar sites, etc.) is combined with conventional prediction models to achieve a lower level of confidence with performance predictions.

This is an elaboration of a scheme that provides an engineer with a choice to select an appropriate resistance factor that suits a specific site condition (Phoon et al. 1995). The degree of site understanding is characterized by the coefficient of variation (COV) of the design property and three COV tiers are proposed for reliability calibration as shown in Tables 6.1 and 6.2. Phoon and Kulhawy (2008) presented different COV tiers for different design properties (Table 3.7). A similar approach was adopted by Paikowsky et al. (2004) in their reliability calibration of resistance factors for deep foundations. It appears that site variability is divided into low (COV < 25%), medium (25% < COV < 40%), and high (COV > 40%). For CAN/CSAS614:2014, the approach to introduce the level of site and model understanding into simplified RBD is essentially to use Monte Carlo simulations, modeling the ground as a spatially varying random field, and carry out a virtual site investigation, design, and construction of the geotechnical system. The detailed calibration steps are described by Fenton et al. (2016):

1. Assume a trial resistance factor for design check for a geotechnical system (e.g., shallow foundation) and limit state (e.g., ULS),
2. Produce a realization from a random field of ground properties,
3. Conduct a virtual site investigation by sampling the realization at some location. The distance between the sampling location and the geotechnical system is related to the level of site and model understanding,
4. Design the geotechnical system using the characteristic geotechnical parameters determined from the sampled locations in step 3 and the resistance factor in step 1.
5. Virtually install the design from step 4 on (or in) the random field realization generated in step 2,
6. Employ an accurate and unbiased model (e.g., the finite element method) to determine if performance is unsatisfactory (exceedance of limit state),
7. Repeat a large number of times, recording the number of failures (Monte Carlo simulation).
8. The probability of failure is then estimated as the number of failures divided by the number of trials in step 7. If this probability exceeds the target value, the resistance factor adjusted downwards and vice-versa. The process is repeated until a resistance factor is obtained that produces designs satisfying the target probability of failure.

Several examples of this calibration method are reported in Fenton et al. (2008) (ULS design of shallow foundation); Fenton and Naghibi (2011) (ULS design of deep foundations in cohesionless soils); Naghibi and Fenton (2011) (ULS design of deep foundations in cohesive soils); Fenton et al. (2005a) (SLS design of shallow foundations); Naghibi et al. (2014) (SLS design of deep foundations); Fenton et al. (2005b) (ULS design of retaining walls).

6.3 ISSUE OF VARIABLE COEFFICIENT OF VARIATION

As highlighted previously, simplified geotechnical RBD formats are harder to calibrate than those in structural engineering, because these formats must cover a wide range of COVs resulting from different soil property evaluation methodologies (Phoon 2015). The following simple example is adopted to demonstrate the issue of variable COV in a concrete way. The COV of a soil parameter is not a constant in the sense that it depends on site variability, measurement error, and transformation (regression) error arising from converting field data to the design soil parameter (Phoon and Kulhawy 1999). It can vary in a wide range, say between 0.1 and 0.7 for the undrained shear strength of a clay. The purpose of this section is to illustrate in a simple way that the widely used simplified RBD format based on *constant* partial factors cannot achieve the same reliability level, even approximately, for scenarios with variable COVs. Consider a friction pile with axial resistance Q and subjected to axial load

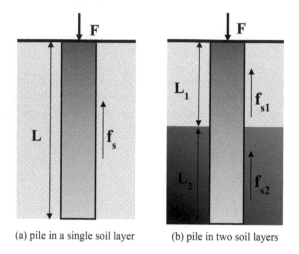

(a) pile in a single soil layer   (b) pile in two soil layers

Figure 6.6 Illustrative pile design examples (Ching et al. 2015, Figure 1).

F (Fig. 6.6a). The axial resistance Q is provided by the side resistance (tip resistance is ignored):

$$Q = \pi \times B \times L \times f_s \tag{6.10}$$

where f_s is unit side resistance; B is the pile diameter; L is the total embedment length. The unit side resistance f_s is lognormal with mean $= \mu_{fs}$ and COV $= V_{fs}$, and the axial load F is (independently) lognormal with mean $= \mu_F = 1000\,\mathrm{kN}$ and COV $= V_F = 0.1$. It is clear that Q is lognormal with mean $= \mu_Q = \pi \times B \times L \times \mu_{fs}$ and COV $= V_Q = V_{fs}$. The limit state function is defined to be $G = \ln(Q) - \ln(F)$. In the standard Gaussian space, the limit state function is

$$G(z_Q, z_F) = \lambda_Q + \xi_Q z_Q - \lambda_F - \xi_F z_F$$
$$\xi = \sqrt{\ln(1 + V^2)} \tag{6.11}$$
$$\lambda = \ln(\mu) - 0.5 \times \xi^2$$

where λ and ξ are respectively the mean and standard deviation of the logarithm of the subscripted variable, and (z_Q, z_F) are jointly standard normal.

Two cases would be considered: a calibration case and a validation case. The mean value and COV for the calibration case are (μ_Q, V_Q), and those for the validation case are (μ'_Q, V'_Q). The mean and COV of the load F for both calibration and validation cases are equal $(\mu_F = 1000\,\mathrm{kN}, V_F = 0.1)$. Basically, the calibration case will be used to calibrate the partial (load and resistance) factors to achieve a prescribed target reliability index of β_T. The validation case will be used to examine whether these partial factors indeed produce a design with an actual reliability index β'_A that is reasonably close to β_T.

6.3.1 Partial factors for the calibration case

Consider the calibration case with resistance mean $=\mu_Q$ and resistance COV $=V_Q=0.3$ and also consider $\beta_T=3.0$. One common method for calibrating the partial factors is the first order reliability method (FORM) as reviewed above (Hasofer & Lind 1974). This method first finds the FORM design point, which is the point on the limit state line that is closest to the origin. Direct calculation shows that the FORM design point has the following coordinates:

$$z_Q^* = \frac{-\beta_T\xi_Q}{\sqrt{\xi_Q^2+\xi_F^2}} \qquad z_F^* = \frac{\beta_T\xi_F}{\sqrt{\xi_Q^2+\xi_F^2}} \tag{6.12}$$

The resulting resistance and load factors, denoted by (ψ, η) respectively, are

$$\psi = \exp(\lambda_Q + \xi_Q z_Q^*)/\mu_Q = \exp(-0.5\xi_Q^2 - \beta_T\xi_Q^2/\sqrt{\xi_Q^2+\xi_F^2})$$

$$\eta = \exp(\lambda_F + \xi_F z_F^*)/\mu_F = \exp(-0.5\xi_F^2 + \beta_T\xi_F^2/\sqrt{\xi_Q^2+\xi_F^2}) \tag{6.13}$$

The design equation $\eta\mu_F \le \psi\mu_Q$ is the Load and Resistance Factor Design (LRFD) format. Note that the calibrated (ψ, η) only depend on (ξ_Q, ξ_F), not on (λ_Q, λ_F). This implies that the calibrated (ψ, η) only depend on the COVs (V_Q, V_F), not on the mean values (μ_Q, μ_F). The calibration case is with $V_Q=0.3$ and $V_F=0.1$, hence the resulting partial factors are $\psi=0.416$ and $\eta=1.096$. In this LRFD format, the nominal load and resistance are assumed to be equal to their respective mean values for simplicity. Typically, the nominal load μ_F is given by the structural engineer. The main task for the geotechnical engineer is to find the adequate dimension (B or L) of the geotechnical structure so that μ_Q is sufficiently large, to fulfill $1.096\,\mu_F \le 0.416\mu_Q$.

6.3.2 Actual reliability index for the validation case

It is of interest to know the actual reliability index, denoted by β_A', implied by the partial factors $\psi=0.416$ and $\eta=1.096$ when they are applied to the validation case. Although we have highlighted that $\beta_A' \ne \beta_T$ in all simplified RBD formats, it is important to know the difference particularly for $\beta_A' < \beta_T$ (unconservative design). The same design equation is applied to the validation case. The only difference is that the calibrated factors are applied to the mean resistance (μ_Q') and mean load (μ_F') for the validation case: $1.096 \times \mu_F' \le 0.416 \times \mu_Q'$. Consider the validation case with $\mu_F'=1000$ kN, $L'=20$ m, $\mu_{fs}'=50$ kN/m², and $V_{fs}'=V_Q'=0.5$. The geotechnical engineer needs to determine the pile diameter B' to fulfill the LRFD design equation $1.096 \times \mu_F' \le 0.416 \times \mu_Q'$. This means that

$$1.096 \times \mu_F' \le 0.416 \times \pi \times B' \times L' \times \mu_{fs}' \tag{6.14}$$

The resulting B' is 0.838 m at the least. The actual failure probability $(p_{f,A}')$ for this design size of $B'=0.838$ m can be determined using Monte Carlo simulation (MCS) (sample size, $n=10^6$):

$$p_{f,A}' = P(\pi B'L'f_s' < F') = 0.0372 \tag{6.15}$$

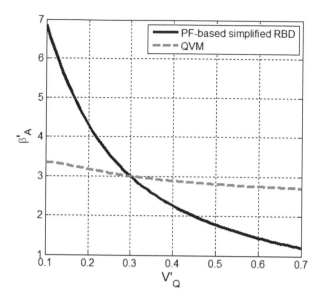

Figure 6.7 Relationship between β'_A and V'_Q.

The corresponding actual reliability index $\beta'_A = 1.78$, significantly less than the target value, $\beta_T = 3.0$. This $\beta'_A = 1.78$ is for the case with $V'_Q = 0.5$. Now consider V'_Q varying between 0.1 and 0.7 for the validation case. The actual reliability index β'_A under various V'_Q is plotted as the solid line in Fig. 6.7. The actual reliability index (β'_A) may be as high as 6.86 (actual failure probability $p'_{f,A} = 3.4 \times 10^{-12}$) when $V'_Q = 0.1$ and as low as 1.21 ($p'_{f,A} = 0.11$) when $V'_Q = 0.7$. It is clear that a uniform reliability level is not achieved by the calibrated partial factors $\psi = 0.416$ and $\eta = 1.096$. In particular, unconservative designs could be produced.

6.4   ISSUE OF VARIABLE SOIL PROFILES

Another difficulty encountered in the calibration of simplified geotechnical RBD formats is the presence of layered soil profiles. This issue does not surface in structural engineering. The examples shown in Fig. 6.6 are constructed to illustrate in a simple way that the widely used simplified RBD method based on *constant* partial factors cannot achieve the same reliability level, even approximately, for these scenarios. It does not matter how numerical values are assigned to the partial factors, say empirical re-distribution of the global factor of safety or rigorous calibration using reliability. The application of a single numerical value for each partial factor regardless of the subsurface profile imposes a fundamental limit on the ability of the design code to achieve a uniform reliability level over the range of scenarios covered by the code. A realistic deep foundation design code will have to cover layered soil profiles.

Consider the first scenario in Fig. 6.6a with $\mu'_F = 1000\,\mathrm{kN}$, $L' = 20\,\mathrm{m}$, $\mu'_{fs} = 50\,\mathrm{kN/m^2}$, and $V'_{fs} = V'_Q = 0.5$. Suppose that an engineer needs to design for the diameter B using constant partial factors $\psi = 0.416$ and $\eta = 1.096$. We have seen in the previous section that the resulting B' is 0.838 m, and the corresponding actual reliability index $\beta'_A = 1.78$. Now consider the second scenario in Fig. 6.6b (two soil layers now) with $\mu'_F = 1000\,\mathrm{kN}$, $L'_1 = L'_2 = 10\,\mathrm{m}$, $\mu'_{fs1} = \mu'_{fs2} = 50\,\mathrm{kN/m^2}$, and $V'_{fs1} = V'_{fs2} = 0.5$. Note that the second scenario has the same total length as the first one. Moreover, the mean value and COV for the second scenario are the same as those for the first one. The only difference is that there are two independent soil layers in the second scenario. With the partial factors $\psi = 0.416$ and $\eta = 1.096$, it is clear that resulting B' is still 0.838 m. The actual failure probability can be determined using MCS $(n = 10^6)$:

$$p'_{f,A} = P(\pi B' L'_1 f'_{s1} + \pi B' L'_2 f'_{s2} < F) = 4.9 \times 10^{-3} \qquad (6.16)$$

where f_{s1} and f_{s2} are independent lognormal random variables with mean = $50\,\mathrm{kN/m^2}$ and COV = 0.5. The actual reliability index β'_A is 2.58. With the same partial factors $\psi = 0.416$ and $\eta = 1.096$ applied to both scenarios, the reliability indices differ by a factor of 1.44 ($\beta'_A = 1.78$ for one layer soil versus $\beta'_A = 2.58$ for two-layer soil), but it is more accurate to note that the probabilities of failure differ by a factor of 7.6 ($p'_{f,A} = 0.0372$ for one layer soil versus versus $p'_{f,A} = 4.9 \times 10^{-3}$ for two-layer soil). In fact, one can show that the actual reliability index for another scenario with five independent soil layers with equal thicknesses further increases to 3.95! It is evident that this partial-factor-based (PF-based) simplified RBD format cannot produce uniform reliability indices over these two scenarios unless the partial factors can be adjusted.

Ching et al. (2015) shows that the issue of soil profile is connected to the issue of variable redundancy. The first scenario in Fig. 6.6a has less redundancy than the second one in Fig. 6.6b. The first scenario has only one soil layer, hence it requires more caution in selecting the design value for f_s, denoted by $f_{s,d}$, than the second scenario with two soil layers. This is because the consequence for selecting an erroneous $f_{s,d}$ for the first scenario is large – there is only a single soil layer providing the resistance (no redundancy). On the other hand, the consequence for selecting an erroneous $f_{s1,d}$ for the second scenario is mitigated by the presence of a second supporting soil layer (more redundancy). It is rather unlikely for the errors in both layers to be identical in sign and in magnitude. To achieve the same reliability level, the $f_{s,d}$ value in the first scenario should be selected in a more conservative way. Namely, a smaller partial factor should be used for the first scenario than for the second scenario. This makes sense intuitively even in the absence of more rigorous probabilistic argument. If a constant partial factor is adopted, the resulting β'_A will not be the same. This is why the PF-based simplified RBD method cannot produce a uniform reliability index when it is applied to problems involving variable redundancy. This issue of variable redundancy is not limited to pile problems. It appears in other geotechnical design problems as well. However, the issue can be explained in a more physically intuitive way for piles, because the degree of redundancy is visually linked to the number of supporting soil layers in a concrete way rather than appearing as a mathematical abstraction. As we will see later in a gravity retaining wall example, this issue of variable redundancy can exist in problems that does not involve a variable number of soil layers.

6.5   QUANTILE VALUE METHOD (QVM)

Ching and Phoon (2011, 2013) developed a quantile-based simplified RBD method that is more robust than the PF-based simplified RBD method. This quantile-based method was referred to as the Quantile Value Method (QVM) in Ching and Phoon (2013). The authors showed that QVM is robust under the presence of variable COV of soil parameters. First, the random variables are classified as stabilizing or destabilizing according to their effects on G. A random variable is stabilizing (or destabilizing) if the increase of this random variable will increase (or decrease) G. The basic idea of the QVM is to reduce any stabilizing random variable (e.g., soil strength) to its ε quantile (ε is small) to obtain its design value, but to increase any destabilizing random variable (e.g., load) to its $1 - \varepsilon$ quantile to obtain its design value. The parameter ε is called the probability threshold, and a *constant* ε is applied to both types of random variables: taking ε quantiles for stabilizing variables and $1 - \varepsilon$ quantiles for destabilizing variables. Then, an engineer can design the size of the geotechnical structure based on these quantile-based design values. It is useful to note that a *constant* ε is equivalent to applying a variable partial factor that changes according to the COV of the design parameter. If X is a lognormal stabilizing random variable, its QVM design value is its ε quantile:

$$X_d = \exp[\lambda_X + \xi_X \Phi^{-1}(\varepsilon)] \tag{6.17}$$

This is equivalent to applying a variable partial factor that changes according to the COV:

$$X_d = \gamma_X \mu_X \qquad \gamma_X = \exp[\lambda_X + \xi_X \Phi^{-1}(\varepsilon)]/\mu_X = \exp[-0.5\xi_X^2 + \xi_X \Phi^{-1}(\varepsilon)] \tag{6.18}$$

The equivalent partial factor depends on ξ_X, which in turn depends on the COV of X.

6.5.1   Robustness of QVM against variable COV

Let us now demonstrate the use of QVM for the pile example in Fig. 6.6a. Again, there are a calibration case and a validation case. For the calibration case, $L = 20\,m$, $\mu_{fs} = 50\,kN/m^2$, $V_Q = V_{fs} = 0.3$, $\mu_F = 1000\,kN$, $V_F = 0.1$, and $\beta_T = 3.0$. Ching & Phoon (2011) derived the relationship between ε and β_T. For the calibration case, this relationship reduces to (Ching and Phoon 2013)

$$\varepsilon = \Phi\left(\frac{-\beta_T\sqrt{\xi_Q^2 + \xi_F^2}}{\xi_Q + \xi_F}\right) \tag{6.19}$$

The calibrated ε value is 9.02×10^{-3}. If one adopts the usual linearization ($\xi_Q^2 + \xi_F^2)^{0.5} \approx 0.7(\xi_Q + \xi_F)$, Eq. (6.19) reduces to an even simpler form: $\varepsilon \approx \Phi(-0.7 \times \beta_T)$.

It is of interest to know the actual reliability index β'_A implied by the calibrated $\varepsilon = 9.02 \times 10^{-3}$ when it is applied to the validation case. Consider the validation case with $\mu'_F = 1000\,\text{kN}$, $L' = 20\,\text{m}$, $\mu'_{fs} = 50\,\text{kN/m}^2$, and $V'_{fs} = V'_Q = 0.5$. The design value for Q', denoted by Q'_d, is its 9.02×10^{-3} quantile:

$$Q'_d = \pi \times B' \times L' \times (9.02 \times 10^{-3}\,\text{quantile of } f'_s) \tag{6.20}$$

Because f'_s is lognormal with mean $= \mu_{fs} = 50\,\text{kN/m}^2$ and COV $= V_{fs} = 0.5$, the 9.02×10^{-3} quantile of f'_s can be calculated analytically:

$$(9.02 \times 10^{-3}\,\text{quantile of } f'_s) = \exp[\lambda_{fs} + \xi_{fs}\Phi^{-1}(9.02 \times 10^{-3})]$$
$$= 14.63\,\text{kN/m}^2 \tag{6.21}$$

The design value for F', denoted by F'_d, is its $(1 - 9.02 \times 10^{-3})$ quantile:

$$(1 - 9.02 \times 10^{-3}\,\text{quantile of } F') = \exp[\lambda_F + \xi_F\Phi^{-1}(1 - 9.02 \times 10^{-3})]$$
$$= 1259.8\,\text{kN} \tag{6.22}$$

The geotechnical engineer needs to find the adequate diameter B' so that Q'_d is sufficiently large, to fulfill $Q'_d = F'_d$ at the least. It is easy to determine that the resulting B' is 1.37 m. The actual failure probability $p'_{f,A}$ for this design size of $B' = 1.37$ m can be determined using MCS ($n = 10^6$) using Eq. (6.15). The resulting $p'_{f,A}$ is 2.54×10^{-3}, and $\beta'_A = 2.80$, which is fairly close to $\beta_T = 3.0$. This $\beta'_A = 2.80$ is for the case with $V'_Q = 0.5$. Now consider V'_Q varying between 0.1 and 0.7 for the validation case. The actual reliability index β'_A under various V'_Q is plotted as the dashed line in Fig. 6.7. β'_A closely follows 3.0. The largest departure from 3.0 occurs at the two extremes: $\beta'_A = 3.34$ for $V'_Q = 0.1$ and $\beta'_A = 2.71$ for $V'_Q = 0.7$. The actual reliability level is fairly uniform, compared to that for the PF-based simplified RBD format (solid line in Fig. 6.7).

6.5.2 Pad foundation supported on boulder clay

The pad foundation example adopted herein was originally developed by the European Technical Committee 10 (ETC 10) of the International Society of Soil Mechanics and Geotechnical Engineering. The focus is on the ultimate limit state (ULS) requirement, i.e. the total vertical load cannot exceed the total resistance. The details for this example can be found in Ching et al. (2014). With a target reliability index β_T of 3.2, Ching et al. (2014) showed that the calibrated ε value for QVM is 0.0083, whereas the calibrated partial factor for the undrained shear strength (s_u) depends on the mean and COV of s_u, as shown in Fig. 6.8. Note that for QVM, a single ε value is calibrated, whereas for the partial factor approach, multiple partial factors are calibrated: one partial factor is calibrated for each partition shown in Fig. 6.8. It is clear that the calibrated partial factor decreases as the COV of s_u increases. This partitioned partial factor approach is consistent to the MRFD approach discussed in Section 6.2. The MRFD approach

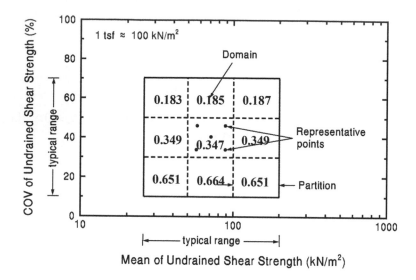

Figure 6.8 Calibrated partial factors for various partitions.

that adopts partitioned partial factors is deemed to be more robust than the LRFD approach that adopts a single universal partial factor.

It is further assumed in Ching et al. (2014) that the mean and COV of s_u for a future design case (validation case) depend on site investigation effort. Section 4.6.1 showed how the mean and COV of s_u can be updated based on OCR (overconsolidation ratio), $q_t - \sigma_v$ (net cone resistance), and N_{60} (SPT N corrected for energy efficiency) information. A variety of site investigation efforts are produced by systematically changing: (a) the number of test types and (b) the test precision. Four scenarios are considered for the number of test types: (T1) only the range of N_{60} is known; (T2) the ranges for N_{60} and $q_t - \sigma_v$ are both known; (T3) the ranges for N_{60} and OCR are both known; and (T4) the ranges for N_{60}, $q_t - \sigma_v$, and OCR are all known. Scenario T1 is considered as the basic case with the least effort, while T4 contains the most information in terms of number of test types. Five scenarios (P0 to P4) are considered for test precision and they are summarized in Table 6.3. The ranges in Table 6.3 represent the bounds for OCR, N_{60}, and $q_t - \sigma_v$ based on the assumption that more precise information on each test type measurement is available perhaps by increasing the number of tests and boreholes. P0 means no site-specific tests are conducted and information bounds are purely estimated from general literature appropriate for "clay". P4 means that sufficiently extensive tests (e.g., multiple CPT soundings or boreholes) are conducted to narrow the ranges. Table 6.4 shows how the updated mean and COV for s_u depend on the site investigation efforts. It is clear that the COV decreases as the precision increases (narrow information bound) and also decreases as the number of test types increases.

Consider a future scenario with test type = T2 and precision = P3. Namely, bounds for N_{60} and $q_t - \sigma_v$ are known, and the bounds are [6, 10] and [1030 kN/m², 1450 kN/m²], respectively. According to Table 6.4, the mean and COV of s_u are

Table 6.3 Characterization of site information: (a) function of test type and precision and (b) information bounds for various precision scenarios (Table 6 in Ching et al. 2014, with permission from ASCE).

		Test type scenario			
Precision scenario		T1	T2	T3	T4
P0	Zero precision	N_{60}	N_{60} & $q_t - \sigma_v$	N_{60} & OCR	$N_{60}, q_t - \sigma_v$ & OCR
P1	Poor precision	N_{60}	N_{60} & $q_t - \sigma_v$	N_{60} & OCR	$N_{60}, q_t - \sigma_v$ & OCR
P2	:	N_{60}	N_{60} & $q_t - \sigma_v$	N_{60} & OCR	$N_{60}, q_t - \sigma'_v$ & OCR
P3	:	N_{60}	N_{60} & $q_t - \sigma_v$	N_{60} & OCR	$N_{60}, q_t - \sigma_v$ & OCR
P4	Excellent precision	N_{60}	N_{60} & $q_t - \sigma_v$	N_{60} & OCR	$N_{60}, q_t - \sigma_v$ & OCR

		Information bounds based on site investigation		
Precision scenario		OCR	N_{60}	$q_t - \sigma_v$ (kN/m²)
P0	Zero precision	[1, 50]	[0, 100]	[200, 6000]
P1	Poor precision	[5, 25]	[3, 18]	[730, 2040]
P2	:	[7.5, 16.7]	[5, 12]	[940, 1580]
P3	:	[8.5, 14.6]	[6, 10]	[1030, 1450]
P4	Excellent precision	[9.5, 13.1]	[7, 9]	[1100, 1350]

Table 6.4 Updated mean and COV for s_u under various scenarios (modified from Table 7 in Ching et al. 2014, with permission from ASCE).

	T1	T2	T3	T4
Mean (kN/m²) (COV) [partial factor]				
P0	115.8 (0.35) [0.349]			
P1	111.3 (0.32) [0.349]	107.3 (0.27) [0.651]	111.3 (0.32) [0.349]	107.3 (0.27) [0.651]
P2	111.4 (0.31) [0.349]	105.7 (0.25) [0.651]	102.4 (0.25) [0.651]	100.4 (0.21) [0.651]
P3	111.3 (0.31) [0.349]	**105.5 (0.24)** [0.651]	99.9 (0.24) [0.664]	99.0 (0.20) [0.664]
P4	112.0 (0.31) [0.349]	105.3 (0.24) [0.651]	99.5 (0.23) [0.664]	98.1 (0.19) [0.664]

105.5 kN/m² and 0.24, respectively (the underlined cell in the table). For this particular scenario, QVM can be adopted for the simplified RBD, and the calibrated ε value is 0.0083 (note that ε = 0.0083 even if the future scenario is with different test type and precision). Because s_u is assumed lognormal, this means that the design value of s_u should be reduced to its 0.0083 quantile:

$$\text{Design value of } s_u = \exp[\lambda_{su} + \xi_{su} \times \Phi^{-1}(0.0083)] \qquad (6.23)$$

where λ_{su} and ξ_{su} are the mean and standard deviation of $\ln(s_u)$:

$$\xi_{su} = \sqrt{\ln(1 + V_{su}^2)} = \sqrt{\ln(1 + 0.24^2)} = 0.237$$
$$\lambda_{su} = \ln(\mu_{su}) - 0.5 \times \xi_{su}^2 = \ln(105.5) - 0.5 \times \xi_{su}^2 = 4.631 \tag{6.24}$$

The resulting design value of s_u for QVM is $58.2\,kN/m^2$. Based on this design value, it is concluded that the dimension (width B) of the pad foundation needs to be at least 2.92 m. The actual reliability index, denoted by β_A, for B = 2.92 m can be determined using reliability analysis (e.g., Monte Carlo simulation). The resulting β_A is equal to 3.22, fairly close to the target reliability index $\beta_T = 3.2$. Table 6.5 shows the final design B sizes and corresponding β_A values for all Tm-Pn scenarios. The underlined cell is the result for the aforementioned T2-P3 scenario. The required B in Table 6.5 shows reasonable trend – the largest for P0 and the smallest for T4-P4. Moreover, the actual reliability index β_A for QVM fairly close to the target value 3.2.

The partitioned partial factor approach can be applied to the same T2-P3 scenario. Table 6.4 shows that the calibrated partial factor for this scenario is 0.651. This means that the design value of s_u should be reduced to 0.651 × mean value = 0.651 ×105.5 = 68.68 kN/m^2. Based on this design value, it is concluded that the dimension (width B) of the pad foundation needs to be at least 2.69 m. The actual reliability index (β_A) for B = 2.69 m is equal to 2.59, somewhat different from the target reliability index $\beta_T = 3.2$. Table 6.6 shows the final design B sizes and corresponding β_A values for all Tm-Pn scenarios with the partitioned partial factor approach. The

Table 6.5 Final design B size/β_A for QVM (modified from Table 10 in Ching et al. 2014, with permission from ASCE).

	T1	T2	T3	T4
P0	3.21 m/2.96			
P1	3.14 m/3.05	3.02 m/3.16	3.14 m/3.04	3.01 m/3.16
P2	3.10 m/3.04	2.94 m/3.20	3.02 m/3.21	2.89 m/3.26
P3	3.10 m/3.08	**2.92 m/3.22**	2.99 m/3.21	2.84 m/3.28
P4	3.08 m/3.01	2.91 m/3.27	2.97 m/3.29	2.84 m/3.21

Table 6.6 Final design B size/β_A for the MRFD approach with partitioned partial factors (modified from Table 11 in Ching et al. 2014, with permission from ASCE).

	T1	T2	T3	T4
P0	3.51 m/3.55			
P1	3.58 m/3.92	2.67 m/2.30	3.58 m/3.88	2.67 m/2.30
P2	3.58 m/4.00	2.69 m/2.54	2.73 m/2.49	2.76 m/2.90
P3	3.58 m/4.08	**2.69 m/2.59**	2.74 m/2.58	2.75 m/3.00
P4	3.57 m/3.93	2.69 m/2.65	2.74 m/2.67	2.76 m/2.99

required B in general shows the reasonable trend – larger for P0 and smaller for T4-P4, but there are some unexpected results, such as T2-P1 having the smallest required B. The actual reliability index β_A is in general around the target value 3.2 with some occasional large deviations from 3.2, e.g., for T2-P1, $\beta_A = 2.30$.

It is evident that QVM is effective in linking the site investigation efforts to design savings (required B reduces with increasing site investigation efforts) and produces β_A that are fairly close to the target value 3.2 in this pad foundation example (Table 6.5). This is not an easy task at all – $\varepsilon = 0.0083$ is NOT calibrated with respect to any specific design case with any particular site investigation effort. In general, the MRFD approach with partitioned partial factors can link the site investigation efforts to design savings, but it is less effective than QVM. It is expected that the LRFD approach with a single universal partial factor will be further less effective than MRFD. The ability to link to site investigation efforts is an obvious advantage for QVM.

6.6 EFFECTIVE RANDOM DIMENSION

Unfortunately, Ching et al. (2015) showed that QVM is not robust against variable redundancy, either. Nonetheless, they discovered an interesting relationship between the probability threshold ε and failure probability p_f. This relationship opens up a possibility to improve the robustness of QVM against variable redundancy. To demonstrate this, consider again the pile examples in Fig. 6.6 with total depth $L = 20$ m, but now the load $F = 1000$ kN is deterministic. Now consider four such examples with the number of independent layers being $n_L = 1$ to 4. Moreover, all layers have the same thickness, e.g., for $n_L = 4$, the thickness of each layer is 5 m. A large n_L is associated with more redundancy. The unit side resistance f_s for each layer is lognormal with mean $\mu_{fs} = 50$ kN/m^2 and COV $V_{fs} = 0.3$. Suppose a constant $\varepsilon = 0.01$ is adopted for QVM. The procedure presented above (Eqs. 6.20 to 6.22) is used to find the resulting B', and the actual failure probability $p'_{f,A}$ and actual reliability index β'_A for this resulting B' are evaluated through MCS. Table 6.7 shows the results. It is remarkable that β'_A increases significantly as n_L increases, indicating that QVM is not robust against variable redundancy. Although not shown here, the PF-based simplified RBD method suffers from the same degree of non-robustness. More interestingly, the following relationship holds approximately (see the rightmost column in Table 6.7):

$$\left(\beta_A / \beta_\varepsilon\right)^2 \approx n_L \qquad (6.25)$$

Table 6.7 QVM design results for four pile examples with different numbers of soil layers.

n_L	ε	β_ε	B'	p'_{fA}	β'_A	$(\beta'_A/\beta_\varepsilon)^2$
1	0.01	2.326	0.658 m	0.01	2.326	1
2	0.01	2.326	0.658 m	3.71×10^{-4}	3.374	2.10
3	0.01	2.326	0.658 m	1.68×10^{-5}	4.148	3.18
4	0.01	2.326	0.658 m	7.00×10^{-7}	4.825	4.30

Table 6.8 QVM design results (considering n_L) for the four pile examples.

n_L	β_ε	ε	$f_{s,d}$	B'	$p'_{f,A}$	β'_A
1	3.00	0.0013	19.85 kN/m²	0.802 m	0.0013	3.00
2	2.12	0.0169	25.69 kN/m²	0.620 m	0.0010	3.08
3	1.73	0.0416	28.80 kN/m²	0.553 m	8.52×10^{-4}	3.14
4	1.50	0.0668	30.83 kN/m²	0.516 m	7.51×10^{-7}	3.18

where $\beta_\varepsilon = -\Phi^{-1}(\varepsilon)$ can be viewed as the reliability index of the input side resistances, because the probability of a side resistance being less than its design value is ε (design value $f_{s,d}$ is taken to be the ε quantile of f_s in QVM).

The observation that $(\beta_A/\beta_\varepsilon)^2 \approx n_L$ suggests the following steps for RBD:

1. Estimate n_L.
2. Given the estimated n_L and the target reliability index β_T, determine $\beta_\varepsilon = \beta_T/n_L^{0.5}$. Further determine $\varepsilon = \Phi(-\beta_\varepsilon)$.
3. Determine the design values of the random variables using QVM. This involves reducing all stabilizing random variables to their ε quantiles to obtain their design values and increasing all destabilizing random variable to their $1 - \varepsilon$ quantiles.
4. Based on the design values obtained in Step 3, the size of the geotechnical structure is obtained by solving $G = 0$.

Now let us apply these steps to the four pile examples with $n_L = 1$ to 4 with $\beta_T = 3.0$. Table 6.8 shows the design results. It is clear that now ε changes with n_L (the third column): ε increases with increasing n_L. This means that a "bolder" design value of $f_{s,d}$ can be adopted for a case with more redundancy (the fourth column). The resulting B' is smaller for a case with more redundancy (the fifth column), and yet the actual reliability index β'_A is still satisfactory (close to the target value $\beta_T = 3.0$; the rightmost column). It is important to point out here that unit side resistance f_s for each layer is assumed to be statistically independent of other unit side resistances. Compared to those in Table 6.7, the actual reliability indices β'_A in Table 6.8 are now significantly more uniform.

In Ching et al. (2015), they proposed the concept of "effective random dimension" (ERD) to characterize the degree of redundancy for limit state functions involving linear sums of normal random variables (possibly correlated). They showed that a simple closed-form formula for ERD exists for linear sums of standard normal random variables. The significance of ERD is the effective number of *independent* standard normal random variables that affect the limit state function. Moreover, they showed that

$$\left(\beta_A/\beta_\varepsilon\right)^2 = \text{ERD} \tag{6.26}$$

for limit state functions involving linear sums of normal random variables. By comparing Eq. (6.26) with Eq. (6.25.), it is clear that ERD has the same physical meaning as the number of soil layers (n_L) for the pile examples in Tables 6.7 and 6.8. Ching et al. (2015) further showed that $(\beta_A/\beta_\varepsilon)^2$ for a general nonlinear limit state function (possibly involving some correlated non-normal random variables) can also be used to characterize its degree of redundancy, to quantify the effective number of *independent* random variables. Therefore, Eq. (6.26) is applicable to general nonlinear limit state functions as well. ERD can be estimated by the characteristics of the problem at hand. Because ERD is dimensionless, it is expected that ERD is dependent on some dimensionless parameters governing the limit state function of interest. To construct the relationship between ERD and these dimensionless parameters, a collection of "calibration cases" are generated. To make the relationship generic, these calibration cases must cover sufficiently diverse design scenarios. For each calibration case, ERD can be determined by $(\beta_A/\beta_\varepsilon)^2$, where $\beta_\varepsilon = -\Phi(\varepsilon)$, whereas β_A is determined by MCS. The relationship between ERD and the governing dimensionless parameters can then be constructed using regression. Once this regression equation is obtained, ERD can be estimated using these dimensionless parameters. This calibration exercise involving MCS is carried out by the code developer, not the practitioner who is applying the code. Note that a similar calibration exercise is also needed for existing simplified RBD formats (Phoon et al. 2013).

6.6.1 Gravity retaining wall

The issue of variable redundancy is not limited to geotechnical structures embedded in a variable number of soil layers, such as the friction pile example studied above. The following gravity retaining wall example (Fig. 6.9) shows that this variable redundancy issue can arise even if the geotechnical structure is embedded in a *fixed* number of soil layers. The retaining wall has a total height of H and a base width B. There is a surcharge pressure q on the retained ground level. The water table is assumed to be at the ground level at the toe of the wall and it is $h_w = \lambda \times H$ above the base of the wall at the heel of the wall. The foundation sand layer has submerged unit weight γ'_s and effective friction angle ϕ'_s, whereas the backfill sand above the water table has dry unit weight γ_d and submerged unit weight $\gamma' = (1+\omega)\gamma_d - \gamma_w$ below the water table (ω is the water content; γ_w is the water unit weight), and its effective friction angle is ϕ'. Sliding failure is typical for this cantilever wall type, because it is quite light. Therefore, the limit state function for the sliding failure is considered in this example.

The issue of variable redundancy not only exists in the pile example with a variable number of soil layers, but also exists in the current gravity retaining wall example, even though this current example involves a fixed number of soil layers. Ching et al. (2015) showed that by solely change the statistics (mean and COV) of the surcharge q, ERD can change from 2.20 to 3.15. Based on 1000 randomly selected calibration cases, Ching et al. (2015) showed that ERD can be effectively estimated using the following equation based on some dimensionless parameters.

$$ERD \approx 3.25 - 0.42(B/H)^2 - 1.19\lambda^2 - 1.04(B/H) + 0.71\lambda$$
$$+ 0.44 \cdot \ln[\mu_q/(\mu_{\gamma d} \cdot H \cdot K_a)]$$
$$+ 1.46 \cdot V_q - 1.87 \cdot V_{\phi'_s} + 1.35 \cdot V_{\phi'} - 0.13\rho - 0.03\mu_{\phi'} \qquad (6.27)$$

where $K_a = \tan^2(45° - \mu_{\phi'}/2)$; $\mu_{\gamma d}$, $\mu_{\phi'}$, and μ_q are the mean values for γ_d, ϕ', and q, respectively; $V_{\phi'}$, $V_{\phi'_s}$, and V_q are the COVs for ϕ', ϕ'_s, and q, respectively. Note that this

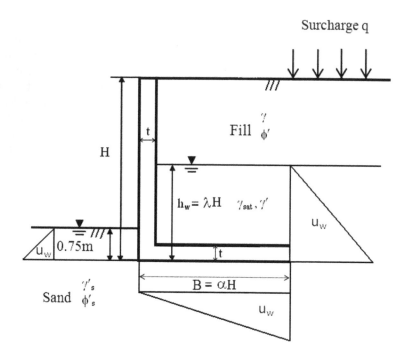

Figure 6.9 Gravity retaining wall example (Ching et al. 2015, Figure 5).

ERD equation is developed by the code developer as part of the reliability calibration exercise, not the practitioner who is applying the code. A new QVM procedure that considers the degree of redundancy, called the ERD-QVM, is proposed by Ching et al. (2015):

1. Estimate ERD using Eq. (6.27).
2. Given the estimated ERD and the target reliability index β_T, determine $\beta_\varepsilon = \beta_T/ERD^{0.5}$. Further determine $\varepsilon = \Phi(-\beta_\varepsilon)$.
3. Determine the design values of the random variables using QVM. This involves reducing all stabilizing random variables to their ε quantiles to obtain their design values and increasing all destabilizing random variable to their $1-\varepsilon$ quantiles. Conceptually, this step is the same as computing a design value by dividing a characteristic value by a partial factor (for a stabilizing variable) or by multiplying a characteristic value by a partial factor (for a destabilizing variable). The outcome is a numerical value that will be substituted into an algebraic design check.
4. Based on the design values obtained in Step 3, the size of the geotechnical structure is obtained by solving $G = 0$. For the current gravity retaining wall example, the size is the base width B. With $\beta_T = 3.0$, Ching et al. (2015) showed that the original QVM with $\varepsilon = 0.0265$ performs satisfactorily: the actual reliability indices β'_A for 1000 validation cases fall into a range between 2.43 and 3.72. Nonetheless, the ERD-QVM further improves the performance: β'_A ranges from 2.64 to 3.44.

6.7 CONCLUSIONS

Simplified RBD formats are expected to dominate geotechnical design codes in the next few decades. They are very attractive to practitioners, because it is possible to enjoy the advantages of reliability-based design without performing reliability analysis. In fact, the computational effort between applying simplified RBD format and applying the factor of safety format is similar, because both design checks are algebraic.

While it is understandable for geotechnical RBD to adopt structural LRFD concepts at its initial stage of development over the past decades, the authors believe that it is timely for the geotechnical design code community to look into how we can improve our state of practice to cater to the distinctive needs of geotechnical engineering practice. Section D.5 in ISO2394:2015 clarifies that the "key goal in geotechnical RBD is to achieve a more uniform level of reliability than that implied in existing allowable stress design". This goal is not explicitly recognized in many existing implementations of simplified geotechnical RBD formats. The advantage of adopting RBD over existing allowable stress design is largely nullified if the prescribed target reliability index cannot be achieved consistently over the full range of design scenarios within the ambit of the design code. This is not a pedantic issue. In fact, it goes to the heart of geotechnical engineering practice that must cater to diverse local site conditions and diverse local practices that grew and adapted over the years to suit these conditions. One obvious example is that the COVs of geotechnical parameters can vary over a wide range, because diverse property evaluation methodologies exist to cater to these diverse practice and site conditions (refer to Section D.1, ISO2394:2015). A simplified geotechnical RBD format that meets the diverse needs of geotechnical engineering practice is not available. Existing LRFD or comparable simplified RBD formats are limited in more than one substantial aspect. For examples, these simplified RBD formats do not allow room for the geotechnical engineer to exercise judgment in response to local site conditions and to incorporate local experience. This chapter demonstrates that improved formats (e.g., ERD-QVM) exist that can cater to a more realistic range of design scenarios. Specifically, ERD-QVM can maintain an acceptably uniform level of reliability over a wide range of COVs of geotechnical parameters and a wide range of layered soil profiles. It can achieve this while retaining the simplicity of an algebraic design check similar to the traditional factor of safety format and LRFD. Nonetheless, it has not been tested in more complex design settings. Complex design settings can arise in the mechanical soil-structure interaction sense, such as staged construction in deep excavations. Complex design settings can also arise in the probabilistic sense, such as slopes which are intrinsically system reliability problems. More research is clearly needed, but it is important for geotechnical RBD to be developed with geotechnical needs at the forefront.

ACKNOWLEDGMENTS

The authors are grateful for the valuable comments provided by Professor C. H. Juang on Robust Geotechnical Design (RGD) and Dr W. Gong on Robust LRFD (R-LRFD).

REFERENCES

Akbas, S.O. & Kulhawy, F.H. (2009a) Axial compression of footings in cohesionless soil. I: Load-settlement behavior. *ASCE Journal of Geotechnical and Geoenvironmental Engineering*, 135 (11), 1562–1574.

Akbas, S.O. & Kulhawy, F.H. (2009b) Reliability-based design approach for differential settlement of footings on cohesionless soils. *ASCE Journal of Geotechnical and Geoenvironmental Engineering*, 135 (12), 1779–1788.

Barker, R.M., Duncan, J.M., Rojiani, K.B., Ooi, P.S.K., Tan, C.K. & Kim, S.G. (1991) *Manuals for Design of Bridge Foundations*. NCHRP Report 343. Washington, DC, Transportation Research Board.

Burland, J.B. & Burbidge, M.C. (1985) Settlement of foundations on sand & gravel. *Proceedings of the Institution of Civil Engineers*, 78 (Pt 1), 1325–1381.

CAN/CSAS614:2014. *Canadian Highway Bridge Design Code*. Mississauga, ON, Canadian Standards Association.

Ching, J. & Phoon, K.K. (2011) A quantile-based approach for calibrating reliability-based partial factors. *Structural Safety*, 33, 275–285.

Ching, J. & Phoon, K.K. (2013) Quantile value method versus design value method for calibration of reliability-based geotechnical codes. *Structural Safety*, 44, 47–58.

Ching, J., Phoon, K.K. & Yu, J.W. (2014) Linking site investigation efforts to final design savings with simplified reliability-based design methods. *ASCE Journal of Geotechnical and Geoenvironmental Engineering*, 140 (3), 04013032.

Ching, J., Phoon, K.K. & Yang, J.J. (2015) Role of redundancy in simplified geotechnical reliability-based design – A quantile value method perspective. *Structural Safety*, 55, 37–48.

Criswell, M.E. & Vanderbilt, M. (1987) *Reliability-Based Design of Transmission Line Structures: Methods*. Report EL-4793 (1). Palo Alto, Electric Power Research Institute. 473 pp.

EN 1990:2002. *Eurocode – Basis of Structural Design*. Brussels, European Committee for Standardization (CEN).

ETC 10. Geotechnical design ETC10 Design Example 2.2. *Pad Foundation with Inclined Eccentric Load on Boulder*. Available from: http://www.eurocode7.com/etc10/Example%202.2/index.html.

Fenton, G.A. & Naghibi, M. (2011) Geotechnical resistance factors for ultimate limit state design of deep foundations in frictional soils. *Canadian Geotechnical Journal*, 48 (11), 1742–1756.

Fenton, G.A., Griffiths, D.V. & Cavers, W. (2005a) Resistance factors for settlement design. *Canadian Geotechnical Journal*, 42 (5), 1422–1436.

Fenton, G.A. Griffiths, D.V. & Williams, M.B. (2005b) Reliability of traditional retaining wall design. *Geotechnique*, 55 (1), 55–62.

Fenton, G.A., Griffiths, D.V. & Zhang, X.Y. (2008) Load and resistance factor design of shallow foundations against bearing failure. *Canadian Geotechnical Journal*, 45 (11), 1556–1571.

Fenton, G.A., Naghibi, F., Dundas, D., Bathurst, R.J. & Griffiths, D.V. (2016) Reliability-based geotechnical design in the 2014 Canadian Highway Bridge Design Code. *Canadian Geotechnical Journal*, 53 (2), 236–251.

Gong, W., Wang, L., Juang, C.H., Zhang, J. & Huang, H. (2014) Robust geotechnical design of shield-driven tunnels. *Computers and Geotechnics*, 56, 191–201.

Gong, W., Khoshnevisan, S., Juang, C.H. & Phoon, K.K. (2016) R-LRFD: Load and Resistance Factor Design considering design robustness. *Computers and Geotechnics*, 74, 74–87.

Hasofer, A.M. & Lind, N.C. (1974) Exact and invariant second-moment code format. *ASCE Journal of Engineering Mechanics*, 100 (1), 111–121.

ISO2394:1973/1986/1998/2015. *General Principles on Reliability for Structures*. Geneva, International Organization for Standardization.

Juang, C.H. & Wang, L. (2013) Reliability-based robust geotechnical design of spread foundations using multi-objective genetic algorithm. *Computers and Geotechnics*, 48, 96–106.

Juang, C.H., Wang, L., Liu, Z., Ravichandran, N., Huang, H. & Zhang, J. (2013) Robust geotechnical design of drilled shafts in sand: New design perspective. *ASCE Journal of Geotechnical and Geoenvironmental Engineering*, 139 (12), 2007–2019.

Kulhawy, F.H. & Phoon K.K. (2002) Observations on geotechnical reliability-based design development in North America. In: *Proceedings, International Workshop on Foundation Design Codes and Soil Investigation in View of International Harmonization and Performance Based Design, Tokyo, Japan*. pp. 31–48.

Naghibi, M. & Fenton, G.A. (2011) Geotechnical resistance factors for ultimate limit state design of deep foundations in cohesive soils. *Canadian Geotechnical Journal*, 48 (11), 1729–1741.

Naghibi, F., Fenton, G.A. & Griffiths, D.V. (2014) Serviceability limit state design of deep foundations. *Geotechnique*, 64 (10), 787–799.

Paikowsky, S.G. (2002) Load and Resistance Factor Design (LRFD) for deep foundations. In: *Proceedings, International Workshop on Foundation Design Codes and Soil Investigation in View of International Harmonization and Performance Based Design, Tokyo, Japan*. pp. 59–94.

Paikowsky, S.G. (2004) *Load and Resistance Factor Design (LRFD) for Deep Foundations*. NCHRP Report 507. Washington, DC, Transportation Research Board.

Paikowsky, S.G., Canniff, M.C., Lesny, K., Kisse, A., Amatya, S. & Muganga, R. (2010) *LRFD Design and Construction of Shallow Foundations for Highway Bridge Structures*. NCHRP Report 651. Washington, DC, Transportation Research Board.

Park, G.J., Lee, T.H., Lee, K.H. & Hwang, K.H. (2006) Robust design: An overview. *AIAA Journal*, 44 (1), 181–191.

Phadke, M.S. (1989) *Quality Engineering Using Robust Design*. New Jersey Prentice Hall.

Phoon, K.K. (2015) Reliability of geotechnical structures. In *Proceedings: 15th Asian Regional Conference on Soil Mechanics and Geotechnical Engineering*, Japanese Geotechnical Society Special Publication, 2 (1), 1–9.

Phoon, K.K. & Kulhawy, F.H. (1999) Characterization of geotechnical variability. *Canadian Geotechnical Journal*, 36 (4), 612–624.

Phoon, K.K. & Kulhawy, F.H. (2008) Serviceability limit state reliability-based design. In: Phoon, K.K. (ed.) *Reliability-Based Design in Geotechnical Engineering: Computations and Applications*. London, Taylor & Francis. pp. 344–383.

Phoon, K.K. & Ching, J. (2013) Can we do better than the constant partial factor design format? In: *Modern Geotechnical Design Codes of Practice – Implementation, Application, and Development*. Amsterdam, IOS Press. pp. 295–310.

Phoon, K.K., Kulhawy, F.H. & Grigoriu, M.D. (1995) *Reliability-Based Design of Foundations for Transmission Line Structures*. Report TR-105000. Palo Alto, Electric Power Research Institute.

Phoon, K.K., Kulhawy, F.H. & Grigoriu, M.D. (2003a) Multiple resistance factor design (MRFD) for spread foundations. *ASCE Journal of Geotechnical and Geoenvironmental Engineering*, 129 (9), 807–818.

Phoon, K.K., Kulhawy, F.H. & Grigoriu, M.D. (2003b) Development of a reliability-based design framework for transmission line structure foundations. *ASCE Journal of Geotechnical and Geoenvironmental Engineering*, 129 (9), 798–806.

Phoon, K.K., Becker, D.E., Kulhawy, F.H., Honjo, Y., Ovesen, N.K. & Lo, S.R. (2003c) Why consider reliability analysis in geotechnical limit state design? In: *Proc. International Workshop on Limit State design in Geotechnical Engineering Practice (LSD2003), Cambridge*, CDROM.

Phoon, K.K., Ching, J. & Chen, J.R. (2013) Performance of reliability-based design code formats for foundations in layered soils. *Computers and Structures*, 126, 100–106.

Ravindra, M.K. & Galambos, T.V. (1978) Load and Resistance Factor Design for steel. *ASCE Journal of Structural Division*, 104 (ST9), 1337–1353.

Taguchi, G. (1986) *Introduction to Quality Engineering: Designing Quality into Products and Processes*. White Plains, New York, Quality Resources.

Task Committee on Structural Loadings (1991) *Guidelines for Electrical Transmission Line Structural Loading, Manual and Report on Engineering Practice*. Vol. 74. New York, ASCE.

Wang, Y., Au, S.K. & Kulhawy, F.H. (2011) Expanded reliability-based design approach for drilled shafts. *ASCE Journal of Geotechnical and Geoenvironmental Engineering*, 137 (2), 140–149.

Chapter 7

Direct probability-based design methods

Yu Wang, Timo Schweckendiek, Wenping Gong,
Tengyuan Zhao, and Kok-Kwang Phoon

ABSTRACT

This chapter focuses on recent development of direct probability-based design methods, including the expanded reliability-based design (expanded RBD) method, reliability-based robust geotechnical design (RGD) method, and the new safety standards for flood defenses in the Netherlands which is the first ever national standard that adopts direct (or full) probability-based design methods. One major criticism to the simplified semi-probabilistic RBD format is displacement of sound engineering judgment and lack of flexibility for practitioners. Because the simplified semi-probabilistic RBD format adopts the same trial-and-error approach as traditional allowable stress design (ASD) methods and it is developed to circumvent the need for practitioners to perform probabilistic analysis, these compromises seem unavoidable. An alternative solution to this dilemma is to maintain the engineering judgment and flexibility similar to ASD methods, but at the expense of performing probabilistic analysis using direct probability-based design methods. It is shown that, with the aid of commonly available computers and widely used computer software such as Microsoft Excel, performing Monte Carlo Simulation (MCS)-based probabilistic analysis and design are becoming more and more straightforward and convenient. MCS is already available in some commercial geotechnical software programs. MCS can be comprehended easily as a repetitive computer execution of traditional ASD design calculation, and the reliability analysis background required for performing MCS is substantially reduced. A gravity retaining wall design example is used in this chapter to illustrate the MCS-based design method in Excel.

7.1 INTRODUCTION

The new edition of ISO2394:2015 considers a design process as a decision making process which shall take basis in information concerning the implied risks (See Clause 4.4.1 of ISO2394:2015). Both failure probability and failure consequences (e.g., loss of lives and injuries, damages to the qualities of the environment, and monetary losses) are integral elements of risk. When the consequences of failure vary substantially and explicit quantification of failure consequences is needed, full risk-informed decision making is often performed (See Clause 7 of ISO2394:2015). For example, assessment on dam safety is frequently a risk-informed decision making process because consequences of dam failure certainly vary beyond normal ranges (e.g., Hartford and Baecher 2004). When the consequences of failure and damage are well understood and within normal ranges, reliability-based decision making can be applied instead of full risk-informed decision making. The reliability-based decision making may be directly based on probabilistic analysis and all specified reliability requirements are explicitly checked and satisfied (See Clause 8 of ISO2394:2015). Such an approach is referred to as direct (or full) probability-based design methods. Application of direct probability-based design methods requires the availability of uncertainty models, reliability methods, and expertise in probabilistic analysis, which are not always available in engineering practices. When in addition to the consequences also the failure modes and the uncertainty representation can be categorized and standardized, the design process may be further simplified a semi-probabilistic reliability-based design (RBD) format (See Clause 9 of ISO2394:2015), as discussed in Chapter 6.

The semi-probabilistic RBD methods have formats similar to the traditional allowable stress design (ASD) methods. The factor of safety (FS) in ASD methods is replaced by a combination of load and resistance factors (or material partial factors) in semi-probabilistic RBD methods. The same trial-and-error approach is used in both ASD and semi-probabilistic RBD methods, in which a trial design is proposed and checked against various design requirements, followed by revision of the trial design, if necessary. No probabilistic analysis is needed for practitioners when using semi-probabilistic RBD methods in engineering practice. The probabilistic aspect of semi-probabilistic RBD methods is reflected through some calibration processes during the RBD code development that produces a table of load and resistance factors (or material partial factors) for a given target probability of failure or reliability index. Examples are given in Table 6.1 and 6.2. Practitioners are only required to select appropriate load and resistance factors (or material partial factors) from the provided table during the design, and they are not involved in the code calibration processes or probabilistic analyses. This process precludes practitioners from performing probabilistic analysis and circumvents the need for practitioners to learn how to perform probabilistic analysis. However, it is a double-edged sword.

Because practitioners are not involved in the calibration processes, many assumptions and simplifications (e.g., uncertainty models, including calculation model uncertainty, probability distributions of loads and geotechnical properties, and propagation of these uncertainties) adopted in the calibration processes are frequently unknown to the practitioners. This situation can lead to potential misuse of the load and resistance factors (or material partial factors) that are only valid for the assumptions and simplifications adopted in the calibration processes. In other words, when using the load and resistance factors (or material partial factors), practitioners have to accept all the assumptions and simplifications adopted in the calibrations. More often than not, the assumptions and simplifications underlying the calibration processes are not stated in the design code (at least not stated explicitly). Only the final outcomes of the calibration processes in the form of load and resistance factors are presented. Practitioners may feel uncomfortable to accept these "black box" calibration processes blindly. In addition, practitioners have no flexibility in changing any of these assumptions/simplifications or making their own judgment because recalibrations are necessary when any assumption or simplification is changed. This situation leads to one major criticism of semi-probabilistic RBD methods: displacement of good geotechnical sense and sound engineering judgment which have been long considered as a critical element in geotechnical practice to cater for diverse property, load and resistance evaluation methodologies and diverse site conditions (Bolton 1983; Fleming 1989; Phoon 2008). In fact, existing LRFD codes that recommend a single value for each resistance factor is an example of straitjacketing that do not allow engineers to incorporate uncertainties specific to their sites/design scenarios into the design. It has been argued that engineers could apply their judgment by judicious selection of a cautious estimate of the nominal/characteristic resistance suitable for a particular site, but this approach is identical to what is being done in ASD. The uncertainties presented in Chapter 3 and 4 are not considered explicitly at the site level and Chapter 1 has argued that imposing on engineering judgment *alone* to assess a cautious estimate in the presence of auto- and cross-correlated input parameters and how these parameters manifest themselves in the response is onerous. Adoption of direct probability-based design methods is a possible mitigation to this criticism, particularly when the expertise in probabilistic analysis required in direct probability-based design methods is substantially reduced or practitioners are equipped with sufficient background in probability and statistics. With rapid development of computer technology and commonly available personal computers (PC), it is now possible to use a PC to perform probabilistic analysis with minimal expertise in probabilistic analysis, although having such expertise is always advantageous.

This chapter focuses on recent development of direct probability-based design methods. It starts with a non-exhaustive list of situations in which using direct probability-based design methods is beneficial and necessary, followed by some recently developed direct probability-based design methods, including the expanded reliability-based design (expanded RBD) method (Wang et al. 2011a; Wang 2011, 2013; Wang and Cao 2013a), reliability-based robust geotechnical design (RGD) method (Juang et al. 2013a, b; Juang and Wang 2013), and the new safety standards for flood defenses in the Netherlands (Schweckendiek et al. 2012; Schweckendiek et al. 2015) which is the first ever national standard that adopts direct (or full) probability-based design methods. Then, two important aspects (i.e., system reliability and reliability target) in the implementation of direct probability-based design methods are briefly reviewed. An example of a gravity retaining wall design is used to illustrate the expanded RBD method and its implementation in a widely used Excel spreadsheet platform. Quantification of uncertainty in soil properties from a limited number of site-specific standard penetration test (SPT) and triaxial test results is also performed using Excel spreadsheet and included in this illustration.

7.2 SITUATIONS OF DIRECT PROBABILITY-BASED DESIGN METHODS BEING NECESSARY

Although there are many cases in which semi-probabilistic design methods in a simplified RBD format are sufficient and probabilistic analysis can be bypassed, sometimes direct probability-based design methods are necessary and beneficial. A non-exhaustive list of such situations is provided below:

1. Out of the calibration domain for semi-probabilistic RBD codes
    Chapter 6 has emphasized the importance of clearly stating the salient features of the underlying calibration domain (e.g., range of pile diameters, pile lengths,

statistics of geotechnical parameters) that have been used during the development of semi-probabilistic RBD codes. The load and resistance factors (or material partial factors) are valid only when the design scenario is within the calibration domain. When the design scenario is out of the calibration domain (e.g., the pile diameters, pile lengths or statistics of geotechnical parameters are out of the range of those that have been used in calibration), it is inappropriate to use the load and resistance factors (or material partial factors) from semi-probabilistic RBD codes. In this case, it is necessary and beneficial to use direct probability-based design methods. In addition, most semi-probabilistic RBD codes are only calibrated for a single soil layer profile, it is inappropriate to directly apply the resulting load and resistance factors (or material partial factors) to a multiple soil layer profile, which is very common in geotechnical practice (see Chapter 6). Recalibration and further studies are needed for using semi-probabilistic methods in a multiple soil layer profile, as discussed in Chapter 6. Before completion of the recalibration and further studies, one feasible alternative to this multiple soil layer profile problem is direct probability-based design methods.

2.   Different calculation models

To carter for the diverse property, load and resistance evaluation methodologies and diverse site conditions in geotechnical practice, many different calculation models have been developed and adopted in traditional ASD practice. For example, bearing capacity of foundation can be estimated using many different calculation models. Some of them have theoretical basis, such as Terzaghi's and Vesic's bearing capacity equations, and some of them empirically correlate the foundation bearing capacity with results of some commonly used in-situ tests, such as SPT or cone penetration test (CPT). Models for settlement calculations are even more diverse. There are at least tens of different calculation models for estimating foundation settlement. In traditional ASD practice, practitioners have the flexibility to exercise their sound engineering judgment to decide which calculation model to use and how to use. In contrast, only one "best" calculation model is selected during the calibration and development of semi-probabilistic RBD codes. The load and resistance factors (or material partial factors) are valid only for the pre-selected calculation model. If the selected model is changed, recalibration is needed and the resulting load and resistance factors (material partial factors) are probably different. When using semi-probabilistic RBD codes, practitioners therefore have to stick to the pre-selected calculation models. They cannot exercise their sound engineering judgment or have the flexibility to use a calculation model that, in their opinions, best suits the design scenario and information in hand, if this calculation model is different from the pre-selected one. Direct probability-based design methods may be used when practitioners prefer to use different calculation models.

3.   Different uncertainty models

During the calibration of semi-probabilistic RBD codes, an uncertainty model is developed and adopted, although it is often opaque to practitioners. The uncertainty model generally includes: (i) decision on which variables are considered as uncertain; (ii) probabilistic modelling of the uncertain variables as random variables (e.g., probability distributions of the random variables); and (iii) auto- and cross-correlation structures. The uncertainty model is an integral part of the calibration process. If the uncertainty model changes, the load and resistance factors (or material partial factors) resulted from the calibration process are very likely to change too. In other words, the load and resistance factors (or material partial factors) are valid only for the specific uncertainty model adopted in calibration process. When using semi-probabilistic RBD codes, practitioners therefore have to accept the uncertainty model underlying the calibration of the load and resistance factors (or material partial factors). They cannot exercise their sound engineering judgment or have the flexibility to develop and adopt an uncertainty model that, in their opinions, best suits the design scenario and site-specific information in hand, if this uncertainty model is different from the one adopted in the calibration. For example, most, if not all, existing semi-probabilistic RBD codes consider the foundation allowable settlement as a deterministic value. However, Zhang and Ng (2005) showed that the foundation allowable settlement is indeed highly uncertain. This uncertainty has significant effect on probabilistic analysis and should be considered in foundation design (Wang et al. 2011a). The existing semi-probabilistic RBD codes cannot accommodate this uncertainty unless recalibration is performed with this uncertainty included in the recalibration. Before such a recalibration is available, direct probability-based design methods may be used in this case or other similar situations.

4. Different target failure probability, p_{ft}

Semi-probabilistic RBD codes often only provide load and resistance factors (or material partial factors) for one or a few pre-selected values of target failure probability, p_{ft}. It is therefore difficult for practitioners to adjust the design p_{ft}, unless recalibration is performed to obtain different sets of load and resistance factors (material partial factors) for other p_{ft} values. Adjustment of p_{ft} is beneficial in geotechnical practice because it allows practitioners to exercise their sound engineering judgment and have the flexibility to adjust the p_{ft} value to reflect the importance of the intended geotechnical structures and the consequence of failure. Such adjustment is common in traditional ASD practice where practitioners adjust the design FS to accommodate the different consequences of failure as one important consideration. Because the FS in ASD codes is replaced by a combination of load and resistance factors (or material partial factors) in semi-probabilistic RBD codes, it is difficult for practitioners to apply their experience in FS adjustment from ASD to semi-probabilistic RBD practice. Direct probability-based design methods are beneficial in this case where the p_{ft} value needed in the design is different from the one pre-specified in semi-probabilistic RBD codes. It is recognized that some codes mitigate this situation by allowing structures to be design according to different target failure probabilities associated with different reliability classes (Table B2 in EN1990:2002).

5. Exact value of failure probability is needed

The exact value of failure probability p_f may be needed in some engineering applications, such as quantitative risk assessment and risk-based decision making (See Clauses 7.1–7.5 of ISO2394:2015). Both failure probability and failure consequences (e.g., economic loss or number of fatality) are integral elements of risk. When the risk needs to be assessed quantitatively, the exact value of failure probability is also needed. In this case, direct probability-based design methods are necessary. It is also worthwhile to note that using the load and resistance factors (or material partial factors) in existing semi-probabilistic RBD codes does not guarantee achievement of the target failure probability prescribed in the codes, as discussed in Chapter 6 and some previous studies (e.g., Wang 2011; Wang 2013). When such a guarantee is needed, the failure probability needs to be evaluated explicitly using direct probability-based design methods. We use the term "exact" to sharpen the contrast between the ability of direct methods to achieve any prescribed p_{ft} to any desired accuracy and the ability of simplified methods to achieve only a few pre-selected values of p_{ft} and even for each p_{ft}, the actual p_f achieved is different from p_{ft}. It goes without saying that the accuracy of p_f is limited by available information and achieving anything more "exact" is not necessarily practically meaningful.

6. Correlated load and resistance

Although semi-probabilistic RBD methods have been successfully applied to foundation design where load and resistance are usually independent, it has been less satisfactory when applied to earth retaining structures or slopes (Christian and Baecher 2011; Wang 2013). One major challenge is that, for earth retaining structures and slopes, the load and resistance are usually originated from the same sources (e.g., effective stress of soil) and correlated with each other. It is therefore difficult to decide whether the effective stress of soil or earth pressure should be regarded as a load or resistance. This situation leads to a difficult but frequently asked question in the retaining wall design with Eurocode 7: should passive earth pressure be regarded as a resistance or load (e.g., Bond and Harris 2008, Wang 2013)? The answer to this question obviously has a significant bearing and may result in different designs, because the partial factors are different for resistances and loads. In addition, it is common that loads and resistances are modelled as independent random variables during RBD code calibrations. Although this uncertainty model is generally sufficient for foundations, it violates the fundamental physics for earth retaining structures or slopes because both load and resistance are originated from effective stress of soil and they intrinsically correlate with each other. Some direct probability-based design methods, such as expanded RBD method (Wang et al. 2011a; Wang 2013), can effectively bypass the difficulty in handling the correlated load and resistance. Details are provided in the next section.

7. Serviceability limit state (SLS) design

Most existing semi-probabilistic RBD codes, except a few (e.g., Phoon et al. 1995), only deal with ultimate limit state (ULS) design of geotechnical structures, without considering serviceability limit state (SLS) design. It is obvious both ULS and SLS designs are necessary for geotechnical structures. In the absence of semi-probabilistic RBD codes for SLS design, direct probability-based design methods may be used to fill the gap.

8. Rock Engineering

It may be difficult to apply a semi-probabilistic RBD approach to rock engineering which is typically governed by geometric uncertainties such as orientation of joints that cannot be easily factored in the conventional LRFD way. In addition, the associated failure wedges are more myriad than a bearing capacity mechanism in soil. As mentioned in Chapter 1, the Commission on Evolution of Eurocode 7 hosted by the International Society for Rock Mechanics (https://www.isrm.net/gca/index.php?id=1143) noted that the partial factor approach in Eurocode 7 "is in many ways inappropriate – and in some circumstances inapplicable – to rock engineering."

One common feature in the situations listed above is displacement of sound engineering judgment and lack of flexibility for practitioners. Because semi-probabilistic RBD codes adopt the same trial-and-error approach as traditional ASD methods and it is developed to circumvent the need for practitioners to perform probabilistic analysis, some compromises seem unavoidable, such as displacement of sound engineering judgment and lack of flexibility for practitioners. One possible solution to this trade-off is to maintain the engineering judgment and flexibility similar to ASD methods, but at the expense of performing probabilistic analysis using direct probability-based design methods.

Similar to the semi-probabilistic RBD methods, the objective of direct probability-based design is to find a set of design parameters such that a prescribed target failure probability p_{ft} is achieved. A trial-and-error approach may be used to adjust the design parameters until the p_f of the trial design is smaller than the prescribed p_{ft}. The failure probability for each trial design is evaluated directly and explicitly using reliability analysis methods, such as first order second moment method (FOSM), first or second order reliability method (FORM or SORM), or Monte Carlo simulation (MCS). Although application of direct probability-based design methods is relatively rare for conventional types of geotechnical structures with no exceptional risk or difficult ground or loading conditions (e.g., the Geotechnical Category 2 defined in Eurocode 7, such as spread foundations, raft foundations, pile foundations, retaining walls, excavations, bridge piers and abutments, embankments and earthworks, ground anchors and other tie-back systems), direct probability-based design methods have been commonly used in several geotechnical related engineering fields, such as earthquake engineering, offshore engineering, dam engineering and nuclear engineering (e.g., site selection and foundation design for dam and nuclear facilities). For example, probabilistic seismic hazard analysis (PSHA, Cornell 1968; Reiter 1990) is a routine element in earthquake engineering practice and there is an entire research community devoted to PSHA. Liquefaction potential of soils may be evaluated probabilistically using results of either SPT (e.g., Cetin et al. 2004; Juang et al. 2008, 2013c) or CPT (e.g., Juang et al. 2000, 2006). Offshore pile foundations are frequently designed using probability-based methods (e.g., Tang et al. 1990; Lacasse et al. 2013; Chen and Gilbert 2014).

Since sufficient expertise in reliability analysis is needed for practitioners to use direct probability-based design methods and practitioners are not necessarily experts in reliability analysis, the need to gain new knowledge becomes a major hurdle for adoption of direct probability-based design methods in geotechnical practice. To remove this hurdle, new direct probability-based design methods for geotechnical structures have been recently developed, such as the expanded RBD method (Wang et al. 2011a; Wang 2011; and Kulhawy et al. 2012). Under the expanded RBD method, the burden imposed on practitioners to learn new reliability concepts is substantially reduced. The design process is conceptualized as a systematic sensitivity study in this method, which is common in geotechnical practice and familiar to practitioners. In such a sensitivity study, a large number of design alternatives (or trial designs) are evaluated systematically, and the optimal design with the maximum utility and satisfying the reliability requirements is chosen as the final design. Details of the expanded RBD method are provided in the following section.

7.3 EXPANDED RELIABILITY-BASED DESIGN (EXPANDED RBD) METHOD

The expanded RBD method (Wang et al. 2011a; Wang 2011; and Kulhawy et al. 2012) formulates the geotechnical design process as an expanded reliability problem in which a single run of MCS is used in a PC to address, explicitly and simultaneously, the ULS, SLS, economically-optimized limit state (EOLS, a measure of utility, Wang and Kulhawy 2008; Wang 2009), and reliability requirements of the design. An expanded reliability problem, as described herein, refers to a reliability analysis of a system in which a set of system design parameters are considered artificially as uncertain with probability distributions specified by the user for design exploration purposes. Consider, for example, a pile foundation design in which design parameters are pile depth D and diameter B. Both D and B are treated artificially as discrete uniform random variables. Then, the design process is considered as a process of finding failure probabilities for design alternatives with various combinations of B and D [i.e., conditional probability p(Failure|B, D)] and comparing them with a target probability of failure p_{ft}, which could be a ULS or SLS requirement. A single run of MCS with a total sample number n is performed to evaluate p(Failure|B, D), as illustrated in Figure 7.1. The traditional ASD calculations are repeated n times in the MCS, and it is equivalent to a systematic sensitivity study that contains n different design cases with various input parameters and/or design parameters. The conditional failure probability p(Failure|B, D) is calculated from the MCS results as (Wang et al. 2011a and Wang 2011):

$$p(\text{Failure}|B, D) = \frac{p(B, D|\text{Failure})}{p(B, D)} p_f \qquad (7.1)$$

in which p(B, D|Failure) = conditional joint probability of B and D given failure. Since B and D are independent discrete uniform random variables, p(B,D) in Eq. 7.1 is expressed as:

$$p(B, D) = \frac{1}{n_B n_D} \qquad (7.2)$$

in which n_B and n_D = number of possible discrete values for B and D, respectively. The quantities p(B, D|Failure) and p_f in Eq. 7.1 are estimated using a single run of MCS in a PC.

Monte Carlo Simulation

Monte Carlo Simulation (MCS) is a numerical process of repeatedly calculating a mathematical or empirical operator in which the variables within the operator are random or contain uncertainty with prescribed probability distributions (e.g., Ang

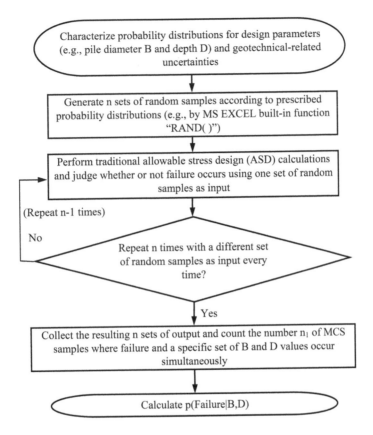

Figure 7.1 Flow chart for Monte Carlo Simulation in expanded RBD method (Modified from Wang et al. 2011a).

and Tang 2007). With the aid of commonly available PC and computer software such as Microsoft Excel, performing MCS is becoming more and more straightforward and convenient. The reliability analysis background required for performing MCS is substantially reduced when using some built-in functions and add-ins in Excel.

Consider the expanded RBD method above. The mathematical operator involves calculation of load (L) and resistance (R) and judgment of whether failure occurs. Failure here does not refer to catastrophic collapse of geotechnical structures, but only refers to events in which the load exceeds resistance (i.e., $L > R$) or some limit states are exceeded. Figure 7.1 shows MCS procedures schematically for the expanded RBD method. The MCS starts with characterization of probability distributions for the design parameters (e.g., B and D for a drilled shaft) and geotechnical-related uncertainties that are considered in the expanded RBD method. In addition to the discrete uniform distributions specified for design parameters (e.g., B and D), proper probability distribution functions are used to model uncertainties that arise in loads, geologic site interpretations, geotechnical properties, and computational models. For example, effective friction angle ϕ' of soil can be modeled by a lognormal distribution (e.g., Phoon et al. 1995; Wang et al. 2011a) or a site-specific probability distribution estimated from site investigation data (e.g., Wang et al. 2015 & 2016). Then, repeated random samples of the uncertain variables (e.g., B, D, and other uncertainties considered) are generated from their respective probability distributions, followed by repeated calculation of L and R and judgment of whether failure occurs using each set of random samples as input. Finally, statistical analysis of the output is performed to estimate p_f and p(B, D|Failure) as:

$$p_f = \frac{n_f}{n} \qquad (7.3)$$

$$p(B, D|\text{Failure}) = \frac{n_1}{n_f} \qquad (7.4)$$

in which n = total number of MCS samples, n_f = number of MCS samples where failure occurs, and n_1 = number of MCS samples where failure and a specific set of B and D values occur simultaneously. Note that a total number $n_B \times n_D$ of n_1 values is obtained from a single run of MCS, and each n_1 value corresponds to a possible combination of B and D. Combining Eqs 7.1–7.4 leads to:

$$p(\text{Failure}|B, D) = \frac{n_B n_D n_1}{n} \qquad (7.5)$$

Note that n_B, n_D and n are pre-specified by practitioners before MCS, and n_1 and n_f are obtained by simply counting the numbers of failure samples for each combination of B and D and the total failure samples in MCS. Estimation of the p(Failure|B, D) using Eq. 7.5 is therefore straightforward.

Figure 7.2 shows an illustration of the p(Failure|B, D) obtained from MCS. Note that the relationships given in Figure 7.2 are variations of p_f as a function of the design parameters B and D that represent different designs. From this perspective, these are results of a sensitivity study on p_f versus the design parameters. Feasible designs can be inferred directly from the figure, and they are those with p(Failure|B, D) $\leq p_{ft}$. The feasible designs satisfy the ULS, SLS, and reliability requirements. Then, the EOLS requirement should be adopted to finalize the design, which will be the one with the minimum construction cost. Wang and Kulhawy (2008) outlined a straightforward optimization process that allows the incorporation of ULS and SLS designs with construction costs to select the most cost-effective geotechnical structures among those being considered. The construction costs for geotechnical structures may be estimated using published, annually-updated, unit cost data, such as Means Building Construction Cost Data (e.g., Means 2007). The construction costs for all feasible designs are calculated as the product of their unit costs and respective design parameters, and the final design is determined by comparing their construction costs. The final design therefore satisfies the ULS, SLS, EOLS, and reliability requirements.

Although MCS has the advantage of conceptual and mathematical simplicity and can be comprehended easily as a repetitive execution of traditional ASD design, there is a question about how many MCS samples are necessary to ensure a desired level of accuracy in the results. As a rule of thumb, the number of MCS samples should be at least ten times greater than the reciprocal of the probability level of interest, i.e., p_{ft}

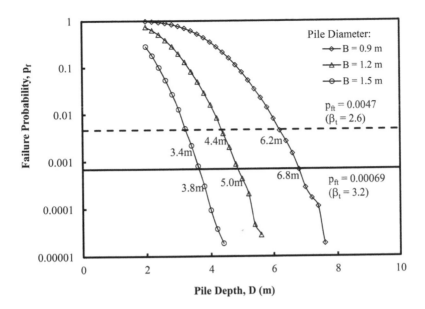

Figure 7.2 An illustration of conditional failure probability from MCS (Modified from Wang et al. 2011a).

(e.g., Roberts and Casella 1999; Wang et al. 2011a). For the expanded RBD method, the minimum number of samples (n_{min}) can be estimated as follows:

$$n_{min} = \frac{10 n_B n_D}{p_{ft}} \qquad (7.6)$$

The n_{min} value estimated from Eq. 7.6 increases rapidly and the computation effort required increases significantly as the number of design parameters and their possible combinations increases. When the computational effort is a concern and improvement of the computational efficiency for MSC is preferred, advanced MCS, such as Subset Simulation (Au and Beck 2001; Au and Wang 2014), may be used together with the expanded RBD method. Subset Simulation can be implemented conveniently in Excel spreadsheet (Au et al. 2010; Wang et al. 2011b). Examples of using expanded RBD method together with Subset Simulation in Excel spreadsheet are referred to Wang and Cao (2013a).

Advantages of expanded RBD method

The expanded RBD method deals rationally with several important characteristics of geotechnical engineering practice, making it perhaps particularly suitable for geotechnical structures. Some advantages of the expanded RBD method are highlighted below:

(1)   The ULS and SLS calculation models in the expanded RBD method are established in the same way as those in traditional ASD methods. Practitioners therefore have the same flexibility to select and use, in their opinions, the "best"

calculation models and make appropriate design assumptions and modifications that best suit the design situation in a particular project. This allows practitioners to exercise their practical good sense and sound engineering judgment, which is always considered as a critical element in geotechnical engineering practice, in a way similar to the traditional ASD methods.

(2)   The uncertainties are modelled explicitly and directly. Practitioners have the flexibility to include uncertainties deemed appropriate and to model the uncertainties in soil and rock properties on a site-specific basis, which is necessary and beneficial because of the site-specific nature of ground properties.

(3)   It decouples reliability assessment from traditional ASD design calculations, and it has a unique capability of dealing rationally with correlated load and resistance (e.g., earth retaining structures and slopes). The expanded RBD method is a repetitive computer execution of traditional ASD design calculations, and the correlation between the load and resistance is implicitly considered in the traditional ASD calculation models. The reliability assessment is performed by MCS, i.e., repetitive computer execution of the traditional ASD calculation models, and the difficulty in handling the correlated load and resistance in semi-probabilistic RBD codes is bypassed in this MCS-based method.

(4)   It is able to properly handle system reliability problems (e.g., multiple failure modes or complex system structure). As far as the multiple failure modes and interaction among various system components are represented properly in the traditional ASD calculation models, the MCS-based reliability assessments in the expanded RBD method is simply a repetitive computer execution of the ASD calculation model. No sophisticated system reliability analysis is required. The process for system reliability problems is largely identical to that for single failure mode or component reliability problems. As discussed in Section 7.6, the majority of geotechnical problems are in fact system reliability problems, and therefore, the expanded RBD method is particularly suitable for geotechnical practice.

(5)   It provides practitioners the flexibility to adjust p_{ft}, without additional calculations, for accommodating specific project needs (see Section 7.7). It also offers additional insight into how the expected performance level changes as the design parameters change.

Finally, the MCS in expanded RBD method is conceptually and mathematically simple (i.e., it is just a repetitive computer execution of the ASD calculations). It can be implemented easily in a spreadsheet environment (e.g., Microsoft Excel). This is particularly convenient for practitioners who frequently perform design calculations using spreadsheets. With rapid development of modern computer technology, thousands of MCS samples can be generated and calculated for conventional foundation designs within seconds. An illustrative example of using expanded RBD method in Excel spreadsheet is provided in Section 7.8.

7.4 RELIABILITY-BASED ROBUST GEOTECHNICAL DESIGN (RGD)

Due to the limited data available, the probability distributions of input parameters and solution models may not be able to be accurately characterized. A new design philosophy called robust geotechnical design (RGD) was recently proposed by Juang et al. (2013a&b) to address such a circumstance. Robust geotechnical design aims to make the response of a geotechnical system robust against, or insensitive to, the variation of uncertain input parameters (referred to as noise factors in RGD). For example, the uncertainties in soil properties are difficult to quantify due to insufficient data, and they may be treated as noise factors in RGD. By systematically changing design parameters, RGD is realized through a multi-objective optimization that explicitly considers all design requirements such as safety, robustness and cost. The results of such an optimization are expressed as a Pareto Front, which is a collection of optimal designs that collectively defines a trade-off relationship between cost and robustness. It is found that the cost generally increases as the robustness improves. Thus, the robustness may be considered as additional conservativeness that is invested in the design and accounts for the hard-to-control (i.e., cannot be easily adjusted by the practitioner) and hard-to-characterize (i.e., the uncertainty is recognized but hard to quantify due to insufficient data) noise factors, such as uncertainty in soil properties. Note that the optimal designs on the Pareto Front meet all the safety requirements. The Pareto Front therefore could be adopted by practitioners to make an informed design decision according to a target cost or robustness.

According to the uncertainty characterization of noise factors, three levels of robust geotechnical design (RGD) could be implemented: (1) site-specific data or knowledge is quite limited and the noise factors could only be characterized by the upper bounds and the lower bounds, the fuzzy set-based RGD (Gong et al. 2014a&b) might be employed; (2) more data availability is achieved and the noise factors may be characterized by probability distributions, however, the statistical information of the corresponding probability distributions (e.g., coefficient of variation and type of distribution) could not be calibrated accurately, the reliability-based RGD (Juang et al. 2013a&b; Juang and Wang 2013; Khoshnevisan et al. 2014) or the sensitivity-based RGD (Gong et al. 2014c; Gong et al. 2016) might be adopted; and (3) sufficient site-specific data or knowledge is available and the statistical information of the probability distributions of noise factors could be accurately characterized, the direct reliability-based method, such as expanded RBD (Wang et al. 2011a; Wang 2011, 2013; Wang and Cao 2013a), can be adopted. The focus of this section is placed on the reliability-based RGD, while the other RGD approaches could be referred to the references listed above.

Within the framework of the reliability-based RGD (Juang et al. 2013a&b; Juang and Wang 2013), the variation of the failure probability of the geotechnical design, caused by the uncertainty in the probability distributions of noise factors, is explicitly considered; and, the essence of reliability-based RGD is to seek an optimal design that simultaneously minimizes the variation of the failure probability (i.e., robust requirement) and the cost (i.e., economic requirement) while meets the target failure probability (i.e., safety requirement). In reference to Juang et al. (2013b), the steps of reliability-based RGD can be summarized in what follows:

(1)   Define the problem of concern and classify all input parameters of the design geotechnical structures into design parameters and noise factors. For the given geotechnical structures, the traditional ASD calculation models can be established.

(2)   Estimate statistics of uncertain parameters, quantify the uncertainty in the statistics of noise factors, and identify the design domain. For geotechnical structures,

key uncertain soil parameters are usually treated as noise factors. The uncertainty in the statistics (e.g., coefficient of variation) of each noise factor may be characterized based upon published literatures and engineering judgment. The design domain generally includes typical ranges of the design parameters. These design parameters might be specified in discrete numbers, leading to a design domain consisting of a finite number (M) of design alternatives.

(3)   Evaluate the robustness of a given design. The system performance of concern of a design alternative, in the context of reliability-based RGD, is the failure probability, and the robustness of a design alternative is therefore measured by the standard deviation of the failure probability. Reliability analysis is performed in this step to evaluate the failure probability and its statistics (i.e., mean and standard) for a given design alternative.

(4)   Repeat Step 3 for all design alternatives specified in Step 2. For each design alternative, the mean and standard deviation of the failure probability are determined.

(5)   Carry out a multi-objective optimization to establish a Pareto Front and choose the most preferred design from the Pareto Front. In this multi-objective optimization, the mean of the failure probability is the design constraint that must be less than the target failure probability, while the standard deviation of the failure probability and the cost are design objectives to be minimized.

Details of the reliability-based RGD methodology and application examples could be referred to a series papers by Juang and his co-workers (e.g., Juang et al. 2013a&b; Juang and Wang 2013).

7.5   THE NEW SAFETY STANDARDS FOR FLOOD DEFENSES IN THE NETHERLANDS

The current requirements of dike safety for flood defenses in the Netherlands are stipulated by the Dutch Flood Defense Act from 1996. Different from most countries, where safety standards and codes of practice refer to design of new structures, the Dutch safety standards employ periodic safety assessments of the existing structures to warrant an appropriate protection level from flooding. When dikes are found to be unsafe in the 6-yearly or 12-yearly assessments, the responsible authorities need to strengthen the structures for achieving the required safety standards. There is a Dutch national dike reinforcement program (HWBP, in Dutch: Hoogwaterbeschermingsprogramma) with a yearly budget of roughly 360 million Euro with the objective of bringing all non-compliant flood defenses up to the standards until roughly the year 2050 (Schweckendiek et al. 2015).

In autumn 2014, the Dutch minister of Infrastructure and the Environment announced that the safety standards established in 1996 will be updated and legally established in 2017. The new safety standards will change the definition for safety standard from an exceedance probability of a normative load event (e.g., a design water level corresponds to an annual exceedance frequency of 1/2,000) to an acceptable annual probability of flooding. The main difference between the current and the new definitions is that, while the current safety levels purely referred to the design hydraulic

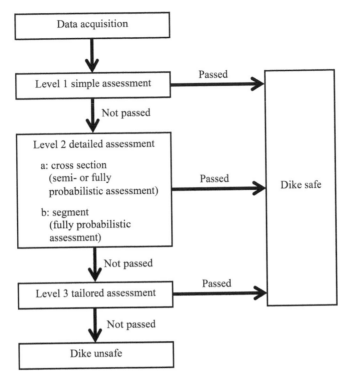

Figure 7.3 Assessment framework and process for the new safety standards in the Netherlands (after Schweckendiek et al. 2015).

load conditions, the new version explicitly includes the resistance of the flood defenses with all associated uncertainties.

The change in safety standard and assessment criterion towards an acceptable probability of flooding has created the need for new safety assessment methods, rules and tools. The national project dedicated to this, WTI-2017, is developing such methods and tools for the first round of the new type of safety assessments starting in 2017. Direct probability-based design methods are adopted, and WTI-2017 aims to facilitate the methods and tools for semi-probabilistic assessments in 2017 and for fully probabilistic analyses of the most important failure mechanisms in 2019 (Schweckendiek et al. 2015). An overview of the safety assessment framework and process is shown in Figure 7.3. The safety assessment is based on evaluation of the relevant failure mechanisms, such as overtopping/overflow, macro instability, and internal erosion/piping. After identifying the relevant failure mechanisms for the dike under investigation, the level 1 simplified assessment is carried out. Level 1 assessments are typically based on the characteristic values of the loading (i.e. water level and wave characteristics, if applicable) and rather easy-to-obtain geometrical parameters. If a failure mechanism cannot be ruled out in level 1, it proceeds to the level 2 detailed assessments. At this level typically physics-based assessment models are used for each failure mechanism, such as limit equilibrium models for slope stability or Sellmeijer criterion for backward internal erosion. Note that the increased level of detail in the assessments demands an increased level of detail in the input data, especially on ground conditions and geotechnical properties.

The most noteworthy feature on level 2 is that the assessment can be semi-probabilistic on a cross section of a dike with characteristic values and partial safety factors (i.e., semi-probabilistic RBD format such as load and resistance factor design) or fully probabilistic, i.e., direct probability-based design methods. The fully probabilistic assessment can be made for one mechanism on a cross section of a dike (level 2a) or for the combination of several mechanisms and sections in one dike segment (level 2b), which is typically in the order of tens of kilometers long. It is explicitly recognized that, compared with fully probabilistic assessments, simplifications made in semi-probabilistic assessments come at the cost of additional conservatism. Therefore, one objective of the WTI-2017 project is to ensure consistency between the semi-probabilistic and the fully probabilistic approaches by calibrating partial safety factors to the target probabilities of failure established in the new safety standards (Schweckendiek et al. 2012; Huber et al. 2015).

If the safety requirements are found to be unsatisfactory after the detailed assessments, level 3 tailored assessments can be considered. In this level, suitable state-of-the-art modeling and/or monitoring techniques may be used. The main consideration to move or not to level 3 is if one expects the extra efforts and investments in data acquisition and modeling to pay off in terms of a sufficiently distinct or more accurate assessment compared to level 2. Finally, if none of the level 1–3 assessments above permits the conclusion that the dike is safe with respect to all failure mechanisms, the dike is considered unsafe and needs to be strengthened.

7.6   SYSTEM RELIABILITY

Although traditional reliability-based design and decision making are primarily applied to elements and individual limit states (e.g., ULS or SLS failure), systems behavior is of concern because systems failure is usually the most serious consequence associated with failure of a structure. It is therefore of interest to assess the probability of system failure following an initial element failure, as highlighted in Clause 8.3 of ISO2394:2015. This is particularly true for geotechnical structures.

Because the majority of geotechnical structures have multiple failure modes, most geotechnical reliability analysis problems are indeed system reliability problems. For example, a simple gravity retaining wall has at least three failure modes: horizontal sliding along the base of the wall, overturning or rotation about the toe of the wall, and bearing capacity failure of the soil beneath the wall. These failure modes tend to interact among each other, because loads and resistances for different failure modes are correlated. For example, self-weight of a gravity retaining wall, which is the major source of resistance against sliding and overturning failure modes, but at the same time, is also a major source of load for bearing capacity failure mode.

A pile foundation for high-rise buildings is often a system of piles, which consists of several pile groups, each group consisting of a few individual piles. The failure of the pile foundation is initiated by yielding of an individual pile within the system. Collapse of the pile system occurs when each of the individual piles within the system collapses and the pile system is unable to accommodate additional load. The evaluation of the reliability of the pile system requires the consideration of the reliability of the individual piles, the pile group effects, and the system effects arising from pile-superstructure interactions (Zhang et al. 2001).

Many geotechnical structures form failure mechanisms in the surrounding soil mass in the ultimate limit state (e.g., slopes, tunnels, deep excavations). For example, a soil slope may have many potential slip surfaces, and each potential slip surface in the soil mass is a failure mode. Therefore, the slope stability problem contains many failure modes, and it is a system reliability problem. This system reliability characteristic can be even more distinct for rock slopes where multiple failure wedges can easily formed. Reliability analysis of slope stability has been performed using FORM. However, only one slip surface (i.e., the so-called most critical slip surface) or a single failure mode is usually used in FORM, and the probability of failure associated with this "most likely" failure mode identified by FORM only provides a lower bound for the system probability of failure. The failure probability from FORM therefore can be significantly underestimated and biased towards the unconservative side (Ching et al. 2009; Wang et al. 2011b; Zhang et al. 2011). Some system reliability methods have been developed recently to address the variation of slip surfaces, particularly when the spatial variability of soil properties is modelled in the analysis (Zhang et al. 2011; Li et al. 2013; Li et al. 2014).

In contrast to a pile foundation where the sliding surface is mostly restricted to the interface between soil and pile, the trajectory of a slip surface in a soil mass is coupled to the specific realization of a random field and can only be determined through numerical analysis. This class of system reliability problems is complex, because of the coupling between mechanics and spatial variability. However, it is not uncommon in geotechnical engineering. MCS-based methods can be used as a viable and unbiased way of estimating system reliability and handling the coupling between mechanics and spatial variability. The multiple failure modes and interaction among various system elements are modeled explicitly in the traditional ASD calculation models, and the spatial variability is modelled separately in the uncertainty models. Then, the MCS-based reliability assessments is simply a repetitive computer execution of the ASD calculation model using samples generated based on the uncertainty models. No sophisticated system reliability analysis is needed. The process for system reliability problems is largely identical to that for single failure mode or element reliability problems.

When the traditional ASD calculation models are complex, the computational time and efforts required for the MCS-based methods might be expensive. In this case, improvement of computational efficiency is preferred, and advanced MCS methods can be used, such as Subset Simulation (Au and Beck 2001; Au and Wang 2014) and important sampling. Both subset simulation and importance sampling have been implemented successfully in the reliability analysis of slope stability (e.g., Ching et al. 2009; Au et al. 2010; Wang et al. 2011b). An Excel-based software package called UPSS (Uncertainty Propagation using Subset Simulation) has been developed to implement Subset Simulation in Excel. UPSS can be obtained from the following web page: https://sites.google.com/site/upssvba/. Examples of using the Excel-based Subset Simulation in foundation and slope stability problems are referred to Wang and Cao (2015).

Table 7.1 Summary of reliability index, probability of failure, and corresponding expected performance level (after U. S. Army Corps of Engineers 1997).

Reliability Index β	Failure Probability $p_f = \Phi(-\beta)$	Expected Performance Level
1.0	0.16	Hazardous
1.5	0.07	Unsatisfactory
2.0	0.023	Poor
2.5	0.006	Below average
3.0	0.001	Above average
4.0	0.00003	Good
5.0	0.0000003	High

Note: $\Phi(\cdot)$ = standard normal cumulative distribution function.

Table 7.2 Summary of target reliability index β_t in several geotechnical RBD codes.

Design Code	ULS β_t	SLS β_t
Electric Power Research Institute (EPRI) multiple resistance and load factor design (MRFD)	3.2	2.6
Canadian Highway Bridge Design Code (CHBDC 2014)	3.1–3.7	2.3–3.1
Canadian National Building Code (NCBC)	3.5	Not available
American Association of State Highway and Transportation Official (AASHTO) foundation design code	2.0–3.5	Not available
Eurocode 7*	4.7	2.9

Note: Reference period of the reliability indices is 1 year, except CHBDC 2014 in which a period of 75 years is used (Fenton et al. 2016); *: Refer to Reliability Class 2 (RC2) in Eurocode.

7.7 RELIABILITY TARGET

As discussed in Clause 8.4 of ISO2394:2015, the target failure probability p_{ft} (or target reliability index β_t) should depend on the consequence and the nature of failure, the economic losses, the social inconvenience, effects to the environment, sustainable use of natural resources and the amount of expense and effort required to reduce the probability of failure. Table 7.1 summarizes various expected performance levels adopted by U. S. Army Corps of Engineers and their corresponding reliability indices and failure probability. The reliability indices for most structural and geotechnical components lie between 1 and 4, corresponding to p_f ranging from about 16% to 0.003%, as shown in Table 7.1. Table 7.2 summarizes target reliability indices recommended in several geotechnical RBD codes (mostly foundation design codes). The reliability target varies for different limit states and in different codes. It is worthy to note that a foundation often consists of many components such as individual piles. The levels of reliability of the entire foundation and the components are often not identical. The components should be so designed that the entire foundation satisfies the reliability target. The selection of reliability target for an individual pile in a pile system has been demonstrated by Zhang et al. (2001).

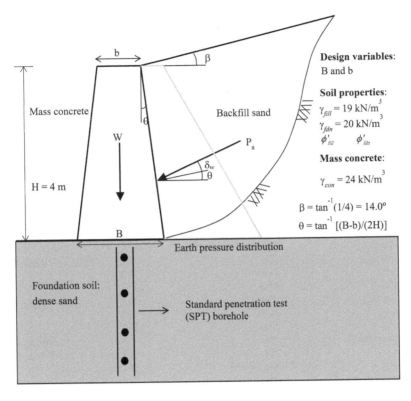

Figure 7.4 Gravity retaining wall design example.

7.8 GRAVITY RETAINING WALL DESIGN EXAMPLE

A gravity retaining wall design example is used in this section to illustrate the expanded RBD method. Figure 7.4 show a design of a mass concrete gravity retaining wall with a symmetrical trapezoid cross section. The wall height is $H = 4\,m$, and the bottom and top wall width are B and b, respectively. B and b are design variables in this example. The slope angle of the backfill soil is β, the angle between the back face of the wall and the vertical is θ. Note that $\theta = \tan^{-1}[(B - b)/(2H)]$ in this example. Site investigation is performed to obtain the soil properties required in the design. A borehole is drilled under the intended retaining wall, and SPT is performed in the borehole. Table 7.3 summaries the corrected SPT N (i.e., $(N_1)_{60}$) values obtained from the tests for foundation soil. Using the soil samples obtained from the borehole, the foundation soil is found to be dense sand with a unit weight $\gamma_{fdn} = 20\,kN/m^3$. In addition, laboratory tests are performed for the coarse-grained backfill soil. The backfill unit weight is $\gamma_{fill} = 19\,kN/m^3$. Two triaxial tests are also carried out for the backfill soil to obtain its effective friction angle, ϕ'_{fill}, as $36.3°$ and $38.6°$, respectively. The unit weight of concrete is $\gamma_{con} = 24\,kN/m^3$. The other geotechnical properties required for the design include effective friction angle of the foundation soil, ϕ'_{fdn},

Table 7.3 Summary of SPT test results for foundation soil.

Depth (m)	Corrected SPT N results, $(N_1)_{60}$
0.5	11.7
2.0	9.8
3.5	11.9
5.0	27.8

Table 7.4 Summary of uncertainty model.

Random Variables	Statistics	Values	Distribution Type
Effective friction angle of backfill soil, ϕ'_{fill}	Mean	37.4°	Figure 7.7 (from
	Standard deviation	4.6°	BEST Excel Add-in)
Effective friction angle of foundation soil, ϕ'_{fdn}	Mean	39.6°	Figure 7.9 (from
	Standard deviation	4.4°	BEST Excel Add-in)
Ratio of δ_w to ϕ'_{fill}, i.e., $r_w = \delta_w/\phi'_{fill}$	Min*	0.5	Triangle distribution
	Peak*	2/3	
	Max*	1.0	
Ratio of δ_b to ϕ'_{fdn}, i.e., $r_b = \delta_b/\phi'_{fdn}$	Min*	0.5	Triangle distribution
	Peak*	2/3	
	Max*	1.0	
Wall bottom width, B	Min*	1.8 m	Discrete uniform distribution
	Max*	3.2 m	
	Increment	0.2 m	
Difference between wall bottom and top width, $x = B - b$	Min*	0.5 m	Discrete uniform distribution
	Max*	1.5 m	
	Increment	1.0 m	

Note: *Min = Minimum, Max = Maximum, Peak = the most probable value.

soil-wall interface friction angle δ_w, and the interface friction angle between the base of the wall and the foundation soil δ_b.

Based on the information above, the expanded RBD method is used to design the gravity retaining wall. Details of the method are illustrated step by step in the following three subsections: uncertainty model, deterministic calculation model (i.e., traditional ASD calcualtion model), and MCS.

Uncertainty model

Table 7.4 summarizes the uncertainty model used in this example. Under the expanded RBD method, B and b are design parameters and they are treated as discrete random variables uniformly distributed over a possible range. Because H = 4 m, a typical B range suggested by Geoguide 1 in Hong Kong (GEO 1993) is 0.5 H–0.7 H or 2.0 m–2.8 m. Therefore, a slightly bigger range of [1.8 m, 3.2 m] with an increment of 0.2 m is adopted in this example. The number of B values $n_B = 8$. To simplify the ASD calculation model (see the next subsection), a new design parameter x = B − b is used to replace b. A possible range of [0.5 m, 1.5 m] with an increment of 1 m is

Table 7.5 Summary of prior knowledge for ϕ'_{fill} and ϕ'_{fdn}.

Random variable Statistic	ϕ'_{fill} (°)		ϕ'_{fdn} (°)	
	Mean	Standard Deviation	Mean	Standard Deviation
Min	25	1.25	34	1.70
Max	45	6.75	45	6.75

adopted for x. In other words, two possible x values of 0.5 m and 1.5 m are considered in this example, and the number of x values $n_x = 2$.

In this example, four geotechnical parameters are considered uncertain and treated as random variables, including ϕ'_{fill}, ϕ'_{fdn}, δ_w, and δ_b. Because δ_w and δ_b are often linked to ϕ'_{fill} and ϕ'_{fdn}, respectively, in geotechnical practice, the ratios between them (i.e., $r_w = \delta_w/\phi'_{fill}$ and $r_b = \delta_b/\phi'_{fdn}$) are treated as random variables. A triangle distribution within a range of [0.5, 1.0] and having a peak value at 2/3 is used to model r_w and r_b. Using the site investigation data mentioned above, Bayesian equivalent sample method (Wang and Cao 2013b, Wang et al. 2016) described in Section 3.9 of Chapter 3 is applied to quantify the uncertainty in ϕ'_{fill} and ϕ'_{fdn}. The Excel-based Bayesian equivalent sample toolkit (BEST) is used in this example. The BEST Excel toolkit can be downloaded freely from https://sites.google.com/site/yuwangcityu/best/1.

The BEST Excel Add-in is used to integrate the site-specific test data with engineering experience and judgment (referred to as prior knowledge in Bayesian methods). Then the integrated knowledge is transformed into a large of number of equivalent sample through Markov chain Monte Carlo simulation. Table 7.5 summarizes the prior knowledge used for ϕ'_{fill} and ϕ'_{fdn} in this example. A uniform distribution is adopted to represent the relatively uninformative prior knowledge (Cao et al. 2016). Only the typical ranges (i.e., the maximum and minimum values) are needed to define the uniform distribution. Note that, because foundation soil is known as dense sand, the minimum value of effective friction angle for the foundation soil is larger than that for the backfill soil.

Two direct measurements of ϕ'_{fill} (i.e., 36.3° and 38.6°) are obtained from laboratory triaxial tests. These two measurement values are integrated with the prior knowledge in Table 7.5 using the User-defined model in BEST, as shown in Figure 7.5. Because ϕ'_{fill} is measured directly, no transformation model is needed. The only input data to the BEST is the observation data (i.e., 36.3° and 38.6°) and prior knowledge in Table 7.5. After the data input, the "Generate" button in Figure 7.5 is clicked to activate the Bayesian equivalent sample generation window in Figure 7.6. 30,000 samples of ϕ'_{fill} are generated in this example. Then, conventional statistical analysis of the 30,000 samples can be performed, such as calculating mean and standard deviation and plotting histogram. Figure 7.7 shows a probability density function (PDF) of the ϕ'_{fill} estimated from histogram.

SPT was performed in the foundation soil, and the test results are summarized in Table 7.3. The BEST program is used to integrate the SPT test results with the prior knowledge in Table 7.5. A transformation model is needed to the correlate SPT N values to the effective friction angle of interest here. As shown in Figure 7.8, the

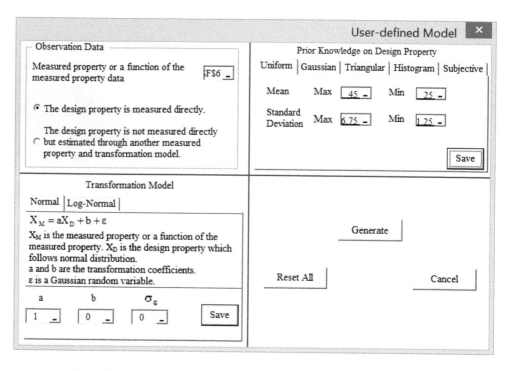

Figure 7.5 The Excel window for quantifying uncertainty in ϕ'_{fill} using BEST.

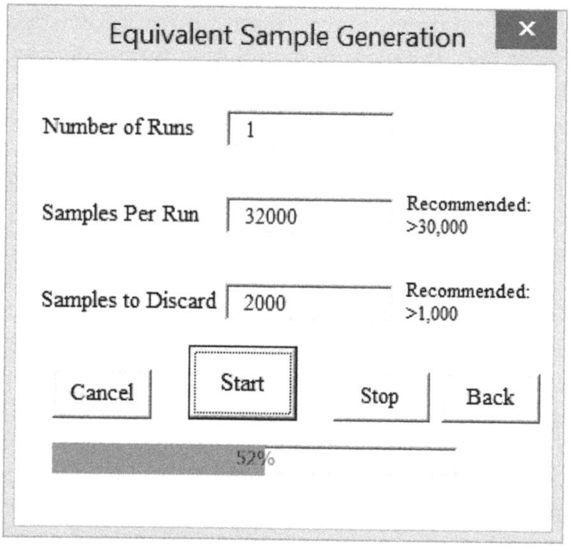

Figure 7.6 The sample generation window of BEST.

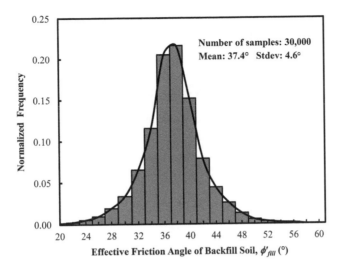

Figure 7.7 Probability density function (PDF) for ϕ'_{fill}.

BEST program contains a built-in model that correlates the corrected SPT N values to effective friction angle using a correlation developed by Ching et al. (2012):

$$\ln(N_1)_{60} = 0.161\phi'_{fdn} + 3.724 + \varepsilon \tag{7.7}$$

where ε represents model uncertainty and follows a normal distribution with a zero mean and standard deviation $\sigma_\varepsilon = 0.496$. Figure 7.8 shows an Excel window when using the BEST program. After specifying the $(N_1)_{60}$ and the prior knowledge, the "Generate" button in Figure 7.8 is clicked to activate the Bayesian equivalent sample generation window in Figure 7.6. 30,000 samples of ϕ'_{fdn} are generated in this example. Then, conventional statistical analysis of the 30,000 samples is performed to obtain mean, standard deviation, and histogram. Figure 7.9 shows the ϕ'_{fdn} PDF estimated from histogram of the 30,000 equivalent samples.

Deterministic calculation model

Traditional ASD calculation model for gravity wall design is adopted as the deterministic calculation model in the expanded RBD method. Practitioners may exercise their sound engineering judgment and select, in their opinions, the most suitable calculation model to best suit the design situation in hands. For example, Coulomb theory of earth pressure is adopted to calculate the active pressure coefficient, K_a, as:

$$K_a = \frac{\cos^2(\phi'_{fill} - \theta)}{\cos^2\theta\cos(\delta_w + \theta)\left[1 + \sqrt{\frac{\sin(\delta_w + \phi'_{fill})\sin(\phi'_{fill} - \beta)}{\cos(\delta_w + \theta)\cos(\theta - \beta)}}\right]^2} \tag{7.8}$$

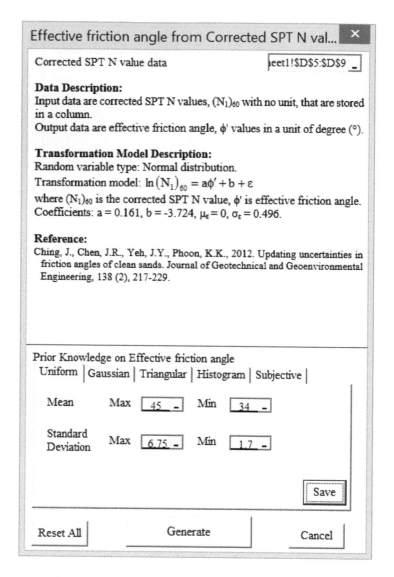

Figure 7.8 The Excel window for quantifying uncertainty in ϕ'_{fdn} using BEST.

The resultant force, P_a, from the active pressure is then expressed:

$$P_a = \frac{1}{2}K_a\gamma_{fill}H^2 \tag{7.9}$$

The horizontal component of P_a is denoted as $P_{a,h}$, and it is calculated as:

$$P_{a,h} = P_a\cos(\delta_w + \theta) \tag{7.10}$$

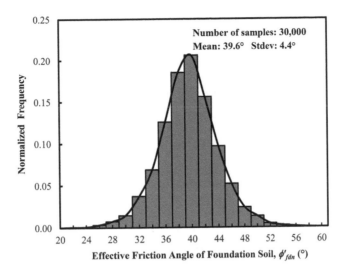

Figure 7.9 Probability density function (PDF) for ϕ'_{fdn}.

The vertical component of P_a is denoted as $P_{a,v}$, and it is calculated as:

$$P_{a,v} = P_a \sin(\delta_w + \theta) \tag{7.11}$$

The weight of the mass concrete wall is denoted as W, and it is calculated as

$$W = \gamma_{con}[(b + B)H/2] \tag{7.12}$$

Three failure modes are considered in this example, including sliding, overturning, and bearing capacity failures. Three factors of safety for sliding, overturning, and bearing capacity failures are denoted as $FS_{sliding}$, $FS_{overturning}$ and FS_{bc}, respectively, and they are calculated using the equations below. When the value of FS is less than 1, the corresponding failure mode occurs. For the sliding failure mode, $FS_{sliding}$ is calculated as:

$$FS_{sliding} = \frac{(P_{a,v} + W)\tan(\delta_b)}{P_{a,h}} \tag{7.13}$$

For the overturning failure mode, $FS_{overturning}$ is calculated as:

$$FS_{overturning} = \frac{M_r}{M_d} = \frac{W\frac{B}{2} + P_{a,v}\left(B - \frac{1}{3}\frac{(B-b)}{2}\right)}{P_{a,h}H/3} \tag{7.14}$$

where M_r is the resisting moment and M_d is the driving moement. For the bearing capacity failure mode, the bearing pressure, q, is calculated as:

$$q = \frac{W + P_{a,v}}{B}\left(1 + \frac{6e}{B}\right) \tag{7.15}$$

where e = eccentricity of the sum of the vertical forces, and it is calculated as:

$$e = \frac{B}{2} - \frac{M_r - M_d}{W + P_{a,v}} \tag{7.16}$$

Because the foundation soil has no effective cohesion and there is no surcharge pressure, the ultimate bearing capacity, q_{ult}, is calculated as (Vesic 1975):

$$q_{ult} = 0.5\gamma_{fdn}B'N_\gamma i_\gamma \tag{7.17}$$

where N_γ = bearing capacity coefficient, i_γ = inclination factors, $B' = B - 2e$ = effective width of foundation base. In addition, N_γ is calculated as:

$$N_\gamma = 2\left(N_q + 1\right)\tan\phi'_{fdn} \tag{7.18}$$

$$N_q = \exp(\pi\tan\phi'_{fdn})\tan^2\left(\frac{\pi}{4} + \frac{\phi'_{fdn}}{2}\right) \tag{7.19}$$

$$i_\gamma = (1 - K_i)^{m_i+1} \tag{7.20}$$

$$K_i = \frac{P_{a,h}}{(W + P_{a,v})} \tag{7.21}$$

$$m_i = 2 \tag{7.22}$$

The FS_{bc} is calculated as:

$$FS_{bc} = \frac{q_{ult}}{q} \tag{7.23}$$

In addition to the checking of the FS_{bc} value using Eq. 7.23, a checking on the e value is performed to ensure that the minimum q value is larger than zero. When the absolute value of e is larger than B/6 (i.e., $|e| > B/6$), the bearing capacity failure mode also occurs and the FS_{bc} value is set as "0".

For a given set of B, b, ϕ'_{fill}, ϕ'_{fdn}, δ_w, and δ_b values, the calculation model described above can be easily implemented in an Excel spreadsheet, as illustrated by Figure 7.10. The Row "1" in Figure 7.10 lists some constant parameters, including the wall height H and respective unit weight of concrete γ_{con}, backfill soil γ_{fill}, and foundation soil γ_{fdn}. Columns "A" to "G" are used to define the input parameters that are treated as random variables, including the design parameters (i.e., B, B − b, and b) and geotechnical parameters (i.e., ϕ'_{fill}, ϕ'_{fdn}, r_w and r_b). Using the information provided in Row "1" and Columns "A" to "G", deterministic calculations are performed in Columns "H" to "AB" using Eqs 7.8–7.23. Judgment of whether or not failure occurs is performed in Columns "AC" to "AE", using an "IF" function in Excel. If a FS is less than 1, failure occurs and "1" is assigned to the corresponding cell. Otherwise, "0" is assigned to the cells. For example, the syntax in Cell "AC8" is "=IF(Z8 > 1,"0","1")". It is

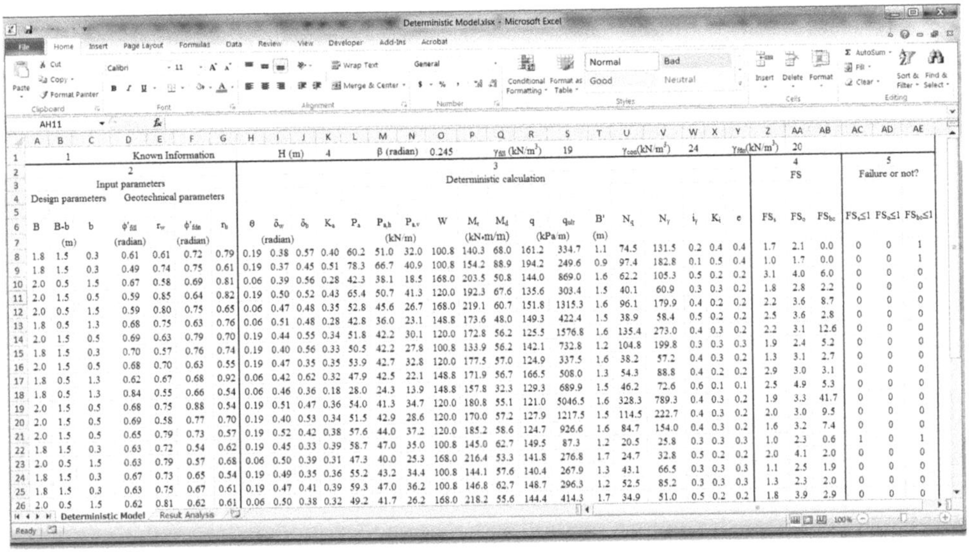

Deterministic Model.xlsx - Microsoft Excel

1 Known Information: H (m) 4; β (radian) 0.245; γ_{fill} (kN/m³) 19; γ_{con} (kN/m³) 24; γ_{fdn} (kN/m³) 20

2 Input parameters (Design parameters; Geotechnical parameters) | 3 Deterministic calculation | 4 FS | 5 Failure or not?

Row	B	B-b	b	ϕ'_{fill}	r_w	ϕ'_{fdn}	r_b	θ	δ_w	δ_b	K_a	P_a	$P_{a,h}$	$P_{a,v}$	W	M_r	M_d	q	q_{ult}	B'	N_q	N_γ	i_γ	K_i	e	FS_s	FS_o	FS_{bc}	$FS_s \le 1$	$FS_o \le 1$	$FS_{bc} \le 1$
7		(m)		(radian)		(radian)			(radian)				(kN/m)			(kN·m/m)		(kPa/m)		(m)											
8	1.8	1.5	0.3	0.61	0.61	0.72	0.79	0.19	0.38	0.57	0.40	60.2	51.0	32.0	100.8	140.3	68.0	161.2	334.7	1.1	74.5	131.5	0.2	0.4	0.4	1.7	2.1	0.0	0	0	1
9	1.8	1.5	0.3	0.49	0.74	0.75	0.61	0.19	0.37	0.45	0.51	78.3	66.7	40.9	100.8	154.2	88.9	194.2	249.6	0.9	97.4	182.8	0.1	0.5	0.4	1.0	1.7	0.0	0	0	1
10	2.0	0.5	1.5	0.67	0.58	0.69	0.81	0.06	0.39	0.56	0.28	42.3	38.1	18.5	168.0	203.5	50.8	144.0	869.0	1.6	62.2	105.3	0.5	0.2	0.2	3.1	4.0	6.0	0	0	0
11	2.0	1.5	0.5	0.59	0.85	0.64	0.82	0.19	0.50	0.52	0.43	65.4	50.7	41.3	120.0	192.3	67.6	135.6	303.4	1.5	40.1	60.9	0.3	0.3	0.2	1.8	2.8	2.2	0	0	0
12	2.0	0.5	1.5	0.59	0.80	0.75	0.65	0.06	0.47	0.48	0.35	52.8	45.6	26.7	168.0	219.1	60.7	151.8	1315.3	1.6	96.1	179.9	0.4	0.2	0.2	2.2	3.6	8.7	0	0	0
13	1.8	0.5	1.3	0.68	0.75	0.63	0.76	0.06	0.51	0.48	0.28	42.8	36.0	23.1	148.8	173.6	48.0	149.3	422.4	1.5	38.9	58.4	0.5	0.2	0.2	2.5	3.6	2.8	0	0	0
14	2.0	1.5	0.5	0.69	0.63	0.79	0.70	0.19	0.44	0.55	0.34	51.8	42.2	30.1	120.0	172.8	56.2	125.5	1576.8	1.6	135.4	273.0	0.4	0.3	0.2	2.2	3.1	12.6	0	0	0
15	1.8	1.5	0.3	0.70	0.57	0.76	0.74	0.19	0.40	0.56	0.33	50.5	42.2	27.8	100.8	133.9	56.2	142.1	732.8	1.2	104.8	199.8	0.3	0.3	0.3	1.9	2.4	5.2	0	0	0
16	2.0	1.5	0.5	0.68	0.70	0.63	0.55	0.19	0.47	0.35	0.35	53.9	42.7	32.8	120.0	177.5	57.0	124.9	337.5	1.6	38.2	57.2	0.4	0.3	0.2	1.3	3.1	2.7	0	0	0
17	1.8	0.5	1.3	0.62	0.67	0.68	0.92	0.06	0.42	0.62	0.32	47.9	42.5	22.1	148.8	171.9	56.7	166.5	508.0	1.3	54.3	88.8	0.4	0.2	0.2	2.9	3.0	3.1	0	0	0
18	1.8	0.5	1.3	0.84	0.55	0.66	0.54	0.06	0.46	0.36	0.18	28.0	24.3	13.9	148.8	157.8	32.3	129.3	689.9	1.5	46.2	72.6	0.6	0.1	0.1	2.5	4.9	5.3	0	0	0
19	2.0	1.5	0.5	0.68	0.75	0.88	0.54	0.19	0.51	0.47	0.36	54.0	41.3	34.7	120.0	180.8	55.1	121.0	5046.5	1.6	328.3	789.3	0.4	0.3	0.2	1.9	3.3	41.7	0	0	0
20	2.0	1.5	0.5	0.69	0.58	0.77	0.70	0.19	0.40	0.53	0.34	51.5	42.9	28.6	120.0	170.0	57.2	127.9	1217.5	1.5	114.5	222.7	0.4	0.3	0.2	2.0	3.0	9.5	0	0	0
21	2.0	1.5	0.5	0.65	0.79	0.73	0.57	0.19	0.52	0.42	0.38	57.6	44.0	37.2	120.0	185.2	58.6	124.7	926.6	1.6	84.7	154.0	0.4	0.3	0.2	1.6	3.2	7.4	0	0	0
22	1.8	1.5	0.3	0.63	0.72	0.54	0.62	0.19	0.45	0.33	0.39	58.7	47.0	35.0	100.8	145.0	62.7	149.5	87.3	1.2	20.5	25.8	0.3	0.3	0.3	1.0	2.3	0.6	1	0	1
23	2.0	0.5	1.5	0.63	0.79	0.57	0.68	0.06	0.50	0.39	0.31	47.3	40.0	25.3	168.0	216.4	53.3	141.8	276.8	1.7	24.7	32.8	0.5	0.2	0.2	2.0	4.1	2.0	0	0	0
24	1.8	1.5	0.3	0.67	0.73	0.65	0.54	0.19	0.49	0.35	0.36	55.2	43.2	34.4	100.8	144.1	57.6	140.4	267.9	1.3	43.1	66.5	0.3	0.3	0.3	1.1	2.5	1.9	0	0	0
25	1.8	1.5	0.3	0.63	0.75	0.67	0.61	0.19	0.47	0.41	0.39	59.3	47.0	36.2	100.8	146.8	62.7	148.7	296.3	1.2	52.5	85.2	0.3	0.3	0.3	1.3	2.3	2.0	0	0	0
26	2.0	0.5	1.5	0.62	0.81	0.62	0.61	0.06	0.50	0.38	0.32	49.2	41.7	26.2	168.0	218.2	55.6	144.4	414.3	1.7	34.9	51.0	0.5	0.2	0.2	1.8	3.9	2.9	0	0	0

Deterministic Model | Result Analysis

Figure 7.10 Illustration of deterministic calculation model in an Excel spreadsheet.

worthwhile to note that, starting from Row "8", each row in Figure 7.10 is a repetitive execution of the deterministic calculation model described above. In other words, each row in Figure 7.10 is a MCS sample using a combination of different B, B – b, ϕ'_{fill}, ϕ'_{fdn}, r_w, and r_b values generated in accordance with the uncertainty model described above.

MCS and expanded RBD method

The MCS is a repetitive computer execution of the deterministic calculation model described above (i.e., a row starting from Row "8" in Figure 7.10). As noted in Figure 7.1, the simulations start with generation of random samples using the probability distributions specified in the uncertainty model (see Table 7.4). Two independent design parameters (i.e., B and x = B – b) and four independent geotechnical parameters (i.e., ϕ'_{fill}, ϕ'_{fdn}, r_w, and r_b) are considered as random variables. Random samples of the uniformly distributed discrete random variables B and x may be easily generated using an Excel function called "randbetween". For example, the discrete B and x = B – b samples can be generated using the following syntax "=0.2*RANDBETWEEN(9,12)" and "=RANDBETWEEN(1,2)-0.5", respectively. The b value is then calculated as b = B – x.

Random samples of ϕ'_{fill}, ϕ'_{fdn}, r_w, and r_b can be generated in Excel using an inverse transformation method. Excel has a built-in random number generator (i.e., "rand()") for a continuous random variable U_i that uniformly distributes between [0, 1]. Because U_i is uniformly distributed between 0 and 1, it can be interpreted as a probability. To generate samples y_i of a non-uniform random variable Y, simply set $U_i = P[Y < y_i]$ = CDF of Y and calculate the value of y_i from U_i (i.e., $y_i = CDF^{-1}(U_i)$, where CDF^{-1} is an inverse CDF function for Y). Then, the y_i values obtained are random samples of Y. For example, r_w and r_b in this design example follows a triangle distribution with the min, max and peak value of 0.5, 1.0, and 2/3, respectively. Their CDF function is expressed as (e.g., $Y = r_w$ or r_b):

$$U_i = \text{CDF}(y_i) = \begin{cases} 0 & y_i < 0.5 \\ 12(y_i - 0.5)^2 & 0.5 \leqslant y_i \leqslant 2/3 \\ 1 - 6(1 - y_i)^2 & 2/3 < y_i \leqslant 1 \\ 1 & 1 < y_i \end{cases} \tag{7.24}$$

Then, the inverse CDF function is expressed as:

$$y_i = \text{CDF}^{-1}(U_i) = \begin{cases} 0.5 + \sqrt{\dfrac{U_i}{12}} & U_i \leqslant \frac{1}{3} \\ 1 - \sqrt{\dfrac{1 - U_i}{6}} & U_i > \frac{1}{3} \end{cases} \tag{7.25}$$

Note that U_i is generated in Excel using the built-in function "rand()". Random samples of r_w or r_b are generated by implementing Eq. 7.25 with U_i as an input in Excel through an "IF" statement.

The inverse transformation method can also be used to generate random samples of ϕ'_{fill} and ϕ'_{fdn}. As described early in this section, the uncertainty in ϕ'_{fill} and ϕ'_{fdn} is quantified by 30,000 equivalent samples, respectively. The PDFs (see Figures 7.7 and 7.9) and CDFs of ϕ'_{fill} and ϕ'_{fdn} are developed from the respective 30,000 equivalent samples, and their CDFs are empirical CDFs that might not be able to be expressed analytically as an equation (e.g., similar to Eq. 7.25 for r_w or r_b). To generate random samples of ϕ'_{fill} (or ϕ'_{fdn}), the 30,000 equivalent samples are firstly sorted in an increasing order of ϕ'_{fill} (or ϕ'_{fdn}) using the "sort" function in Excel. Then, a new column with an integer from "1" to "30000" is added to denote the ranking of each equivalent sample of ϕ'_{fill} (or ϕ'_{fdn}) after sorting. Finally two Excel built-in functions, "RANDBETWEEN(1,30000)" for generating random integers from 1 to 30,000 and "VLOOKUP" for finding the ϕ'_{fill} (or ϕ'_{fdn}) value corresponding the random integer generated, are used together to generate random samples of ϕ'_{fill} (or ϕ'_{fdn}) from the ranking column and the column with the 30,000 equivalent samples.

After random samples of B, $x = B - b$, ϕ'_{fill}, ϕ'_{fdn}, r_w, and r_b are generated, they are inserted into Columns "A" to "G" in Figure 7.10 to perform deterministic calculations repetitively. In this design example, the numbers of B and x values considered are $n_B = 8$ and $n_x = 2$, respectively. If a $p_{ft} = 0.001$ is adopted in the analysis, using Eq. 7.6 leads to a minimum MCS sample number $n_{min} = 160,000$. In this example, $n = 1,600,000$ (i.e., about 100,000 samples for each combination of B and B − b values) is adopted to ensure good accuracy of the MCS results. It takes only several minutes for a PC with an Intel Core i7-2600 CPU @3.40 GHz and 16.0 GB RAM to perform such an MCS run.

After the MCS, statistical analysis is performed by simply counting the number of failure samples, and the failure probability is the ratio of the failure sample number to the total sample number in each combination of B and B − b values. Figure 7.11 shows the results from the expanded RBD method, which is a variation of failure probability as a function of B and B − b values. For a target failure probability of 0.001 (i.e., $\beta_t = 3.09$, see the horizontal solid line in Figure 7.11), feasible designs includes those with B ≥ 2.8 m and B − b = 0.5 m and those with B ≥ 3.0 m and B − b = 1.5 m. If the cross section area of the retaining wall is used as an index of construction cost for

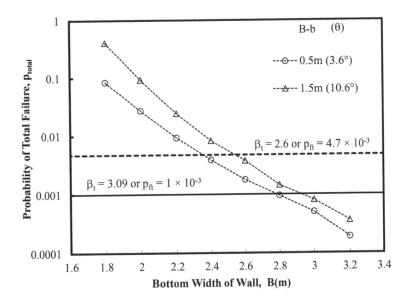

Figure 7.11 Results from the expanded RBD method.

the EOLS requirement, the final design is $B = 3.0$ m and $B - b = 1.5$ m (or $b = 1.5$ m) with the minimum cross section area of 9 m² and $p_f = 0.0008$.

Result analysis

The expanded RBD method allows practitioners to adjust the target failure probability easily without additional computational efforts. For example, if the target failure probability is adjusted from 0.001 to 0.0047 (i.e., $\beta_t = 2.6$, see the horizontal dashed line in Figure 7.11), then, the feasible designs include those with $B \geq 2.4$ m and $B - b = 0.5$ m and those with $B \geq 2.6$ m and $B - b = 1.5$ m. The final design is $B = 2.6$ m and $B - b = 1.5$ m (or $b = 1.1$ m) with the minimum cross section area of 7.4 m² and $p_f = 0.00375$. As the target failure probability increases or the safety requirements become less stringent, the construction cost, as measured by the cross section area of the retaining wall, also decreases. It is a trade-off between safety and cost. The expanded RBD method provides practitioners with a quantitative relationship between the risk and cost associated with various design alternatives and enables practitioners to have a risk-informed decision making on the final design.

Three failure modes (i.e., sliding, overturning and bearing capacity failure) are considered in the gravity retaining wall design example. It is a system reliability problem with multiple failure modes. Additional insights into the interaction between different failure modes can be obtained from the expanded RBD method. Figure 7.12 shows the failure probabilities for sliding, overturning and bearing capacity failure modes, respectively. As B increases, the occurrence probability of three failure modes all tends to decreases. As shown in Figure 7.12(b), the probability for overturning failure is relatively small and has little contribution to the overall failure probability of the retaining wall. When $p_{ft} = 0.001$ and $B-b = 0.5$ m, the most feasible design is $B = 2.8$ m (see Figure 7.11). For this particular feasible design, about 100,000 MCS samples are

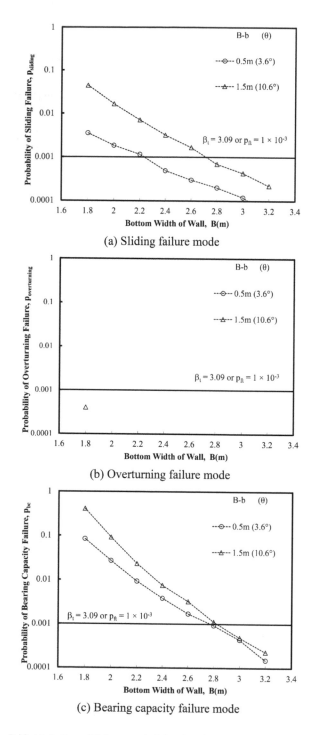

(a) Sliding failure mode

(b) Overturning failure mode

(c) Bearing capacity failure mode

Figure 7.12 Variation of failure probability for three different failure modes.

generated in the expanded RBD method. The number of failure samples is 95, and the total failure probability is 0.00095. Among these 95 failure samples, bearing capacity and sliding failure occurs in 93 and 20 samples, respectively. Overturning failure mode is not observed for this feasible design. Therefore, the failure probability for sliding, overturning and bearing capacity failure modes is 0.0002, 0, and 0.00093, respectively. In addition, among the 20 sliding failure samples, bearing capacity failure mode also occurs in 18 of them. Therefore, bearing capacity failure mode is the dominant failure mode that controls this feasible design.

In contrast, when $p_{ft} = 0.001$ and $B - b = 1.5\,m$, the most feasible design is $B = 3.0\,m$ (see Figure 7.11). Among about 100,000 MCS samples generated for this particular feasible design, the number of failure samples is 80, and the total failure probability is 0.0008. Among these 80 failure samples, bearing capacity and sliding failure mode occurs in 49 and 43 samples, respectively. No overturning failure mode is observed for this feasible design. Therefore, the failure probability for sliding, over-turning and bearing capacity failure modes is 0.00043, 0, and 0.00049, respectively. In addition, sliding and bearing capacity failure modes occur together in 12 samples. Both sliding and bearing capacity failure modes contribute significantly to the total failure probability and play an important role in the design. In other words, this particular feasible design is governed by both sliding and bearing capacity failure modes. This is different from the feasible design with $B - b = 0.5\,m$ and $B = 2.8\,m$ discussed in the previous paragraph. Through detailed analysis of the failure samples generated in the expanded RBD method, practitioners are able to identify the failure mode(s) that con-trols the design and obtain insights into the interaction between different failure modes.

7.9 CONCLUDING REMARKS AND FUTURE WORK

Although most existing RBD codes are in a simplified semi-probabilistic RBD format, sometimes direct probability-based design methods are beneficial and necessary. One major criticism to the simplified semi-probabilistic RBD format is displacement of sound engineering judgment and lack of flexibility for practitioners, which has been long recognized as an important and critical element in geotechnical practice. Because the simplified semi-probabilistic RBD format adopts the same trial-and-error approach as traditional ASD methods and it is developed to circumvent the need for practi-tioners to perform probabilistic analysis, these compromises seem unavoidable. One alternative solution to this trade-off is to maintain the engineering judgment and flexi-bility similar to ASD methods, but at the expenses of performing probabilistic analysis using direct probability-based design methods. With the aid of commonly available PC and widely used computer software such as Microsoft Excel, performing MCS-based probabilistic analysis and design are becoming more and more straightforward and convenient. MCS has the advantages of conceptual and mathematical simplicity and can be comprehended easily as a repetitive computer execution of traditional ASD design calculation. The reliability analysis background required for performing MCS is substantially reduced when using some built-in functions and add-ins in Excel. In addition, the MCS-based design process can be conceptualized as a systematic sensitiv-ity study, in which a large number of design alternatives (or trial designs) are evaluated systematically and the optimal design satisfying the reliability requirements and with the maximum utility is chosen as the final design.

ACKNOWLEDGEMENTS

The authors would like to thank Professor C. Hsein. Juang at Clemson University, USA for reviewing a draft of this chapter. The work described in this chapter was supported by grants from the Research Grants Council of the Hong Kong Special Administrative Region, China (Project No. 9042172 (CityU 11200115) and Project No. 8779012 (T22-603/15N)). The financial supports are gratefully acknowledged.

REFERENCES

Ang, A.H.-S. and Tang, W.H. (2007) *Probability Concepts in Engineering: Emphasis on Applications to Civil and Environmental Engineering*, John Wiley & Sons, New York.

Au, S.K. & Beck, J.L. (2001) Estimation of small failure probabilities in high dimensions by Subset Simulation. *Probabilistic Engineering Mechanics*, 16 (4), 263–277.

Au, S.K. & Wang, Y. (2014) *Engineering Risk Assessment with Subset Simulation*. Singapore, John Wiley & Sons. ISBN: 978-1118398043. 300 pp.

Au, S.K., Cao, Z. & Wang, Y. (2010) Implementing advanced Monte Carlo simulation under spreadsheet environment. *Structural Safety*, 32, 281–292.

Bolton, M.D. (1983) Eurocodes and the geotechnical engineer. *Ground Engineering*, 16 (3), 17–31.

Bond, A. & Harris, A. (2008) *Decoding Eurocode 7*. London and New York, Taylor & Francis.

Cao, Z., Wang, Y. & Li, D. (2016) Quantification of prior knowledge in geotechnical site characterization. *Engineering Geology*, 203, 107–116.

Cetin, K.O., Seed, R.B., Der Kiureghian, A., Tokimatsu, K., Harder, L.F., Kayen, R.E. & Moss, R.E.S. (2004) Standard penetration test-based probabilistic and deterministic assessment of seismic soil liquefaction potential. *ASCE Journal of Geotechnical and Geoenvironmental Engineering*, 130 (12), 1314–1340.

Chen, J. & Gilbert, R. (2014) Insights into the performance reliability of offshore piles based on experience in hurricanes. In: *From Soil Behavior Fundamentals to Innovations in Geotechnical Engineering*. pp. 283–292.

Ching, J., Phoon, K.K. & Hu, Y.G. (2009) Efficient evaluation of reliability for slopes with circular slip surfaces using importance sampling. *ASCE Journal of Geotechnical and Geoenvironmental Engineering*, 135 (6), 768–777.

Ching, J., Chen, J.R., Yeh, J.Y. & Phoon, K.K. (2012) Updating uncertainties in friction angles of clean sands. *ASCE Journal of Geotechnical and Geoenvironmental Engineering*, 138 (2), 217–229.

Christian, J.T. & Baecher, G.B. (2011) Unresolved problems in geotechnical risk and reliability. *Geotechnical Risk Assessment and Management, Geotechnical Special Publication No. 224*, 50–63.

Cornell, C.A. (1968) Engineering seismic risk analysis. *Bulletin of the Seismological Society of America*, 58 (5), 1583–1606.

EN 1990:2002. Eurocode – Basis of structural design. European Committee for Standardization (CEN), Brussels, Belgium.

Fenton, G.A., Naghibi, F., Dundas, D., Bathurst, R.J. & Griffiths, D.V. (2016) Reliability-based geotechnical design in 2014 Canadian Highway Bridge Design Code. *Canadian Geotechnical Journal*, 53 (2), 236–251.

Fleming, W.G.K. (1989) Limit state in soil mechanics and use of partial factors. *Ground Engineering*, 22 (7), 34–35.

Geotechnical Engineering Office (GEO) (1993) *Geoguide 1: Guide to Retaining Wall Design*. 2nd edition. Hong Kong, The Government of Hong Kong Special Administration Region.

Gong, W., Wang, L., Juang, C.H., Zhang, J. & Huang, H. (2014a) Robust geotechnical design of shield-driven tunnels. *Computers and Geotechnics*, 56, 191–201.

Gong, W., Wang, L., Khoshnevisan, S., Juang, C.H., Huang, H. & Zhang, J. (2014b) Robust geotechnical design of earth slopes using fuzzy sets. *ASCE Journal of Geotechnical and Geoenvironmental Engineering*, 141 (1), 04014084.

Gong, W., Khoshnevisan, S. & Juang, C.H. (2014c) Gradient-based design robustness measure for robust geotechnical design. *Canadian Geotechnical Journal*, 51 (11), 1331–1342.

Gong, W., Khoshnevisan, S., Juang, C.H. & Phoon, K.K. (2016) R-LRFD: Load and resistance factor design considering design robustness. *Computers and Geotechnics*, 74, 74–87.

Hartford, D.N.D. & Baecher, G.B. (2004) *Risk and Uncertainty in Dam Safety*. London, Thomas Telford Publishing.

Huber, M., Teixeira, A. & Schweckendiek, T. (2015) Effects of system behaviour in the calibration of partial safety factors. In: *Proceedings of the 5. Symposium zur Sicherung von Dämmen, Deichen und Stauanlagen, Siegen, Germany, 19–20 February, 2015*.

ISO 2394:1973/1986/1998/2015. *General Principles on Reliability for Structures*. Geneva, International Organization for Standardization.

Juang, C.H. & Wang, L. (2013) Reliability-based robust geotechnical design of spread foundations using multi-objective genetic algorithm. *Computers and Geotechnics*, 48, 96–106.

Juang, C.H., Chen, C.J., Rosowsky, D.V. & Tang, W.H. (2000) CPT-based liquefaction analysis, Part 2: Reliability for design. *Geotechnique*, 50 (5), 593–599.

Juang, C.H., Fang, S.Y. & Khor, E.H. (2006) First-order reliability method for probabilistic liquefaction triggering analysis using CPT. *ASCE Journal of Geotechnical and Geoenvironmental Engineering*, 132 (3), 337–350.

Juang, C.H., Fang, S.Y. & Li, D.K. (2008) Reliability analysis of liquefaction potential of soils using standard penetration test. Chapter 13. In: Phoon, K.K. (ed.) *Reliability-Based Design in Geotechnical Engineering*. London and New York, Taylor & Francis.

Juang, C.H., Wang, L., Liu, Z., Ravichandran, N., Huang, H. & Zhang, J. (2013a) Robust geotechnical design of drilled shafts in sand: New design perspective. *ASCE Journal of Geotechnical and Geoenvironmental Engineering*, 139 (12), 2007–2019.

Juang, C.H., Wang, L., Khoshnevisan, S. & Atamturktur, S. (2013b) Robust geotechnical design – Methodology and applications. *Journal of GeoEngineering*, 8 (3), 71–81.

Juang, C.H., Ching, J. & Luo, Z. (2013c) Assessing SPT-based probabilistic models for liquefaction potential evaluation: A 10-year update. *Georisk: Assessment and Management of Risk for Engineered Systems and Geohazards*, 7 (3), 137–150.

Khoshnevisan, S., Gong, W., Juang, C.H. & Atamturktur, S. (2014) Efficient robust geotechnical design of drilled shafts in clay using a spreadsheet. *ASCE Journal of Geotechnical and Geoenvironmental Engineering*, 141 (2), 04014092.

Kulhawy, F.H., Phoon, K.K. & Wang, Y. (2012) Reliability-based design of foundations – A modern view. In: Rollins, K. & Zekkos, D. (eds.) *Geotechnical Engineering State of the Art and Practice (GSP 226)*. Reston, ASCE. pp. 102–121.

Lacasse, S., Nadim, F., Langford, T., Knudsen, S., Yetginer, G.L., Guttormsen, T.R. & Eide, A. (2013) Model uncertainty in axial pile capacity design methods. In: *Offshore Technology Conference, 6–9 May, 2013, Houston, Texas, USA*.

Li, L., Wang, Y., Cao, Z.J. & Chu, X. (2013) Risk de-aggregation and system reliability analysis of slope stability using representative slip surfaces. *Computers and Geotechnics*, 53, 95–105.

Li, L., Wang, Y. & Cao, Z.J. (2014) Probabilistic slope stability analysis by risk aggregation. *Engineering Geology*, 176, 57–65.

Means (2007) *2008 RS Means Building Construction Cost Data*. Kingston, MA, R.S. Means Co.

Phoon, K.K. (2008) Numerical recipes for reliability analysis – A primer. In: Phoon, K.K. (ed.) *Reliability-Based Design in Geotechnical Engineering: Computations and Applications*. London, Taylor & Francis. pp. 1–75.

Phoon, K.K., Kulhawy, F.H. & Grigoriu, M.D. (1995) *Reliability-Based Design of Foundations for Transmission Line Structures*. Report TR-105000 Palo Alto, Electric Power Research Institute.

Reiter, L. (1990) *Earthquake Hazard Analysis*. New York, Columbia University Press. 254 pp.

Robert, C. & Casella, G. (2004) Monte Carlo Statistical Methods, Springer.

Schweckendiek, T., Vrouwenvelder, A.C.W.M., Calle, E.O.F., Kanning, W. & Jongejan, R.B. (2012) Target reliabilities and partial factors for flood defenses in the Netherlands. In: Arnold, P., Fenton, G.A., Hicks, M.A. & Schweckendiek, T. (eds.) *Modern Geotechnical Codes of Practice – Code Development and Calibration*. London, Taylor and Francis. pp. 311–328.

Schweckendiek, T., Slomp, R. & Knoeff, H. (2015) New safety standards and assessment tools in the Netherlands. In: *Proceedings of the 5. Siegener Symposium "Sicherung von Dämmen, Deichen und Stauanlagen", Siegen, Germany, 19–20 February, 2015*.

Tang, W.H., Woodford, D.L. & Pelletier, J.H. (1990) Performance reliability of offshore piles. In: *Offshore Technology Conference, 5/7/1990, Houston, Texas, USA*.

U.S. Army Corps of Engineers (1997) Introduction to Probability & Reliability Methods for Geotechnical Engineering. Washington, DC, Engineering Technical Letter 1110-2-547, Dept of Army.

Vesić, A.S. (1975) Bearing capacity of shallow foundations, Chapter 3. In: Winterkorn, H.F. & Fang, H.Y. (eds.) *Foundation Engineering Handbook*. New York, Van Nostrand Reinhold. pp. 121–147.

Wang, Y. (2009) Reliability-based economic design optimization of spread foundations. *ASCE Journal of Geotechnical and Geoenvironmental Engineering*, 135 (7), 954–959.

Wang, Y. (2011) Reliability-based design of spread foundations by Monte Carlo Simulations. *Geotechnique*, 61 (8), 677–685.

Wang, Y. (2013) MCS-based probabilistic design of embedded sheet pile walls. *Georisk*, 7 (3), 151–162.

Wang, Y. & Kulhawy, F.H. (2008) Economic design optimization of foundations. *ASCE Journal of Geotechnical and Geoenvironmental Engineering*, 134 (8), 1097–1105.

Wang, Y. & Cao, Z. (2013a) Expanded reliability-based design of piles in spatially variable soil using efficient Monte Carlo simulations. *Soils and Foundations*, 53 (6), 820–834.

Wang, Y. & Cao, Z.J. (2013b) Probabilistic characterization of Young's modulus of soil using equivalent samples. *Engineering Geology*, 159, 106–118.

Wang, Y. & Cao, Z. (2015) Practical reliability analysis and design by Monte Carlo Simulations in spreadsheet. Chapter 7. In: Phoon, K.K. & Ching, J. (eds.) *Risk and Reliability in Geotechnical Engineering*. Leiden, CRC Press. pp. 301–335.

Wang, Y., Au, S.K. & Kulhawy, F.H. (2011a) Expanded reliability-based design approach for drilled shafts. *ASCE Journal of Geotechnical and Geoenvironmental Engineering*, 137 (2), 140–149.

Wang, Y., Cao, Z.J. & Au, S.K. (2011b) Practical reliability analysis of slope stability by advanced Monte Carlo Simulations in spreadsheet. *Canadian Geotechnical Journal*, 48 (1), 162–172.

Wang, Y., Zhao, T. & Cao, Z. (2015) Site-specific probability distribution of geotechnical properties. *Computers and Geotechnics*, 70, 159–168.

Wang, Y., Cao, Z. & Li, D. (2016) Bayesian perspective on geotechnical variability and site characterization. *Engineering Geology*, 203, 117–125.

Zhang, J., Zhang, L.M. & Tang, W.H. (2011) New methods for system reliability analysis of soil slopes. *Canadian Geotechnical Journal*, 48 (7), 1138–1148.

Zhang, L.M. & Ng, A.M.Y. (2005) Probabilistic limiting tolerable displacement for serviceability limit state design of foundations. *Geotechnique*, 55 (2), 151–161.

Zhang, L.M., Tang, W.H. & Ng, C.W.W. (2001) Reliability of axially loaded driven pile groups. *ASCE Journal of Geotechnical and Geoenvironmental Engineering*, 127 (12), 1051–1060.

Index

Milton Keynes UK
Ingram Content Group UK Ltd.
UKHW051853071024
449327UK00025B/1943